Abner Shimony is an eminent philosopher and theoretical physicist, best known for contributions to experiments on foundations of quantum mechanics, notably the polarization correlation test of Bell's Inequality and thereby of the family of local hidden-variables theories. *Search for a Naturalistic World View* consists of essays written over a period of four decades and aims at linking the natural sciences, especially physics, biology, and psychology, with epistemology and metaphysics. It will serve professionals and students in all of these disciplines.

Volume I, *Scientific Method and Epistemology,* advocates an "integral epistemology" combining conceptual analysis with results of empirical science. It proposes a version of scientific realism that emphasizes causal relations between physical and mental events and rejects a physicalist account of mentality. It offers a "tempered personalist" version of scientific methodology, which supplements Bayesianism with *a posteriori* principles distilled from exemplary cognitive achievements. It defends the general reliability, corrigibility, and progressiveness of empirical knowledge against relativism and skepticism.

Volume II, *Natural Science and Metaphysics,* widely illustrates "experimental metaphysics." Quantum-mechanical studies argue that potentiality, chance, probability, entanglement, and nonlocality are objective features of the physical world. The variety of relations between wholes and parts is explored in complex systems. One essay proposes that in spite of abundant phenomena of natural selection, there exists no principle of natural selection. A defense is given of the reality and objectivity of transiency. A final section consists of historical, speculative, and experimental studies of the mind–body problem.

"Abner Shimony's work makes signal contributions to every subject his hand has touched in general epistemology and philosophy of science: probability theory, foundations of quantum mechanics, scientific inference, measurement, holism, and many more. He was a pioneer in the transformation of Bayesian statistics into a core resource for philosophical reflection of scientific methodology, as well as in the experimental and theoretical exploration of Bell's inequalities for the possibility of hidden variables in quantum mechanics. His writings are invaluable as both challenge and resource for current research."

Bas van Fraassen, Princeton University

Search for a naturalistic world view
Scientific method and epistemology

Search for a naturalistic world view

VOLUME I

Scientific method and epistemology

ABNER SHIMONY

Professor of Philosophy and Physics
Boston University

CAMBRIDGE
UNIVERSITY PRESS

Published by the Press Syndicate of the University of Cambridge
The Pitt Building, Trumpington Street, Cambridge CB2 1RP
40 West 20th Street, New York, NY 10011-4211, USA
10 Stamford Road, Oakleigh, Victoria 3166, Australia

First published 1993

Printed in the United States of America

Library of Congress Cataloging-in-Publication Data
Shimony, Abner.

Search for a naturalistic world view / Abner Shimony.

p.   cm.

Includes index.

Contents: v. 1, Scientific method and epistemology –
v. 2, Natural science and metaphysics.

ISBN 0-521-37352-2 (v. 1), – ISBN 0-521-37744-7 (v. 1 : pbk.)
ISBN 0-521-37353-0 (v. 2), – ISBN 0-521-37745-5 (v. 2 : pbk.)

1. Science – Methodology.   2. Science – Philosophy.   3. Knowledge,
Theory of.   4. Metaphysics.   I. Title.
Q175.S53483   1993
501 – dc20   92-38168
                                              CIP

A catalog record for this book is available from the British Library.

Volume 1
ISBN 0-521-37352-2 hardback
ISBN 0-521-37744-7 paperback

Volume 2
ISBN 0-521-37353-0 hardback
ISBN 0-521-37745-5 paperback

To Dora Farber Shimony,
with love and gratitude

the pillar of cloud by day, and the pillar of fire by night
—Exodus

Contents

Acknowledgments

The pleasantest duty in compiling this collection of essays is to acknowledge a number of people who influenced them in one way or another, by formative ideas, inspiration, discipline, suggestions for research, criticism, and encouragement. I have enjoyed the friendship of three generations, that of teachers, that of fellow students and colleagues, and that of my students. In thanking all of them publicly I hope not only to fulfill my personal obligation to them, but to provide an elucidation to readers that could hardly be accomplished in another way. Although I subjectively regard my world view as a coherent synthesis of many elements, I have sufficient detachment to realize that the reader could easily be surprised and baffled by certain tensions, certain contrapositions of thesis and antithesis, and certain shifts of method in my expositions. I wish the reader not to regard these as inconsistencies or idiosyncrasies, but instead to recognize that complexity is inevitable in a world view to which many diverse influences have contributed.

The philosophers from whom I learned most during my undergraduate years at Yale (1944–48) were Frederic Fitch, Paul Weiss, and Robert Calhoun. Their ideas and styles of thought exercised a direct influence on me, and in addition they introduced me to three profound philosophers: Alfred North Whitehead, Charles S. Peirce, and Kurt Gödel.

Fitch was diffident in manner but rigorous in reasoning. He had developed an elegant formulation of predicate calculus and some demonstrably consistent and mathematically rich systems of logic. But what most impressed me was his application of logic to some traditional philosophical problems: possibility and necessity, the ontological status of mathematical entities, inductive inference, and the existence of God. In one seminar, I recall, he demonstrated the existence of God (in the sense of a proposition that logically implies all true propositions) and then shyly said that if he were now allowed to strengthen his premises slightly he could demonstrate monotheism. I was already dimly aware of the division of professional philosophy into analytic and speculative subdisciplines, but the example of Fitch consciously or unconsciously strengthened my resolution not to accede to this bifurcation.

Weiss was ebullient and enormously energetic. He is the only philosopher of my personal acquaintance who has deliberately set out to construct a "system." I was greatly impressed by Weiss's book *Reality,* and even though later I became very critical of his mode of philosophizing I never abandoned certain theses of that book: especially, the interdependence of the criteria of coherence and correspondence, the meshing of epistemology and ontology, and the role of adumbration in perception. I am also grateful to Weiss for his encouragement to be daring and to attempt some original (even if grossly premature) ventures with difficult philosophical problems.

Calhoun was the most charismatic person I had ever encountered – deeply religious, politically courageous, dignified without any self-aggrandizement, and wonderfully eloquent. His introductory course in history of philosophy was legendary, and I learned the fascination of the history of ideas more from him than from anyone else. He mainly taught in the Divinity School, and to hear more of his lectures I attended his course on the history of Christian doctrine, which interested me primarily because of his presentation. Once while returning from his lecture I was reviewing in my mind the *homoousia–homoiousia* controversy and was thrown from my bicycle, which taught me (as others had learned previously) that theological disputes can be hazardous to one's health.

Weiss had been Whitehead's graduate student, Calhoun regarded Whitehead as the greatest twentieth-century philosopher, and Fitch was an expert on his and Russell's *Principia Mathematica*. As a result, I studied Whitehead more seriously than any other philosopher. His speculation that the ultimate concrete entities in the universe are protomental seemed to me to be the obvious solution to the mind–body problem, and I still regard it as a deep idea that has not been sufficiently explored. I was impressed (but later with many reservations) by the liaisons he established between his philosophy of organism and modern physical theory. I was convinced then (though not now) that Whitehead's theory of prehensions provided an answer to Hume's skeptical doubts about induction. Whitehead's philosophical methodology, which combines phenomenology and the hypothetico-deductive method (Chap. 1 of the first part of *Process and Reality,* Chap. 15 of *Adventures of Ideas*) seemed then to be admirable and still does.

Weiss had been co-editor with Charles Hartshorne of the first six volumes of the Harvard edition of Peirce's papers. I read Peirce avidly and assented to almost everything that I understood of his semiotics, phenomenology, scientific methodology, pragmatism, critical common-sensism, and evolutionary metaphysics. Peirce's mixture of logical toughness, immersion in the history and practice of the natural sciences, and metaphysical speculation was inspiring to me then and continues to be so.

I heard about Gödel's incompleteness theorem from Fitch without studying it in detail, but we did devote a semester to reading his *Consistency of the Continuum Hypothesis*. This reading helped me to appreciate the philosophical passages in Gödel's articles "Russell's Mathematical Logic" and "What Is Cantor's Continuum Hypothesis?" I was convinced that the propositions of mathematics could not be regarded as tautologies (as held by Ramsey and Wittgenstein) but are analytically true because of internal relations among concepts, that mathematical entities have a Platonic mode of existence, that impredicative definitions are not inherently vicious, that human intuition into mathematical relations is penetrating but "astigmatic," and that the hypothetico-deductive method is an appropriate tool in the foundations of mathematics. Although I claim no expertise on philosophy of mathematics, I continue to be convinced by these theses of Gödel, as well as by most of the reports of his ideas on logic and mathematics in Hao Wang's *Beyond Analytic Philosophy* and *Reflections on Gödel*.

Whitehead, Peirce, and Gödel were all monumental figures in the history of logic, who greatly influenced various branches of twentieth-century analytic philosophy. It was indicative of the unusual character of philosophy at Yale University in the forties that an assiduous student could be somewhat acquainted with the work of these three masters and yet almost entirely unaware of the movements which they affected. Although I did read Wittgenstein's *Tractatus* with Rulon Wells, I barely heard of the Vienna Circle and vaguely recall dismissing what I did hear of it as "nominalism." The news that reached me of the therapeutic linguistic philosophy of later Wittgenstein, Ryle, and Austin seemed to me incredible, and I conflated them with Korzybski and Hayakawa.

Finally, I should mention another kind of influence at Yale. Although I did not take Henry Margenau's course in philosophy of physics, I did take a course in mechanics with him and was made aware that there are some important connections between physics and philosophy. This awareness was strengthened by conversations with Adolf Grünbaum, who arrived as a graduate student at the end of my undergraduate career. Their influence became effective after a few years of delay.

When I went to the University of Chicago (1948–49), Rudolf Carnap influenced me more than any other faculty member. He was an immensely impressive man, who had raised the level of rigor in the fields of logic, analytic philosophy, semantics, and philosophy of science, and he seemed to have at his command an inventory of all the important argumentation in the many fields of his research. I did not become his disciple, mainly because of the conviction of the significance of metaphysical statements that I had contracted from my former teachers and from the writings of Whitehead, Peirce, and Gödel. Carnap did not demand discipleship, however, and he was generous with his time and encouragement, although he must have been perplexed by my peculiar combination of interests in logic, mathematics, and metaphysics. Aiming at the standards of clarity in expression and argument that Carnap set was a most valuable discipline. His formulation of the problems of probability and induction constituted the framework within which my doctoral research was conducted; and even though I returned to Yale for this work and he left Chicago for the Institute for Advanced Study, he continued to give me excellent criticism and advice.

At Chicago Richard McKeon gave unique instruction in reading philosophical texts, with attention to organization and to details; and he made comparisons of philosophical systems, asserting their "incommensurability" (to use a later term) due to shifts of meanings of crucial terms and differences in modes of reasoning. Reading with John William Lenz, who had been a student of McKeon earlier, was very helpful for learning McKeon's techniques away from the tension of the classroom. I never accepted the thesis of incommensurability, but was disciplined in my later work by attention to his theses and analyses.

After I returned to Yale John Myhill was on the philosophy faculty for two years, teaching splendid courses in proof theory and recursive function theory. I am mainly grateful to him, however, for discouraging my ambition in mathematical logic, therefore effectively exiling me to the world of concrete existence.

In 1955 I went to Princeton to seek a second doctorate in physics. The faculty was extraordinary, and the entire experience was strenuous and revealing. I am most deeply grateful to Eugene Wigner, who directed my doctoral dissertation on statistical mechanics and encouraged my later work on foundations of quantum mechanics. The preponderance of the physics community at that time accepted some variant of the Copenhagen interpretation of quantum mechanics and believed that satisfactory solutions had already been given to the measurement problem, the problem of Einstein–Podolsky–Rosen, and other conceptual difficulties. My decision to devote much research effort to these problems would have been emotionally more difficult without Wigner's authority as one of the great pioneers and masters of quantum mechanics. Equally important was the inspiration of witnessing Wigner's immersion in scientific research and of hearing him say that when one understands a phenomenon one has "an elevated feeling."

I also learned much from the classes of John Archibald Wheeler, from later conversations with him, and from his papers. There is no other living physicist who combines to the same degree the daring to make far-reaching speculations ("geometrodynamics," "superspace," "charge without charge," "mass without mass," "law without law," etc.) with control of the mathematical apparatus needed for inferring consequences of speculations and making connections with experiment. I do not know whether he would accept the denomination "experimental metaphysician," but he has been an inspiration for me to do so.

Two other great physicists were de facto my teachers without official status. While I was teaching at MIT (1959–68), I took Laszlo Tisza's course on statistical thermodynamics, and I had valuable conversations with him then and later regarding the physics of open systems, the proper ontology of microphysics, the methodology of physical theory, and other matters. The other was John S. Bell, whose theorem on the impossibility of a "local" hidden-variables interpretation of quantum mechanics is regarded by many people, including myself, as the most profound discovery in natural philosophy in our generation, and who was working at the time of his death on the stochastic modification of quantum mechanics, which he felt to be a viable and promising idea. Although he was not formally trained in philosophy, he possessed philosophical virtues more purely than any other person of my acquaintance: intense desire for deep understanding, breadth of perspective, great analytic power, strong endowment of common sense, and complete intellectual honesty. Bell not only posed some of the most important research problems in the foundations of quantum mechanics but set the standards for work in the field.

The person whose influence is most pervasive throughout these essays is not a teacher but a close friend and collaborator since our student days at the University of Chicago, Howard Stein. Because of the similarities in our background and interests, and probably also because of the exchange of ideas over a period of more than four decades, we have arrived at similar opinions about the indispensability of mathematics, physics, conceptual analysis, history of science, and history of ideas to philosophy generally, and especially to philosophy of science. I have drawn upon his expertise in all of these, especially in mathematics and history of science, where his knowledge is much deeper and more detailed than mine. In addition to recognizing all these elements, one must properly combine them, and Stein's philosophical studies of episodes in the history of physics from Galileo and Newton to the present are unparalleled models in this respect. Finally, he has meticulously read a large number of my essays and has given such penetrating criticism that I have come to regard him as a second intellectual conscience.

I have greatly benefited since 1968 from the opportunities provided by a joint appointment in Philosophy and Physics at Boston University and from the stimulation of colleagues in both departments. All have been tolerant of my cross-disciplinary teaching and research, and some have even accepted my thesis that there is no sharp boundary between the fields. There have been continuing lively discussions of naturalistic epistemology, foundations of quantum mechanics, and reductionism. I am particularly indebted to Robert Cohen, Marx Wartofsky, John Stachel, Charles Willis, George Zimmerman, Armand Siegel, Michael Martin, Judson Webb, Joseph Agassi, James Hullett, Milič Čapek, Jaakko Hintikka, and Sahotra Sarkar.

Long-forgotten conversations with innumerable friends and students have surely left their imprint on these essays in ways that I cannot reconstruct. However, I can at least acknowledge my indebtedness to those I remember: Joel Lebowitz, Bernard d'Espagnat, Richard Jeffrey, Carl Hempel, Andrew Frenkel, Stanley Tennenbaum, Robert Palter, Kenneth Friedman, Penha Dias, Nicolas Gisin, Philip Pearle, Donald Campbell, Don Howard, Martin Klein, David Mermin, Sylvan Schweber, David Bohm, Yakir Aharonov, Herbert Bernstein, Andre Mirabelli, Hyman Hartman, Hans Primas, Paul Teller, Georges Lochak, Michael Redhead, Gordon Fleming, and Louis Mink.

I am indebted to my collaborators in research – Michael Horne, John Clauser, Richard Holt, Anton Zeilinger, and Daniel Greenberger – for their expertise in quantum mechanics, atomic phyics, and optical and neutron interferometry, but much more for sharing the excitement of the hunt.

So far I have only thanked those who affected my professional training and research, but there were earlier influences. Especially the formation I received from Morris and Dora Shimony was permanent and generated a feeling toward the natural world that preserves a kind of religious sentiment. I hope that the arguments and analyses of these essays do not conceal the sense of wonder that animates them.

PART A
The dialectic of subject and object

1
*Integral epistemology**

This essay is both an appreciation of the epistemological contributions of Donald Campbell and a statement of an epistemological program which is different from his in several respects.

In a lecture to the Boston Colloquium for the Philosophy of Science in 1977, he said:

What I am doing is "descriptive, contingent, synthetic epistemology." . . . I make a sharp distinction between the task and permissible tools of descriptive epistemology on the one hand and traditional, pure, analytic, logical epistemology on the other. Descriptive epistemology is a part of science rather than philosophy, as that distinction used to be drawn by philosophers. It is science of science, scientific theory of knowledge, were those terms not too pretentious for the present state of the art.

While I want descriptive epistemology to deal with normative issues, with validity, truth, justification of knowledge – that is, to be epistemology – descriptive epistemology can only do so at the cost of presumptions about the nature of the world and thus beg the traditional epistemologist's question [Campbell, 1977a, p. 1].

Campbell's resolute restriction of his investigations to descriptive epistemology is both his great strength and his weakness. It is his strength because it frees him, at one stroke, from the slow-paced type of inquiry that dominates the literature of analytic epistemology: for example, "When I see a tomato there is much that I can doubt" (Price, 1932, p. 33). His investigations lead out of the study into the open air. There is a wonderful sweep in his survey of the stages of cognitive development (Campbell, 1974, pp. 422–434). In the perspective that Campbell offers, nature is in no way subservient to humans; however, because of the sequence of adaptations to nature which occurred in the human phylogeny, we are able to achieve something approaching objective knowledge.

The weakness in Campbell's program is that the traditional problems of analytic epistemology continue to be haunting, especially when answers

This work originally appeared in M. Brewer and B. Collins (eds.), *Scientific Inquiry and the Social Sciences,* San Francisco: Jossey-Bass Inc., Publishers, 1981. Reprinted by permission of the publisher.

*The research for this essay was supported in part by the National Science Foundation.

to some of them are postulated as preconditions for his own investigations. He postulates an ontology of independent real objects: that there are entities in the universe which do not depend for their existence upon their being perceived or known by human beings. He accepts without argumentation the correspondence theory of truth: that a sentence is true if and only if the state of affairs which it expresses is the case, so that the truth of a sentence does not depend upon its being believed, or upon the utility of believing it, or upon the existence of evidence supporting it. He postulates a causal theory of perception: that one can understand the content of perceptual experience only by taking into account the causal relations which link the perceiver to objects existing independently of the perceiver's faculties. If Campbell's descriptive epistemology entirely abstained from normative questions, then there would be nothing wrong in principle with such postulation. It would be analogous to putting one's trust in the mathematicians and assuming the correctness of useful mathematical theorems without checking the proofs oneself. Since, however, Campbell wishes to deal with normative questions, especially with the justification of knowledge, he cannot avoid the problem of justifying his postulates.

This essay will propose an *integral epistemology,* in which certain methods of descriptive epistemology (which Campbell espouses) and certain methods of analytic epistemology (from which he abstains) are combined for the purpose of rationally assessing claims to human knowledge. It is anticipated that the results of scientific investigation about human beings will shed light on the reliability of human cognition, and reciprocally that adequate justification can be given for the presuppositions of scientific investigations. The proposed integral epistemology is unequivocally naturalistic, following Campbell not only in his general thesis that a necessary condition for understanding human cognition is to see man's place in nature but also in his insistence that detailed attention to the sciences is indispensable for solving epistemological problems. The proposed approach differs from his in its envisagement of a dialectical structure of epistemology and in its resort to methodological, decision-theoretical, and semantic analysis.

A disclaimer should be made at the outset with regard to novelty, not just because of the usual obligation to acknowledge intellectual indebtedness but because an integral epistemology is a synthesis by its conception. A naturalistic view of human knowledge is at least as old as Aristotle's *De Anima,* though it has been greatly expanded by applications of the theory of evolution; and the thesis that epistemology has a dialectical structure goes back, of course, to Plato. Sustained attempts to incorporate naturalistic epistemology into a dialectical framework are, however, uncommon. Among classical philosophers, Peirce seems to come closest to the integral epistemology which I envisage, and the contemporary philosophers who approach it most closely – notably, Quine (1969) and Rescher (1977) – acknowledge their affinity to him. Rescher explicitly characterizes his epistemology as dialectical, but Quine (1974, p. 137) does not seem to like the word, in spite of his espousal of a philosophical methodology which seems to me unequivocally dialectical. A dialectical naturalism is also outlined in Shimony (1970).

SOME DIALECTICAL THESES

A dialectical method offers a reasonable resolution to a difficulty which is intrinsic to the enterprise of epistemology. If epistemology is a comprehensive study of human cognition, then the question of the validity of its own procedure falls within its domain. How, then, does the enterprise rationally begin? One radical type of answer to this difficulty is to identify incontrovertible initial propositions, guaranteed, for example, by clarity and distinctness, by self-certification, or by intuition; a radical answer at the opposite extreme is skepticism. In the work of a number of philosophers who call themselves "dialectical," one finds several characteristic features which permit them to navigate with control between these two extremes. Tentative suppositions are accepted at the beginning of inquiry, but they are subject to criticism and may be revised and refined on examination. Fundamental principles are, then, the end rather than the beginning of inquiry. The terms in initial suppositions are often vague, clarification of meaning being inseparable from the assessment of propositions. Most important, the dialectic is open, with no foregone conclusions and no suppositions that are so entrenched that they cannot be critically evaluated. In particular, the openness applies to the dialectical method itself. Inquiry may show that the dialectical method is in need of supplementation or that it is only a temporary expedient on the way to a deeper method.

Because it is characteristic of dialectical methods to examine the initial suppositions of inquiry and to reflect on the dialectical process itself, there is a widespread opinion that these methods entail a commitment to a coherence theory of truth – the theory that the truth of a sentence consists in its properly fitting into a system of assertions. The opinion is correct concerning some historically important versions of dialectic – notably, Hegel's – but it is not correct as a generalization. The intergral epistemology envisaged in this paper agrees with Campbell's recommendation "to accept the *correspondence meaning of truth and goal of science,* and to acknowledge *coherence* as the major but still fallible *symptom of truth*" (1977a, p. 20).

In the subsequent discussion of connections between description and justification, four procedural or organizational theses will be deployed. For the most part, these theses are implicit in the general features of dialectical methods which have already been noted.

(1) *Commonsense judgments about ordinary matters of fact have a prima facie credibility, which should not be discounted without clear positive reasons.* The distinctions between truth and falsity, between reliability and unreliability, between appearance and reality are habitually made in ordinary discourse, which consists largely of expressions of the kinds of judgments mentioned in thesis 1. Perhaps anticipations of these distinctions can even be found in stages of individual development before ordinary discourse is mastered, but one can hardly doubt that a secure achievement of a commonsense view of the world is a precondition to any kind of sophisticated examination of these distinctions. Consequently, a sweeping skepticism about the reliability of commonsense judgments undermines the rough-hewn contrast between, for instance, true and false judgments or real and apparent properties of things, with the danger that the dialectical process of refining these epistemological distinctions will be aborted at an early stage. It must be emphasized that the presumption of truth for the specified class of judgments does not imply a full commitment to them.

(2) *The way of inquiry should not be blocked.* This thesis is Peirce's "supreme maxim of philosophizing" ([1897] 1931, p. 56). The thesis is implicit in the openness of the dialectic, which was noted previously, but it is a reminder that openness is not preserved without some effort. In a later section (headed "Evolutionary Considerations"), this thesis will be applied in the course of evaluating criteria of meaningfulness.

(3) *Epistemology and knowledge of the world (a vague phrase, intended to refer both to the sciences and to metaphysics) should mesh and complement each other.* This thesis really has the status of a desideratum, for there is no a priori guarantee that we are capable of "closing the circle," in the sense of giving an objective account of the knowing subject as an entity in the world in a manner that adequately explains its cognitive capacities. Nevertheless, thesis 3 is reasonable, and its failure to hold in a particular philosophical system is prima facie evidence that something is amiss, unless convincing reasons are given why it should fail (as Kant, of course, tries to do in the *Critique of Pure Reason*).

(4) *A vindicatory argument – that is, an argument to the effect that a certain method, M, will yield good approximations to the truth in a certain domain if any method will, so that nothing is lost and possibly something is gained by the use of M – is an acceptable form of epistemological justification.* This thesis has a decision-theoretical character. It is similar to recommending a strategy that performs as well as or better than all

other strategies in a game. The thesis has been used by Hans Reichenbach, Herbert Feigl, Wesley Salmon, and others in discussions of the justification of inductive inference. It provides, incidentally, a good example of a somewhat vague concept that becomes clarified as a result of dialectical examination. In common sense there is a vague concept of justification, the explication of which has been a major philosophical problem. The retreat, under the critical scrutiny of Hume and others, from attempts to provide a justification in a strong sense for inductive inference has constituted a remarkable extended dialogue, and the resort to vindication has been a way of salvaging something from the epistemological optimism of an earlier epoch. An important warning is that, despite its reasonableness as stated, thesis 4 is difficult to apply in actual epistemological argumentation, because one seldom can be confident that the condition "nothing is lost and possibly something is gained" is satisfied.

Although these four theses are obviously in need of further discussion, it may be more illuminating to aim at clarification in the course of applying them to concrete epistemological problems than to analyze them further in an abstract way. The following three sections will examine the epistemological relevance of three different bodies of factual information about human cognition, together with some nondescriptive considerations. The section headed "Empirical Psychology and the Reliability of Perceptual Judgments" will discuss the reliability of perceptual judgments in the light of psychological studies of perception and cognition. The section on "Evolutionary Considerations" will consider the implications of evolutionary biology for the scope of human knowledge. And the section headed "Inductive Inferences and Vindicatory Arguments" will outline a program for the justification of inductive inference, involving among other things some reliance on the history of science (which is entitled to be recognized as part of descriptive epistemology). It must be emphasized in advance that answers to epistemological problems cannot be anticipated simply as corollaries of scientific results. An appropriate deployment of factual information is impossible without analysis, if only because of the considerable differences in the questions typically posed by scientists and by epistemologists. This warning is not a revocation of claims for the power of a naturalistic point of view but is, rather, a reassertion of the conception of an integral epistemology, in which factual information is indispensable but not self-sufficient.

EMPIRICAL PSYCHOLOGY AND THE RELIABILITY
OF PERCEPTUAL JUDGMENTS

An index of the significance of empirical psychology for epistemological purposes is the breadth of the overview that it provides of some of the traditional debates of analytic epistemology. The strong points on opposite sides of debates are often seen to fall into place; and, even more impressively, the naturalistic point of view is often completely at ease with conclusions which analytic epistemologists seem to have reached reluctantly, with a sense of abnegation and retrenchment. I shall illustrate these claims by examining the epistemological problem of the reliability of perceptual judgments. (Another illustration, concerning the extent to which observation is theory laden, is given in Shimony, 1977.)

Within empiricism there has been a major debate (Swartz, 1965) between sense data theorists (such as Price and C. I. Lewis) and their critics (such as Firth and Quinton). The issue can be posed more generally than it usually is: Are there intermediate steps between the physical stimulation of the sense organs and the final perceptual judgment which justify or support the perceiver's final judgment? The candidates for the intermediate steps are not restricted to sense data. Few, if any, authors dissent from the proposition that there are unconscious causal links, at least partly neural in character, in the total process of perception, but there is disagreement about the significance of these links for claims of knowledge.

Among the arguments presented by sense data theorists, there is one that generically favors epistemic mediation, without specifying its character. The argument is that a perceptual judgment, of the sort that can be expressed in a statement about physical objects, has a content that could not be summarized in any finite body of observational data; the judgment implies, for example, what would be seen if the identified object were viewed under normal conditions from any of an infinite number of vantage points. But a judgment of such strength must be an extrapolation from the actual observational base. That sense data are the entities that mediate between physical stimuli and final perceptual judgments can be demonstrated, according to sense data theorists, in an indirect way by consideration of illusions. Since an imitation often is classified perceptually as a real object of a certain kind, something common must be presented to the observer in the two cases. A reasonable candidate for the common element is *appearance,* which is interpreted as a network of sense data.

This reasoning has frequently been criticized. Quinton (1965), for example, argues that the identification of the locution "This appears to be ϕ" with "There is a ϕ sense datum" is mistaken. He claims that the primary usage of "appears" is not to signal a definite claim to a meager bit of knowledge about the state of the subject but, rather, to signal a tentative belief in a proposition about ordinary objects. He does, however, admit that there is a secondary sense of "appears," which is applicable when persons turn their attention away from things and toward their own awareness – as, for example, in painting impressionistically or taking audiological tests. This possibility of switching to what he calls a "phenomenological frame of mind," however uncommon its occurrence may be, seems at first to support the claim that sense data are revealed directly by introspection. The revelation occurs when one performs upon oneself the operation called "perceptual reduction," in which one attempts to eliminate all elements of experience that are not unequivocally "given" – in particular, all conceptual contributions (Firth, 1965, p. 235).

The critics of sense data theories insist, however, that the phenomenological frame of mind is incompatible with the normal object-perceiving state and that characteristics discerned in the former by introspection cannot legitimately be imputed to the latter (Firth, 1965, pp. 236–237; Quinton, 1965, p. 507). Firth (1965, p. 214) asserts, furthermore, that "sensations do not occur as constituents of perceptions, but at most only as complete and individual states of mind." And Quinton (1965, p. 510) says, "When we transfer our attention from objects to experience, an enormously richer awareness is obtained. We then suppose that we were in fact having experiences of as complex and detailed a kind while attending to the objects, although we were unaware of the complexity and detail. This move is not inference supported by recollection, but a convention."

It follows from this sharp dichotomizing of mental states that perceptual judgments cannot be interpreted as the results of unconscious inference from premises concerning sense data. This conclusion does not require the rejection of unconscious mental processes. Rather, it rests on a sharp distinction between causes and reasons. Whatever the causal importance of the unconscious processes, they cannot be supposed to have the status of premises in the epistemic system of a subject who is operating in the normal mode of perception. Quinton (1965, p. 525) softens this conclusion somewhat: "Perception is an intelligent activity (not an infallible reflex). . . . But not all intellectual processes are types of reasoning." Unfortunately, however, he provides no explanatory details. I suggest that one of the important ways in which empirical psychology can illuminate traditional epistemological debates is by providing some detailed information about the "intelligent activity" at which Quinton hints.

I shall now list five propositions, expressing experimental results and theoretical ideas of recent psychology, which are relevant to the problem of epistemic mediation in perception. It should be noted in advance that the five propositions are far from being straightforward matters of fact. Each of them contains one or more terms – like "information" in proposition 1 and "expectancies" in proposition 4 – which should be clarified by more conceptual analysis than there will be space for in this paper. That these propositions are drawn from different schools is not merely opportunistic, for I am convinced by a synoptic view (see Neisser, 1976, p. 24) which combines elements from information-processing theories, from hypothetico-deductive models (especially Bruner and Gregory), from Campbell's (1956) analysis of perception as substitute trial and error, from theories which emphasize the richness of stimulus information (J. and E. Gibson), and from developmental psychology (especially Piaget).

(1) There are stages in the process of perception which are fairly well defined temporally and describable in such terms as the selection, transformation, encoding, storing, and use of information. Especially relevant for the debate between sense data theorists and their critics is the experimental evidence for a very brief period of iconic storage, in which an image is presented as a whole, as if offered for further examination, and a longer period, in which features are analyzed and the resulting information is stored in a short-term memory (Neisser, 1967, pp. 89–90, p. 301).

(2) The experimental data concerning visual and auditory pattern recognition favor a hierarchically arranged feature-analysis mechanism, rather than, for example, a template mechanism (Neisser, 1967, ch. 3).

(3) Feature analysis is not merely a matter of making decisions but requires a constructive activity of synthesizing a perceptual object: "The mechanisms of visual imagination are continuous with those of visual perception" (Neisser, 1967, pp. 94–95).

(4) The perceptual process is one of active search, in two different ways. First, there is an internal search for appropriate features and appropriate concepts (Bruner, 1957). Second, there is a search of the environment for additional stimulus information to confirm or disconfirm expectancies aroused by earlier perceptual activity (Gibson, 1950; Campbell, 1956; Neisser, 1976).

(5) The generic trait of search for significant stimulus information is innate, since it is exhibited in newborn infants (Bower, 1966). However, the ability to search with increased control, purposefulness, and sensitivity to salient features of the physical stimuli is a skill, which is developed naturally by a growing child interacting with the environment but which can be greatly enhanced by training (Gibson, 1969).

The epistemological importance of these results and theoretical ideas goes beyond adjudicating between the sense data theorists and their critics, though one can score points for and against both sides. It seems to count against the sense data theorists that the stage of perception which is closest to the physical stimulus and least modified by subjective processing is epistemically labile: "After a tachistoscopic exposure in Sperling's . . . experiment, the subject feels that he 'saw' all the letters, but he cannot remember most of them. The uncoded ones slip away even as he tries to grasp them, leaving no trace behind. Compared with the firm clarity of the few letters he really remembers, they have only a marginal claim to being 'conscious' at all" (Neisser, 1967, p. 301). On the other hand, the occurrence of a stage of feature detection as part of normal perception of objects is a point in favor of the *general* position of the sense data theorists that epistemic mediation occurs in perception, with the qualification that the identification of a feature may not be a conscious occurrence. In spite of this qualm, however, the evidence for a temporal fine structure to perception at least indicates greater continuity between the "phenomenological frame of mind" and the normal state of object perception than the critics of sense data theories admitted.

What empirical psychology is beginning to provide is a view of perception as an intelligent activity in a way that none of the opposing schools within empiricism explicitly formulated (though the quotation from Quinton shows that some epistemologists vaguely foresaw the possibility of an extension of the conception of rational process beyond ordinary inference). In this view, the remarkable reliability of perception under normal circumstances – reliability in the sense of agreement on the whole between any confident perceptual judgment and further judgments based on reexamination of the scene; in the sense of consilience among normal observers with similar access to a scene; and in the sense of helpful guidance in practical activity – is the result of an immense amount of critical scrutiny. The scrutiny occurs on two entirely different time scales. One is the time scale of seconds, on which whatever may be called an act of perception is consummated. (In speaking of this time scale, I make no commitment to the view that perception occurs in discrete units; this view has been justly criticized by Gibson and Neisser, and earlier by Dewey; see Ratner, 1939, p. 958 and p. 962.) It is the scale on which both the internal search for relevant features of the stimulus information occurs, and on which there is search of the environment for information which confirms or disconfirms expectations. The other is the time scale of years, during which skills in perceptual search have been developed, to some extent by neural maturation but certainly in large part as a result of trial and error.

The fact that perception involves search, with resulting confirmation or disconfirmation, and often with consequent refinement of expectancies on the short time scale and of search strategy on the long one, can be construed as a vindication of those epistemologists who maintain that perception is an epistemically mediated process. The price of the vindication, however, is a considerable transformation of the conception of mediation. Furthermore, those epistemologists who maintain the opposite view regarding mediation can find some comfort in the fact that perception, like other skills which are improved by practice, has the property

that its analytic components are usually smoothly integrated and suppressed from consciousness.

One cannot stop at this point in adjudicating, however, since there
is at least one other constituent in the "intelligence" of perception. The
smooth integration which normally occurs when expectations are fulfilled
can be interrupted by disconfirmations or by various kinds of challenges,
such as the one posed by a painter who is trying to replicate an optical
effect or by an epistemologist who is trying to understand the structure of
his knowledge. The ease of such interruption and the ease of resumption
of the integrated sequence are further evidence that the phenomenological frame of mind and the normal state of object perception are not as
disjoint as the critics of sense data theories maintained. (It may be relevant to note that a common pedagogical technique for developing skills
other than perception is to exhibit the elements of a temporal sequence
separately – for instance, musical phrases or parts of a tennis stroke –
and then to coach the student into smoothly integrating them into a larger
unity.) The possibility of consciously verifying whether it is appropriate
to attribute a certain feature to the stimulus information and of consciously searching the environment for confirming or disconfirming information concerning a tentative identification is one of the major ways
in which perception is distinguished from reflex activities. In this way a
kind of control is exercised by the subject as a whole – whatever that may
mean – over the perceptual process, with some similarity to the control
exercised in inference and problem solving, which are considered exemplary instances of intelligent activity.

This is a good place to point out that my invocation of empirical psychology for epistemological purposes differs in important respects from
Quine's. According to Quine (1969, pp. 82–83), "Epistemology . . . studies a natural phenomenon, *viz.*, a physical human subject. This human
subject is accorded a certain experimentally controlled input – certain
patterns of irradiation in assorted frequencies, for instance – and in the
fullness of time the subject delivers as output a description of the three-
dimensional external world and its history." Quine regards the observer
as a black box whose perceptual processes are reliable if his reports are
correct and his behavioral responses adequate, as judged by third persons who take for granted the scientific description of the physical world.
This way of construing naturalistic epistemology is inseparable from a
commitment to behaviorism in psychology and seems to me excessively
narrow.

The cognitive psychology which I have invoked, in order to show that
the reliability of ordinary perceptual judgments is well grounded in the
critical scrutiny exercised by normal subjects, aims at "getting inside the

black box," at least in the minimal sense of offering a "subpersonal theory" (see Dennett, 1978) analogous to microphysics, and perhaps in the further sense of linkage with introspection. In giving an account of the structure of my knowledge, I wish to be free to switch back and forth between a first- and third-person point of view. The ability to take a third-person point of view of myself is not an achievement of epistemological sophistication but, rather, the by-product of socialization and of the mastery of a public language; and it is, of course, a most remarkable thing, which we are far from completely understanding. The ability to take a first-person point of view is so fundamental that it would seem not to need a defense, except for the challenge to its scientific respectability by various forms of behaviorism; and the best defense on the same plane as the challenge is to refer to a cognitive psychology, which is a well-articulated and fruitful alternative to behaviorism.

EVOLUTIONARY CONSIDERATIONS

The discussion of the "closing of the circle" of epistemology and knowledge of the world can be carried out at different levels, with different degrees of fineness of description. At one level – the level at which one studies a human being functionally, as an organism interacting with its environment and endowed with strategies for pursuing vital goals – the closing of the circle is quite well established. We can see, from a variety of psychological studies, *that* human cognitive strategies are effective for practical purposes and, to some extent, *how* they are effective. It is possible to discuss these questions without inquiring into the genetic question of how human beings came to be endowed with such effective cognitive strategies and with apparatus for carrying them out. The genetic question obviously lurks in the background, however; indeed, one reason why it was not obtrusively disturbing is that almost all readers, and certainly all who follow Campbell, take for granted an evolutionary answer. *Some* answer to this question – either the evolutionary one or a surrogate – is necessary if we are to know where human beings fit in nature and therefore to accomplish the "closing of the circle" at a level deeper than that of practical functioning. (I should add, however, in view of the difficulty posed by the mind–body problem, that an evolutionary answer to the genetic question does not ipso facto "close the circle" at the deepest level. Instead, we are confronted with the problem in a new guise: how is it that creatures endowed with the consciousness that we know ourselves to possess could have evolved from ancestors apparently not so endowed?)

Another reason why the question of the origin of human cognitive faculties did not arise obtrusively in the previous section is that we were concerned with the reliability of perception, which is a practical matter, or certainly can be so understood, given the customary usage of the word *reliability.* A perceptual judgment is reliable if on the whole it is confirmed by further experience, if on the whole it agrees with the judgments of other well-placed and competent observers, and if on the whole it can serve as a basis for actions which favor the interests of the perceiver. That perception is reliable in this sense is a commonplace, shared by a number of philosophers, including some who are not naturalistic. In particular, philosophies which offer radically different answers to the genetic question may concur regarding the reliability of perception in the sense stated. Berkeley, for example, repeatedly insists that his immaterialism does not undermine the commonsense view of the world, and indeed he claims to provide a better explanation of the accessibility of the commonsense world to knowledge by perception than do the advocates of a mysterious material substratum.

Nevertheless, different answers to the genetic question are linked with different epistemological positions, but the linkage concerns other problems than those considered so far; for example, the extent to which perception reveals properties of material objects which are independent of the cognitive relation, and the extent to which the structure of experience is imposed by the mind of the subject. I agree with Campbell that, in assessing the various solutions offered to these problems, one must take into account the overwhelming evidence for an evolutionary history of human cognitive faculties. This evidence makes very implausible the Kantian thesis that the mind legislates to nature. Furthermore, the evolutionary point of view permits a naturalistic explanation of the pervasiveness of the concepts of substance and causality in experience, contrary to Kant's thesis that we can recognize the operations of the understanding but cannot give a causal explanation for these operations. (According to Kant, an attempt to give such a causal explanation would be an instance of applying the category of causality to the things in themselves.) As Lorenz has argued, with Campbell's enthusiastic concurrence (1974, pp. 441–447), a lineage of animals which evolves in a physical environment inhabited by relatively stable macroscopic objects and governed by macroscopic regularities would benefit from a propensity to apply the categories of substance and causality to the sensory content of experience. The general lines of this argument are independent of the detailed nature of the propensity – whether it is ontogenetically innate, or requires neurological maturation, or requires some empirical input to consummate the imprinting. The evolutionary point of view suggests a less rigid role of the categories of the understanding than the one to which Kant is committed because of his theory of the *constitutive* function of the understanding.

Insofar as flexibility in applying the categories seems to be required not only to account for the heterogeneous intermingling of regularity and randomness in ordinary experience but also to fit the picture of the world provided by modern physics, the evidence seems to favor a naturalistic rather than a transcendental explanation of a priori elements in experience.

Our concern with the structure of epistemology, however, should make us wary of an easy victory for naturalism. It may reasonably be objected that a question has been begged when the theory of evolution and other theories with which it is linked, such as historical geology and parts of the physical sciences, are accorded a realistic interpretation. If scientific theories are interpreted nonrealistically, merely as means for "economy of thought" or as instruments for anticipating experience of the ordinary world or even one's personal sensations, then their significance for adjudicating epistemological arguments becomes negligible. Consequently, the appeal to evolutionary and other advanced scientific theories is illegitimate in epistemology without an antecedent analytic argument that these theories should be interpreted realistically. The objection can then continue that strong arguments raised by Berkeley, Hume, Mill, James, and the positivists suggest that a realistic interpretation of scientific theories is either meaningless, or else possibly meaningful but in principle undemonstrable by human beings, or else indistinguishable in cognitive content from sufficiently sophisticated phenomenalistic interpretations.

This issue is intricate, and evidently I cannot do justice to it in a part of a section of a paper which aims only at giving an overview. Fortunately, I feel that I can respectably abstain from a detailed discussion of the issue, since Skagestad examines it in this volume and I have written about it elsewhere (Shimony, 1971). A rough outline of my position is the following.

The arguments against realistic interpretations depend on the adoption of quite narrow criteria of meaningfulness and of evidential support. There are no strict and compelling reasons for the adoption of these narrow criteria, and there is a general metaphilosophical or dialectical reason for adopting broader criteria at least at the outset of inquiry: specifically, Peirce's maxim that the way of inquiry should not be blocked (thesis 2, presented earlier). Given broader criteria, one has the possibility of bringing relevant evidence to bear on realism, instrumentalism, and other rival interpretations of scientific theories. The possibility is also kept open that the initially broad criteria of meaningfulness and evidential support will be tightened as a consequence of inquiry, perhaps for reasons of scientific fruitfulness or of coherence (the "closing of the circle" of epistemology and knowledge of the world). If such tightening should occur, it would be nonarbitrary, in contrast to an initial stringency of criteria, which has the effect of hedging a favored philosophical position against evidence. The final step in the argument is that when realism, instrumentalism, and phenomenalism are placed on the same footing as candidates for the interpretation of scientific theories, none being excluded and none favored a priori, then realism is overwhelmingly supported a posteriori.

Needless to say, the foregoing analysis is excessively condensed, but at least it illustrates a major claim of integral epistemology: that naturalistic, analytic, and dialectical considerations must be intertwined if one is to do justice to the peculiar difficulties of the central problems of epistemology.

INDUCTIVE INFERENCE AND VINDICATORY ARGUMENTS

Concern for the structure of epistemology requires that at least a few statements be made about inductive inference, even though they will be inadequate for a complicated subject. Campbell explicitly states that one of the presuppositions of his descriptive epistemology is the validity of inductive inference, and he accepts Hume's skepticism with regard to the possibility of justifying this presumption. One might have expected that he would attempt, as some other naturalistic epistemologists have done, to justify this presupposition by evolutionary considerations, along the following line: (1) Something like inductive inference has been practiced throughout the evolutionary process. (2) In particular, long before the articulation of the scientific method, the higher animals and primitive humans developed (by trial and error) the strategy of proposing hypotheses and testing them by observations. (3) The success in the biological sense of a population which has adopted this strategy shows that it is a good strategy to employ in the world as it actually is constituted, and this fact justifies inductive inference. Actually, Campbell does assert (1) and (2), but he refrains from (3). He does so, I believe, because he realizes that the transition to (3) is itself an instance of inductive inference, the validity of which is under examination. In other words, Hume's criticism of the attempt to justify inductive inference by its past successes applies equally well to arguments which appeal to evolutionary evidence.

If Hume is not to have the last words, several strategies are open. One is to make full use of the apparatus of probability theory and decision theory, which themselves are defensible on analytic or a priori grounds. Another is to concede that any kind of inductive justification of induction involves some circularity but to reason that some types of circularity are nonvicious, in the sense that they do not foreclose the issue under examination and do not hedge a favored position against adverse evidence. This strategy is clearly in the spirit of the dialectical approach to epistemology advocated as a meta-theory in an earlier section (headed

'Some Dialectical Theses') and followed elsewhere in this paper. The third strategy is to employ a vindicatory argument, as sketched in thesis 4 of that section. My own opinion is that all these strategies are needed and must be used in tandem. I am not satisfied that any methodologist has worked out the details of how the meshing of these strategies is to be accomplished, but I have advocated a program along the following lines, which seems to me to be in the right direction (Shimony, 1970, 1976).

Briefly, I believe that the theory of inductive inference has an a priori component, consisting of personal probability theory and decision theory, and making free use of deductive logic. The personal probability evaluation should be qualified by a maxim of critical open-mindedness, which is called "the tempering condition"; that is, give sufficient prior probability to any seriously proposed hypothesis that it may achieve high posterior probability if favored by the data.

A vindicatory argument can be given as follows for this maxim. If we are not optimistic about human ability to make good conjectures, then seriously proposed hypotheses are coordinate with frivolously proposed ones and with merely logically possible ones that no one has ever singled out for mention. But concerning any nontrivial matter of interest, particularly concerning generalities, infinitely many rival hypotheses are logically possible, and it is hopeless to explore all of these. Our only hope of approaching the truth concerning nontrivial matters is to be optimistic about human abductive power, in the sense of permitting a seriously proposed hypothesis to be accepted after experimental scrutiny. This optimism precludes no particular method for assigning an order to hypotheses, since the proposal of an order would itself be a special case of abduction. Consequently, nothing is lost and possibly something is gained – in accordance with the general line of vindicatory arguments – if our subjective probability evaluations are made in accordance with the tempering condition. The tempering condition as stated, however, is manifestly vague, and attempts to eliminate this vagueness by a priori explications of the phrases "seriously proposed" and "sufficient prior probability" are likely to be arbitrary.

A more promising procedure is to use a posteriori considerations in explicating these phrases. By reflecting on exemplary scientific discoveries in the past, one can recognize types of hypotheses which deserve prima facie attention – meaning, of course, no more than a moderate degree of prior probability, remote from full credence. Circularity obviously enters into the proposed epistemological structure, since there is reliance on exemplary scientific achievements, and these in turn could have been identified only by inductive inference. The circularity is nonvicious, since a seriously proposed hypothesis of the past can be subjected to experimental scrutiny and as a result may acquire so low a posterior probability that it is discredited. Furthermore, there is an openness in the procedure in spite of the recourse of past achievement, for a radically new hypothesis can be accorded the status of "seriously proposed" for many reasons; it may, for example, satisfy the standards of exposition of exemplary hypotheses of the past, in spite of the novelty of its content.

This program for justifying inductive inference evidently needs much further clarification and exploration. However, space permits only two last remarks, which bear on the structure of epistemology. The first is to reiterate the indispensability of analytic considerations in an integral epistemology. Without probability theory and decision theory, and without the appeal to a vindicatory argument, there seems to be no alternative to skepticism regarding the justification of induction. The second remark is that what Campbell calls "descriptive epistemology" has a role in the justification of induction, as it does in other epistemological matters. But it appears that the term should be extended, so as to apply not only to results from the natural sciences but also to some historical information. As far as we can reconstruct the reasoning process of prescientific people, we find a mélange of critical and noncritical elements. Along with *modus tollens* and rudimentary probability theory, there is uncritical use of simple inductions and very uncritical reliance on analogies and anthropocentric explanations. If one argues from the biological success of Homo sapiens to the reasonableness of the genetic thought processes of man, then one has an undiscriminating justification of the whole package of primitive thought processes, critical and uncritical together. The a posteriori considerations which permit us to separate out different components of the package and to assess them are considerations concerning not the evolutionary development of the race but, rather, the history of science. The improvement of primitive inductive processes is thus a cultural achievement. It is, however, a very special cultural achievement, since it is the by-product of search for the truth about nature. There is no better illustration of a dialectical structure in epistemology than the refinement of inductive inferences by reflection on exemplary results of science, which themselves were achieved inductively.

REFERENCES

Bower, T. G. R.: 1966, 'The Visual World of Infants.' *Scientific American* **215**(12), 80–92.
Bruner, J. S.: 1957, 'Perceptual Readiness.' *Psychological Review* **64**, 123–152.
Campbell, D. T.: 1956, 'Perception as Substitute Trial and Error.' *Psychological Review* **63**, 330–342.

Campbell, D. T.: 1959, 'Methodological Suggestions from a Comparative Psychology of Knowledge Processes.' *Inquiry* **2**, 152–182.

Campbell, D. T.: 1974, 'Evolutionary Epistemology.' In P. A. Schilpp (Ed.), *The Philosophy of Karl Popper*. La Salle, Ill.: Open Court.

Campbell, D. T.: 1977a, Introductory comments to a lecture on Evolutionary Epistemology to the Boston Colloquium for the Philosophy of Science, February 22.

Campbell, D. T.: 1977b, 'Descriptive Epistemology: Psychological, Sociological, and Evolutionary.' William James Lectures, Harvard University.

Dennett, D. C.: 1978, 'Toward a Cognitive Theory of Consciousness.' In C. W. Savage (Ed.), *Minnesota Studies in the Philosophy of Science*. Vol. 9: *Perception and Cognition: Issues in the Foundations of Psychology*. Minneapolis: University of Minnesota Press, pp. 201–228.

Dretske, F.: 1971, 'Perception from an Evolutionary Point of View.' *Journal of Philosophy* **68**, 584–591.

Firth, R.: 1965, 'Sense-Data and the Percept Theory.' In R. J. Swartz (Ed.), *Perceiving, Sensing, and Knowing*. New York: Doubleday, 1965, pp. 486–496.

Gibson, E. J.: 1969, *Principles of Perceptual Learning and Development*. New York: Appleton-Century-Crofts.

Gibson, J. J.: 1950, *The Perception of the Visual World*. Boston: Houghton Mifflin.

Gregory, R. L.: 1970, *The Intelligent Eye*. New York: McGraw-Hill.

Heffner, J.: 1981, 'Causal Relations in Visual Perception.' *International Philosophical Quarterly,* **21**, 303–332.

Neisser, U.: 1967, *Cognitive Psychology*. New York: Appleton-Century-Crofts.

Neisser, U.: 1976, *Cognition and Reality*. San Francisco: Freeman.

Peirce, C. S.: 1931, 'Notes on Scientific Philosophy.' In C. Hartshorne and P. Weiss (Eds.), *Collected Papers of Charles Sanders Peirce*. Vol. 1. Cambridge, Mass.: Harvard University Press. (Original manuscript 1897.)

Price, H. H.: 1932, *Perception*. London: Methuen.

Quine, W. V.: 1969, 'Epistemology Naturalized.' In *Ontological Relativity and Other Essays*. New York: Columbia University Press, pp. 69–90.

Quine, W. V.: 1974, *The Roots of Reference*. La Salle, Ill.: Open Court.

Quinton, A. M.: 1965, 'The Problem of Perception.' In R. J. Swartz (Ed.), *Perceiving, Sensing, and Knowing*. New York: Doubleday, pp. 497–526.

Ratner, J. (Ed.): 1939, *Intelligence in the Modern World: John Dewey's Philosophy*. New York: Random House.

Rescher, N.: 1977, *Methodological Pragmatism*. Oxford: Blackwell.

Russell, B.: 1921, *The Analysis of Mind*. London: Allen and Unwin.

Shimony, A.: 1970, 'Scientific Inference.' In R. G. Colodny (Ed.), *The Nature and Function of Scientific Theories*. Pittsburgh: University of Pittsburgh Press, pp. 79–172.

Shimony, A.: 1971, 'Perception from an Evolutionary Point of View.' *Journal of Philosophy* **68**, 571–583.

Shimony, A.: 1976, 'Comments on Two Epistemological Theses of Thomas Kuhn.' In R. S. Cohen, P. K. Feyerabend, and M. W. Wartofsky (Eds.), *Essays in Memory of Imre Lakatos*. Dordrecht, Netherlands: Reidel.

Shimony, A.: 1977. 'Is Observation Theory-Laden? A Problem in Naturalistic Epistemology.' In R. G. Colodny (Ed.), *Logic, Laws, and Life: Some Philosophical Complications*. Pittsburgh: University of Pittsburgh Press, pp. 185–208.

Swartz, R. J. (Ed.): 1965, *Perceiving, Sensing, and Knowing*. New York: Doubleday.

COMMENT

At the end of the first section of this paper I note that there are several contemporary philosophers with an affinity to dialectical naturalism, even if they do not use this phrase. I have since then found that Richard Boyd explicitly advocates both components of the phrase. Section 2 of his paper "Scientific Realism and Naturalistic Epistemology" (in *PSA 1980,* ed. P. Asquith and R. Giere, East Lansing, MI: Philosophy of Science Association, 1981, pp. 613–62) is entitled: "Outline of a Naturalistic and Dialectical Version of Scientific Realism."

Three interesting criticisms of the general point of view of this paper are given in *Naturalistic Epistemology: A Symposium of Two Decades,* ed. A. Shimony and D. Nails (Dordrecht: Reidel, 1987). The first is by Paul Sagal, "Naturalistic Epistemology and the Harakiri of Philosophy," pp. 321–32, with a comment by Joseph Agassi, pp. 337–40, and one by me, pp. 333–34. The second is by Joseph Agassi, "Naturalistic Epistemology: the Case of Abner Shimony," pp. 341–51, particularly analyzing the section entitled "Some Dialectical Theses"; my reply is on pp. 352–55. The third is by Marx Wartofsky, "Epistemology Historicized," pp. 357–73, with my reply, pp. 375–77.

2

Reality, causality, and closing the circle

I. PHILOSOPHICAL PERSPECTIVE

The literature of natural science freely employs locutions that prima facie refer to entities – for example, to atoms, the electromagnetic field, a galaxy, black holes, a DNA molecule, a neuron. Scientists speak of properties and states of these entities, of the temporal development of their states, of their interactions with other entities, and of their detection by suitable instruments. Contemporary philosophers of science have intensively discussed metaphysical questions about the ontological status of these putative entities, epistemological questions about cognitive claims regarding them, and semantical questions about the reference of scientific terms and the truth of scientific statements. Although much of the argumentation on these questions is analytically subtle, it is typically not carried out in the context of a broad philosophical world view. This feature of current philosophy of science may be regarded as a virtue by those who are generally suspicious of systems of philosophy, but I dissent.

The purpose of this paper is to discuss various aspects of the problem of realism, especially the ontological status of putative scientific entities, in a broader and I hope more systematic way than usual. I propose to relate the problem of realism to a program which is familiar in systematic philosophies of the past: to understand the knowing subject as an entity in nature and to assess claims to knowledge in the light of this understanding. Such a program aims at the integration of epistemology with the natural sciences and metaphysics. It intends to show how claims to human knowledge of the natural world can be justified, and in turn how the resulting view of the world can account for the cognitive powers of the knowing subject. For brevity I shall refer to this program as "closing the circle." The paradigm is provided by Aristotle, whose *Physics* and *Metaphysics* purport to show that the formal cause of a substance determines its normal behavior, while *On the Soul* purports to demonstrate that the active intellect by its essence has the power to grasp the forms of substances presented by sensory experience, thereby justifying the methodology of intuitive induction of the *Posterior Analytics*.

The integration of Aristotle's philosophy should be recognized as admirable structurally, regardless of the weaknesses of the components. The program of closing the circle is explicit or implicit in the philosophies of Plato, Descartes, Spinoza, Leibniz, Hegel, Peirce, and Whitehead and in variant form in that of Berkeley (who regards minds as entities, but not in what he calls "nature"). Most of the recent philosophers of science who recognize the importance of the program of closing the circle are naturalistic epistemologists, insisting upon the relevance of biological and psychological discoveries to the assessment of claims to knowledge. There are, of course, numerous variants of naturalistic epistemology, and my own is atypical in maintaining that there is an irreducibly mentalistic component in nature, in meshing "first-person" and "third-person" considerations in epistemological analysis, and in doubting the literal applicability of the ideas of Darwinian natural selection to the evolution of science.

It is an open question whether the closing of the circle can be achieved. In Sections II and III some strong considerations will be given for a positive answer, especially against arguments of Kant and Putnam that in principle it is impossible. The analysis of their arguments will require a fairly extensive examination of the concept of causality and of the mind–matter relationship. The position that I arrive at in Section IV concerning ontology would probably be classified as a kind of scientific realism, according to most of the criteria for that affiliation (e.g., Hooker 1987, p. 256), but with reservations that are not found in most formulations of scientific realism.

It is noteworthy that one of the most influential critics of scientific realism, Hilary Putnam, recognizes its relation to the program of closing the circle. He claims to link the impossibility of a "totalistic explanation," which is his pejorative phrase for closing the circle, to the impossibility of the metaphysical realist thesis that the world consists of a fixed totality of mind-independent objects.

I can symphathize with the urge to *know,* to *have* a totalistic explanation which includes the thinker in the act of discovering the totalistic explanation in the totality of what it explains. . . . But I am saying that the project of providing such an explanation has failed.

It has failed not because it was an illegitimate urge – what human pressure would be more worthy of respect than the pressure to *know?* – but because it goes beyond the bounds of any explanation that we have. Saying this is, perhaps, not putting the grand projects of Metaphysics and Epistemology away for good – what another millennium, or another turn in human history as profound as the Renaissance, may bring forth is not for us today to guess – but it is saying that the time has come for a moratorium on Ontology and a moratorium on Epistemology. Or rather, the time has come for a moratorium on the kind of metaphysical speculation that seeks to describe the Furniture of the Universe and to tell us what is Really There and what is only a Human Projection, and for a moratorium on the kind of epistemological speculation that seeks to tell us the One Method by which all beliefs can be appraised. (1990, pp. 117–18)

The thesis in this passage on the limits of human cognition is made more explicit elsewhere, for instance, "What I am saying, then, is that elements of what we call 'language' or 'mind' penetrate so deeply into what we call 'reality' that the very project of representing ourselves as being 'mappers' on something 'language-independent' is fatally compromised from the start" (*ibid.*, p. 28). Putnam's acknowledgment of limitations of human cognitive powers places him in the tradition of Kant, as he recognizes explicitly (e.g., 1990, p. 41 and p. 162). He is not driven to relativism or skepticism, but rather to what he calls "internal realism" (e.g., 1987, p. 7) or "realism with a small 'r'" (*ibid.*). "It *is* a kind of realism, . . . a belief that there is a fact of the matter as to what is rightly assertible for us, as opposed to what is rightly assertible from the God's eye point of view so dear to the classical metaphysical realist" (1983, p. xviii).

If it could be demonstrated that the closing of the circle is in principle an impossible ideal, then there would be momentous consequences for the problem of scientific realism, and more generally for epistemology and metaphysics. A retrenchment of the aims of philosophical inquiry, perhaps along the lines of Putnam's internal realism, might then be a reasonable strategy. There do exist instances of demonstrations that a project is impossible in principle, notably the impossibility of algorithms for solving certain classes of mathematical and logical problems. But these demonstrations depend upon explicitly characterizing the task that is to be performed and the means which are allowed – requirements that clearly are not met in epistemology and metaphysics. The main value of a less than definitive proof of the impossibility of closing the circle is an unintended heuristic: to search for the meshing of epistemology and metaphysics along other avenues than those which have led to an impasse.

This paper is written in the optimistic spirit of Bacon's *Magna Instauratio*. In a famous passage preceding his constructive methodological proposals he warns of "Idols" or false notions that distract the human mind from knowing the truth about nature. Of the four types of Idols – of the Tribe, of the Cave, of the Market-Place, and of the Theatre – it is the first that is most relevant to my investigation:

The Idols of the Tribe have their foundation in human nature itself and in the tribe or race of men. For it is a false assertion that the sense of man is the measure of things. On the contrary, all perceptions as well of the sense as of the mind are according to the measure of the individual and not according to the measure of the universe. And the human understanding is like a false mirror, which, receiving rays irregularly, distorts and discolours the nature of things, by mingling its own nature with it. (1937, Aphorism XLI, pp. 278–79)

Bacon was sanguine that his method of induction would suffice to identify and correct the distorted representation of nature caused by generic imperfections of the human senses and understanding (and also correct the other Idols, which are due to the idiosyncratic distortions of individual minds and the fallacies originating in language and culture). Today the weaknesses of Bacon's inductive logic are well known. Furthermore, as the heirs of profound analysts of human cognitive faculties, like Kant and James, we can hardly avoid the judgment that Bacon naively underestimated the pervasiveness of subjective contributions to human representations. Nevertheless, it may be possible to vindicate Bacon's optimism that the Idols of the Tribe, and other Idols, can be identified and corrected, though the task is sure to require much more than Baconian methodology. The disciplines of neurophysiology, perceptual psychology, cognitive psychology, and related studies, supplemented by the historical study of scientific methodology, have already yielded much information on how "the human understanding . . . distorts and discolors the nature of things," and there may be no subjective source of error that cannot in principle be detected by scientifically studying the cognitive apparatus itself. If the operations of the mind, considered as a natural system but operating under the constraints of a culture, are well understood, this knowledge can be applied to the fine-tuning of the methods of investigation themselves. If so, the program of closing the circle can help to realize the vision of Bacon's instauration.

II. KANT AND THE LIMITS OF CAUSALITY

In the passage by Bacon just quoted, there is an obvious assumption that both the objects studied in nature and the faculties of the knowing subject are causally related to "perceptions as well of the sense as of the mind." The correct scientific method can disentangle the intertwined strands of causality, separating those which cause distortions in the mirror of the understanding from those which produce an accurate reflection of nature. But this ambitious epistemological project is irredeemably flawed if causality itself is one of the Idols. To say that causality is one of the Idols of the Tribe is a thoroughly un-Kantian way of paraphrasing the Kantian doctrine that causation is one of the categories imposed by the understanding upon experience. The paraphrase seems to ascribe a pervasive skepticism to Kant, whereas he is not at all skeptical concerning knowledge within the proper domain of application of the categories. Kant offers a remarkable quid pro quo: in return for acquiescing to transcendental idealism, which renounces theoretical knowledge of things in themselves, he offers "empirical realism," which includes a refutation of Berkeley's immaterialism, an answer to Hume's skeptical doubts on induction, and a rich body of synthetic a priori judgments concerning the phenomenal domain (1929, A370ff).

For the purposes of this paper it is essential to review some arguments against Kant's doctrine of the limited applicability of causality, even though most or all of them are familiar in the literature, because the alleged inapplicability of causality to the things in themselves is an obvious threat to the program of closing the circle. There will be no need, however, to summarize in any detail Kant's doctrine and argumentation concerning causality, or to be drawn into questions of textual exegesis. I shall assume a general acquaintance with Kant's position and shall make free use of the relevant texts in the course of assessing it.

Causality, like all the categories of the understanding, is constitutive of experience by imposing a priori constraints upon appearances (1929, A161–2/B201). The most striking of these rules of constraint is the second Analogy of Experience, "All alterations take place in conformity with the law of the connection of cause and effect" (*ibid.,* A189/B232). Kant's discussion shows that the second Analogy is intended to ensure not only that there must be a temporal relation of cause and effect among appearances, but that this succession is deterministic, so that "a principle of sufficient reason is thus the ground of possible experience" (*ibid.,* A201/B246). In spite of these claims, however, it has been frequently noted that the actual function of the law of connection of cause and effect is regulative – that a law of connection should be sought because it is there to be found (cf. Walsh 1975, p. 99).

So far, however, no decisive objection has been given, because Kant consistently claims that the understanding only provides the form of experience, while sensation supplies the matter (1929, A20/B34), and the details of the law of connection presumably come from empirical data. But when one undogmatically examines the phenomenology of experience, the situation is much more complex than Kant's division of labor between sensation and the understanding would lead one to expect, and the complexities are hard to reconcile with the general view of transcendental idealism. There is a wide spectrum of instantiations of causality: in some (especially involving our own bodies), causal connections are inescapable; in some they are not inescapable and yet are relatively easy to find; in some, all relevant factors are directly observable and nevertheless clever experimentation and analysis are needed in order to disentangle strands; in some, elaborate theories and indirect confirmations are needed in order to reveal causal connections remote from direct observation; in some, chance seems to reign in spite of underpinnings that are known theoretically to be deterministic (situations of chaos); and in some there are deep considerations supporting the nonexistence of a deterministic underpinning. There is, to my knowledge, no natural way to explain in Kantian terms the diversity of this spectrum. But if transcendental idealism is abandoned, there is a natural scenario: that causality is a constraint upon the temporal evolution of things in themselves, a constraint which varies in strictness in different situations and varies in accessibility to the human cognitive system for diverse reasons, such as the strength of interaction, the specialization of human sensory receptors, and the presence of noise.

A different problem concerning causality is raised by Kant's fundamental epistemological principle that the "matter" (or sensuous content) of appearances does not originate in the knowing subject as do the forms of the understanding, but is received by sensation, which is "the effect of an object upon the faculty of representation" (*ibid.*, A20/B34). Since the object referred to in this passage must be something beyond appearances, Kant seems to be applying the concept of causality to the interaction of a thing in itself with the knowing subject, which is another thing in itself. One might try to clear Kant of an apparent contradiction by saying that the word "effect" is loosely used, without the intention of a causal connection. But the lapse of wording is suspiciously appropriate, because constraint is exercised upon the state of the knowing subject by whatever it is that supplies the matter of appearances, and constraint is a kind of causality.

Furthermore, Kant admits the application of the categories, including causality, to empirical psychology, "which would be a kind of *physiology* of inner sense, capable perhaps of explaining the appearances of inner sense" (*ibid.*, A347/B405). He carefully distinguishes the self studied in this way from the transcendental self, which is the subject present in all knowledge, known only through the forms that it imposes upon experience. There is undoubtedly great subtlety in Kant's discussion of the self, for example in his acute criticism of Descartes's claims for the certainty and primacy of self-knowledge (*ibid.*). Nevertheless, there remain baffling questions about the exact relation between the phenomenal and transcendental selves. What is most subversive to Kant's epistemology is that investigation of the phenomenal self by the methods of the cognitive scientists has yielded information that overlaps, and to a considerable extent refines and corrects, the results which Kant's critical method yields concerning the operations of the transcendental self. Piaget's developmental psychology, for instance, seems to cut across the line between the phenomenal and transcendental selves. If the roles of these selves seem to merge, then

Kant's doctrine that the phenomenal self is subject to the categories of the understanding undermines his doctrine that they do not apply to the transcendental self.

Finally, the Kantian restriction of the category of causality to the phenomenal domain is seriously weakened by a generic failure of the fundamental quid pro quo of transcendental idealism. The reward for abandoning all pretense to theoretical knowledge of things in themselves is supposed to be a body of synthetic a priori knowledge, arising from the forms which the intuition and the understanding impose upon experience. Or the point can be inverted, as Kant does in the *Prolegomena* (Kant 1947), by arguing that the undoubted possession of synthetic a priori knowledge in mathematics and pure natural science can be explained only by the attribution of ordering principles to the mind itself. But the developments in mathematics and physics in the last century and a half have undermined the plausibility of any claim to synthetic a priori knowledge. And when this bonus is withdrawn, the motivation for acquiescing to a renunciation of theoretical knowledge of the things in themselves collapses. The Kantian tradition remains influential, but claims of limitations in principle on human knowledge from philosophers who have scrapped the Kantian machinery of space and time as forms of intuition and of causality as a form of the understanding are, to my mind, pallid both in content and in plausibility in comparison with the Kantian original.

If the knowing self is causally connected with other entities, and the phenomena of experience are the effects of interactions governed by causal laws, then the Kantian locution of "transcendental self," connoting that its properties are neither directly nor indirectly revealed by the data of experience, becomes entirely inappropriate. The way is open to hypothetico-deductive inferences about the intricate machinery of cognition, consisting *inter alia* of the means by which physical stimulations of the sensory receptors give rise to neural signals, the mysterious transducers whereby these cause sensations, the operations of memory and imagination, and the apparatus whereby concepts are applied to percepts. Furthermore, the machinery of cognition can be studied both ontogenetically and phylogenetically. In other words, the way is open to a naturalistic study of mind, and the program of closing the circle of epistemology and metaphysics can be pursued by the methods of the natural sciences. There has been a flowering in recent decades of naturalistic and evolutionary epistemologists who are sanguine about the prospects of a causal analysis of cognition (Campbell, Quine, Popper, Vollmer, Wuketits, Hooker, and many others). But they may be prematurely optimistic. The Kantian renunciation of a causal analysis of things in themselves is not the only obstacle to the program of closing the circle. Regardless of the wealth of psychophysical correlations that are found in ordinary life and explored by neurologists, pharmacologists, and psychologists, there may be crucial aspects of the mind–matter relation which are impervious to causal analysis. So claims Putnam, in the argument against a completely naturalistic treatment of mind which will be considered in the following section.

Before proceeding further in the analysis of the knowing subject it is important to comment on the complexity of the concept of causality. If the transcendental philosophy of Kant were correct, then formal principles constraining experience would be known a priori, among them the Second Analogy of Experience (cited previously), which prescribes that temporal alterations of phenomena are deterministic, proceeding according to "rules" or laws (1929, A199/B244). Rejecting the transcendental philosophy opens the possibility that things in themselves alter in time. The laws constraining their temporal alteration are to be found by hypothetico-deductive reasoning, that is, by inferences from evidence which itself is the result of interaction between things in themselves and the knowing subject. Not only the details of these laws, but their general character as well, are to be determined by inference from experience. In other words, once the transcendental philosophy is set aside, one cannot give more than a rough provisional characterization of causality a priori, as a "constraint upon temporal evolution," and even the modalities of constraint remain open to empirical investigation. The development of physics has revealed some remarkable and unforeseen information about these modalities of constraint. The special theory of relativity imposes a limit to the velocity of propagation of influences, by requiring the causal independence of events with spacelike separation. Quantum mechanics in its quotidian application (where difficulties about the reduction of the wave packet are disregarded) treats physical processes stochastically, with the ranges of possible outcomes governed by probability distributions. Furthermore, there is some tension between the quantum-mechanical and relativistic treatments of causality, since entangled quantum states entail causal connections between certain events with spacelike separation, though not in a way that permits superluminal communication; it is generally recognized that the situation is unsatisfactory and calls either for a modification of current physical theory or for deeper conceptual analysis of existing theories (Cushing and McMullin 1989, *passim*). Finally, there are speculations of various kinds concerning the fundamental nature of physical causation. The concept of a temporally invariant general law which constrains temporal contingencies was preserved in the transition from classical to modern physics, but Dirac (1938) was prompted by a remarkable agreement between one number occurring in cosmology and another occurring in elementary particle physics to propose that one or more of the so-called "constants" of nature changes with the age of the universe. More radically, Peirce (1935, pp. 15–16, p. 84) and Wheeler (1983, 1985) have speculated about the evolution of the forms of laws out of an initially chaotic universe. There are also speculations that the primitive type of causality is not a general law but a constraint exercised by one individual entity upon another (e.g., Whitehead's concept of prehension, to be discussed in Section III); law-governed causality is then regarded as something derivative, resulting from the statistics and the cooperation of innumerable instances of primitive causality. The time may not be ripe for judging whether such speculations are susceptible of controlled experimental and analytic investigation, but they are effective reminders of the continuity of modern physics with metaphysics.

One important aspect of causality that is relevant to the concerns of this paper is entirely independent of these deep speculations. It often happens that laws govern a composite entity without explicit reference to its composition. Sometimes one speaks of the "scale of nature" (Tisza 1963, Weisskopf 1979), according to which there are entities of a certain size, composed of well-defined entities of a smaller size, et cetera (but with no commitment to ad infinitum), with laws formulated autonomously at each level. In physics one has the sequence of macroscopic bodies, atoms, electrons and nucleons, and quarks and gluons within a nucleon; in biology there is the sequence of macro-organism, cell, organelles, and constituent macromolecules; and other sciences provide analogous sequences. The relations among laws at a coarse level in the scale of nature and laws on a finer level are so complex and diverse that one can hardly make general statements about them without reservations, but some rough generalizations can be asserted.

(i) Typically, the laws governing phenomena on a coarse level are more accessible to the unaided senses or require less elaborate instruments than those on a fine level, and hence the former are more readily discovered by human investigators.

(ii) Typically, the laws on a coarse level hold in a narrower range of preparations and environmental circumstances than laws at a finer level.

(iii) Typically, the laws on a fine level suffice in principle to explain the very existence of the entities treated on a coarse level (i.e., their characteristic features and their stability), to derive the laws on the coarse level, and to exhibit the limits of validity of these laws. This rough generalization is a commitment to the program of reduction of macrotheories to microtheories, which I have defended elsewhere with certain reservations (Shimony 1985, 1987).

(iv) Typically, the study of composite entities from the standpoint of the laws governing their components is nontrivial, because (to speak quasi-mathematically) very many coupled equations must be solved simultaneously. If reduction is possible in principle (point iii), then there is no "objective emergence" of properties of the composite entity that could not be anticipated by an omniscient calculator on the basis of the laws and initial conditions governing the components, but there may be "epistemic emergence" due to the impracticability of these calculations.

(v) Reducibility in principle of the laws on a coarse level to those on a finer level does not preclude great differences in the causal modality of the laws on the different levels. For instance, statistical laws on a micro-level can suffice to explain the virtually deterministic laws that hold on a coarse level on account of the law of large numbers; and conversely, in chaotic situations effective randomness at a coarse level may be superposed upon a deterministic regime at a finer level, because of extreme sensitivity of long range results to the uncontrollable details of the initial conditions.

A remark is needed in order to avoid the impression that the only kind of hierarchical structure that I recognize in nature is a scale of size. Other crucial parameters provide other types of hierarchies. The most familiar is the distinction between a nonrelativistic and a relativistic regime, according as the velocities of the bodies under consideration in a given frame of reference are much less than the velocity of light or comparable to it. Equally important in elementary particle physics is the scale of energy. In our generation it was discovered by Weinberg, Salam, and Glashow that only at low energies can the theories of the electromagnetic force and of the weak force be treated autonomously with high accuracy; at energies comparable to or greater than the rest energies of the $W^\pm$ and Z_0 particles, a unification of the theories is necessary to account for experimental results. It is anticipated by most elementary particle physicists that the autonomy of the electroweak theory and the theory of the strong force will break down at higher energies, and at even higher energies the theory of gravitation will have to be unified with these.

Finally, it should be noted that throughout the foregoing discussion of causality the common locutions in the natural sciences about entities were freely employed, in a manner that might appear biased towards a realistic interpretation of scientific theory. It is indeed difficult to avoid such locutions without cumbersome substitutes, and an argument could be given that this fact already lends some support to realism. My intention, however, is to hold ontological commitments in abeyance at this stage of the analysis and to turn explicitly to this question in Section IV, after a preparatory discussion of the program of closing the circle.

III. PUTNAM AND THE POSSIBILITY OF A NATURALISTIC THEORY OF MIND

Galileo's distinction between the intrinsic properties of physical entities and the sensory qualities "projected" onto them by the perceiving subject was inseparable from the immense success of physics since the sixteenth century. The set of intrinsic properties was extended as physics developed, so as to include such things as charge distributions and electromagnetic field strengths, but never so as to reintroduce sensory qualities. The most remarkable feature of the picture formulated in terms of the intrinsic physical properties is its causal closure, not just as a matter of theoretical idealization but in actual experimentation and in engineering applications. Whatever predictions, exact or statistical, that can be made regarding later values of the intrinsic properties are based upon laws and initial conditions formulated in terms of them alone. Sensory qualities are indeed indispensable to human observers for inferences about these laws and initial conditions, but once the inferences are made the sensory qualities are irrelevant and superfluous for the subsequent temporal evolution of the intrinsic physical properties. Note that this way of talking about "causal closure" avoids the supposition that the concept is contingent upon Laplacean determinism; what has been said about the self-contained picture of intrinsic physical properties applies as well to stochastic physical theories as to deterministic classical physics.

In spite of the fruitfulness of Galileo's distinction, Putnam (1987, p. 5) follows Husserl in characterizing it as "disastrous," for reasons implicit in the account of causal closure. Sensory qualities could not be projected unless they pertained to the perceiving subject, but this fact already endows them with some kind of ontological status. Furthermore, they are not negligible appendages to scientific investigation, for without them the intrinsic physical properties would be unknown to human investigators. The "disaster" is the impossibility of closing the circle of knower and known within the Galilean framework. It is remarkable that not only the distinction between intrinsic and projected properties but also the appreciation of the consequences of the distinction was anticipated clearly by Democritus. Sextus Empiricus cites Democritus as letting sensation chide the reason by saying, "Miserable mind, will you reject us after receiving the grounds for belief from us? That rejection is ruin for you" (Nahm 1945, p. 209).

Galileo's distinction could be sustained, as Descartes suggested explicitly, by a dualism in which minds have the same ontological status as physical entities. If the two basic types of entities interact causally and mutually, then the causal closure of the physical aspects of the world is lost. Descartes and his followers, however, failed to work out the details of a mind–matter interaction in a way that even partially approached the precision of mathematical physics. Another philosophical possibility is to preserve the causal closure of the physical aspects of the world by a parallelist theory of minds and physical entities, but this theory leaves entirely obscure the correlation of the careers of non-interacting types of entities, as well as the epistemological problem of mental inferences concerning physical bodies.

The only idea that Putnam regards as offering any promise for closing the circle is the functionalist theory of mind, which he proposed early in his career (1975, *passim*). Functionalism is an identity theory, regarding a sensation to be identical with an appropriate brain state (1981, p. 78), but it differs from other identity theories by characterizing the brain state not in manifestly physical terms but rather in terms of a program. A program is an organization to accomplish a certain end, and in principle it can be realized in systems of different material composition, or even in a "disembodied spirit" (*ibid.*, p. 79), though in the brain it is realized by organic molecules. In this way functionalism aims at transcending the framework of materialism while remaining within it. In various expositions of internal realism since the late seventies, however, Putnam has been critical of his own earlier proposal of functionalism (and self-criticism is always interesting – witness *The God that Failed*!)

One of his criticisms is that sensory *qualia* are simply not amenable to functionalist analysis (1981, pp. 79ff). For consider the entire set of programs corresponding to states of consciousness of one human mind, and now suppose that the states of consciousness are transformed systematically with respect to color, so that in all of them red and blue are permuted. This shift could not be accounted for by a shift of the programs that allegedly correspond to states of consciousness, since neither the organization nor the biological function of the set of programs is affected by a systematic shift of colors (as they would be by an unsystematic shift). Consequently, the differences among color *qualia* would dangle without explanation if the only tools for an explanation were these provided by functionalism.

A second criticism of functionalism concerns intentionality (1981, p. 79; 1987, pp. 11–16). Even if physicalism is rescued by a functionalist account of *qualia,* the program of Galileo is incomplete unless the *projection* of sensory qualities onto objects which do not intrinsically possess them is explained. But

Projection is thinking of something as having properties it does not have, but that we can imagine . . . , without being conscious that this is what we are doing. It is thus a species of *thought* – thought about something. Does the familiar

'Objectivist' picture have anything to tell us about thought (or, as philosophers say, about 'intentionality'), that is, about *aboutness?* (1987, pp. 11–12)

Putnam explores the prospects of a functionalist account of intentionality and is pessimistic, for there seems to be no one program, but rather an indefinite number of alternative programs, for achieving a propositional attitude like belief or doubt. Thus he is led to assert that thinking creatures are not only "compositionally plastic," which was part of his earlier conception of functionalism, but "programmatically plastic" (*ibid.*, pp. 14–15). As I understand him, an indefinite set of equally admissible programs is something that must be characterized in mentalistic terms, if at all, in contrast to a single definite program, which is characterized by a specific set of formal rules.

This argument completes Putnam's case against the program of closing the circle (or "totalistic explanation," in his terminology): knowledge of nature cannot be described without recognition of the intentionality of the subject's mental states, but intentionality cannot be explained in a naturalistic theory of mind.

Internal realism, according to Putnam, has the virtue of accommodating without loss the impossibility of a totalistic explanation. It permits him, for example, to maintain both his functionalist theory of mind and his criticism of it.

Today I am still inclined to think that the theory is right; or at least that it is the right *naturalistic* description of the mind/body relation. There are other, 'mentalistic', descriptions of this relation which are also correct, but not reducible to the world-picture we call 'Nature' (indeed the notions of 'rationality', 'truth', and 'reference' *belong* to such a 'mentalistic' version. . . . This fact does not dismay me; for, as Nelson Goodman has emphasized, one of the attractive features of non-realism is that it allows the possibility of alternative right versions of the world.) (1981, p. 79)

Elsewhere (1990, pp. 5–6) Putnam acknowledges an affinity to Niels Bohr's principle of complementarity, according to which we must renounce the unification of different 'complementary pictures' (e.g., wave and particle, space–time description, and causal description) into a single picture of physical reality. In another essay (1988) I have raised some doubts about the coherence of Bohr's (1958) philosophy of complementarity, and similar considerations apply to Putnam's internal realism. The danger of incoherence is evident in the parenthetical clause of the following passage:

it may be possible to see how it can be that what is in one sense the 'same' world (the two versions are deeply related) can be described as consisting of 'tables and chairs' . . . in one version *and* as consisting of space–time regions, particles and fields, etc. in other versions. (1987, p. 20)

One can hardly avoid the question of *how* "the two versions are related" – by what chains of inference, or analogies, or causal connections, or other links? There is some tension within internal realism because of its recognition of a deep relation between the two versions of the world while abstaining – either as a matter of principle or for lack of resources – from exploring this relation. My own conjecture is that Putnam's difficulty at this point is due largely to his presupposition of a materialist view of nature.

Before proceeding to a constructive discussion of a naturalistic theory of mind, I shall comment briefly on another of Putnam's arguments against a totalistic explanation, which is based upon the logic of self-reference. He reviews the semantical paradoxes and arrives at Tarski's conclusion that they arise from allowing a language to contain its own truth-predicate. But a totalistic explanation, including an explanation of the subject who is formulating the explanation, is according to Putnam an epistemic analogue to a semantically closed language and shares the same generic weaknesses (1990, pp. 11–17). He does not spell out in detail the epistemic analogue to the paradox of the liar, however, and I am dubious that a full argument can be given.

First of all, self-reference per se does not entail inconsistency, and Putnam himself points out that in a consistent language rich enough to contain arithmetic it is always possible to construct sentences that refer to themselves. Second, the analogy that Putnam wishes to establish does not hold in the program of closing the circle as I conceive it and as it is articulated by systematic philosophers of the past, like Aristotle. The program aims at a comprehensive set of principles concerning the world, which are applicable to the activity of cognition whereby the principles are known. But a set of principles by itself does not permit the deduction of a sentence analogous to "This sentence is false" regarding a specific act of cognition. Such a deduction would require the supplementation of the envisaged principles by auxiliary sentences; and it would not be unreasonable to exorcize the inconsistency by challenging these auxiliaries, because some of them would presumably be reports of a specific state of mind, which is very elusive.

I propose now to approach the problem of the place of mind in nature in a stepwise manner, following the discussion at the end of Section II of the complexities of causality.

At the coarse level of common sense many causal regularities of a qualitative kind are known widely and reliably. First of all, the knowing subject somehow pervades or inhabits or is intermingled with a body, and the resulting total system, the living body, is capable of a large degree of autonomous behavior; a number of optional motions are available under normal conditions to the living body without specific determination by the environment, such as stretching, flexing, walking, speaking, striking, swallowing, and so forth. Of course the living body is not causally closed, because the environment imposes not only general constraints and opportunities but also sometimes quite specific forces – for example, resistance of a solid obstacle, blows by massive objects, and excitations of the sense organs; and conversely, the autonomous motions of the living body are efficacious for modifying the motions of the physical things of the environment. Also recognized at the coarse level of common sense are the specialized functions of the sense organs for mediating certain causal connections between the environment and the knowing subject – for example, that covering or damaging the eyes interrupts visual sensations. Much more of this kind of phenomenology can be reported, but enough has been said for the purposes of this paper. Beyond all details, what is striking philosophically about the commonsense treatment of the knowing subject is its uninhibited merging of the physical and mental aspects of the living body.

The merging of physical and mental descriptions of the living body is not entirely confined to the level of common sense, however, because investigations at a finer level have discovered and systematized a rich and diverse body of psychophysical correlations. Some of these have wide scope, like Fechner's law of the proportionality of the magnitude of a sensation to the logarithm of the associated physical stimulus (and corrections thereof by Stevens and others). There are some precise special correlations, like that between the frequency of an acoustical vibration and the sensed pitch, and its optical analogue. There are correlations between stimuli of loci of the brain and types of experience; and also between regions of the brain and psychological capacities, as exhibited negatively by observations on patients with brain lesions and traumas. There are correlations between the concentrations of various chemical agents in the brain and emotional states, which have recently been described with increased precision because of the discovery of specialized receptors for the agents. But without any attempt to catalog psychophysical correlations, these few examples suffice to suggest a general observation of great importance: that the level on which the correlations are exhibited is very large compared to that of organic molecules, and for the most part large compared to that of neurons. (There are some exceptions to the second part of this generalization in lower organisms, where a single neuron sometimes controls a single reflex (Changeux 1985, p. 101), but in higher vertebrates a psychological reaction or capacity is characteristically associated with the collective behavior of many neurons.) Consequently, there is a limitation of scale to psychophysical correlation. There is no apparent obstacle to understanding the physical aspects of the living body in terms of the microphysical principles that govern inorganic phenomena, and triumphant statements like the following must be taken very seriously:

Similar methods led from EEG recordings to measures of the electrical activity in single neurons, from studies of the postsynaptic response to the opening of ion channels, itself based on changes in molecular configuration. At each stage, the level of organization is "reduced" to a more elementary level, from the population of cortical neurons to the cell or the synapse, from the nerve impulse or postsynaptic current to changes in single molecules. Each time, a seemingly single, continuous wave is "cut up" into discrete elementary units, which are seen as together accounting for the wave, in the sense that the wave can be fully "reconstituted" from these discrete components. A global activity can thus be reduced to physicochemical properties and can be described in the same terms as those employed by the physicist or the chemist. Admittedly, in practice we do not use the complete chemical structure of the receptor molecule (which is known) to describe the postsynaptic response to acetylcholine, but it is certainly legitimate to do so. (*ibid.*, p. 95)

By contrast, mentality seems to be completely lost in descending to the microscopic level in the scale of nature.

Once again one feels a powerful motivation for a physicalist world view and for a theory of mind, like functionalism, that conforms to physicalism. But again one must remember how difficult it is for physicalism to do justice to *qualia,* intentionality, and other unavoidable aspects of mentality. (See also the discussion of the Phenomenological Principle in Section 3 of "The Transient *Now*," Chapter 18 in Volume II.) Hence there is a heuristic reason for resisting the conclusion of the preceding paragraph and exploring the conjecture that mentality is pervasive in nature at the most fundamental level.

The most systematic version of such a conjecture is A. N. Whitehead's "philosophy of organism," formulated in his *Process and Reality* (1929) and more compactly in Part III of his *Adventures of Ideas* (1933). The ultimate concrete entities in his ontology are "actual occasions," characterized as occasions of experience, where anthropocentrism is avoided by construing "experience" with great latitude. Each occasion has spatial and temporal spread, though Whitehead is vague about the extent of both. The fundamental relation between occasions is "prehension," whereby the experience of an earlier occasion is an ingredient in the experience of one in the process of becoming, a paradigm case being the memory of a past event. The elementary particles of physics are taken to be temporal chains of occasions, while larger physical and organic bodies are highly organized spatiotemporal societies of occasions. The relation of prehension is the ground of causality. In particular, a causal law of physics holds when a vast spatiotemporal region is characterized by a certain pattern, which then constitutes a massive and dominant contribution to new occasions, thus ensuring with near certainty the conformity of the future to the past. All actual occasions share generic features, but in systems which we usually regard as inorganic the constituent occasions are characterized by a low level of experience, a repetitiousness of inherited patterns, and an almost complete absence of significant innovation. Thus Whitehead is proposing a kind of reductionism which is the inverse of that of the physicalists, since his fundamental level is mentalistic in character. His scheme is well adapted to incorporate the ideas of prebiotic and biological evolution in order to account for the emergence of societies endowed with high-level experience:

in bodies that are obviously living, a coördination has been achieved that raises into prominence some functions inherent in the ultimate occasions. For lifeless matter these functionings thwart each other, and average out so as to produce a negligible total effect. In the case of living bodies the coördination intervenes, and the average effect of these intimate functionings has to be taken into account. (1933, ch. 13, sect. VI)

Incidentally, Whitehead's scheme accommodates the two features of mentality that Putnam found most difficult to treat naturalistically: the primitiveness of sensory *qualia* fits an ontology in which occasions of experience are the ultimate entities, and intentionality is treated in a remarkably original way by the doctrine of prehension.

Many weaknesses have been pointed out in the details of Whitehead's philosophy of organism, but the most disturbing feature is his extrapolation of the concept of experience beyond the domain where it is familiar. Admittedly there are good heuristic reasons for some extension. The relations of peripheral to focal vision and of ground to figure show that even conscious experience has gradations; while the latency of memory and skills, the influence of suppressed memories on emotional tone, the wealth of data mined in free association, and so on indicate a continuity between conscious and unconscious levels of experience. The similarities of human to subhuman behavior suggest the appropriateness of attributing some kind of experience to organisms with considerably simpler nervous systems than ours. Nevertheless, one can argue plausibly that all control is lost over the concept of "experience" when it is applied to the occasions of an electron. It is specious to account for the evolution of animals endowed with consciousness from a primordial soup of inorganic constituents by attributing protomentality to the latter, if there are no experimental tests and little if any conceivable meaning of that protomentality. Even a respectful critic like Lovejoy accused Whitehead of not being "an adversary of the dualism with which we are here concerned, but only a dualist with a difference" (1960, p. 169).

There have been a number of suggestions that the thesis of mentality at a basic level in nature can be strengthened and refined by appropriate use of some of the ideas of quantum mechanics (e.g., Stapp 1990, Bohm 1980, Malin 1988, Penrose 1989, Lockwood 1989, Squires 1990). It is argued that some ideas of quantum mechanics are structural and metaphysical in character and hence are not intrinsically limited to material systems.

One idea is potentiality: that when the complete state of a system is given, there are always some properties which are objectively (not just epistemically) indefinite, but these can become definite when the system interacts with an appropriate environment. Potentiality is a modality of being intermediate between full actuality and mere possibility. The concept of potentiality may provide an answer to Lovejoy's criticism of Whitehead, cited above: it could be postulated that the protomentality attributed to all occasions has only a potential relation to consciousness, and the circumstances for actualizing this potentiality include the presence and activation of a nervous system. This proposal would be no more than a verbal dodge, however, unless it could be deployed in psychological experiments analogous to the physical experiments on diffraction and interference where quantum mechanical potentiality is normally exhibited.

Another structural idea of quantum mechanics is entanglement, which is a radically new relation between a composite system and its parts. In a classical system with interacting parts, the failure of each part to be causally closed implies that the dynamical developments of the parts are coupled, but at any moment each part is in a definite state, characterizable without reference to the others. Quantum mechanics, however, permits states of the whole which cannot be factorized into definite states of the parts even at one moment. This nonfactorizability, named "entanglement" by Schrödinger, implies correlation among phenomena with no analog in classical physics. (See "Events and Processes in the Quantum World," Chapter 11 in Volume II.) If the concept of entanglement can be separated from its usual physical context, it could be of immense value in a mentalistic ontology. A high-level mind might be considered to be a composite system, in an entangled state of its many constituent parts. Then the inescapable intuition of common sense that mentality is pervasive throughout the human body could perhaps be given a deep explanation. The conjecture that quantum-mechanical entanglement is essential to mentality may appear to be merely another version of functionalism, since both regard mentality as a collective phenomenon. But there are two essential differences. (i) The first is that quantum-mechanical entanglement essentially involves potentialities, while functionalism does not. In an entangled state of a system with two parts, for example, there is a property A of part I that can be actualized as $a_1, \ldots, a_n$ and there is a property B of part

II that can be actualized as $b_1, ..., b_n$, but if A is actualized as a_k then B is as b_k and conversely. If the state of a system were only a compendium of actual values of properties, as in classical physics, there could be no such entanglement; and when the functionalists discuss the collective state of the nervous system they never depart from the classical conception of a state as a compendium of actualities. (ii) Functionalism is physicalistic, whereas the conjecture that entanglement is essential to mentality need not be. If elementary systems are assumed to be protomental in character, then the correlation of these by quantum mechanical entanglement is not the origin of the nonphysical aspect of nature but only its coordinator and intensifier.

I shall not draw out this discussion of neo-Whiteheadian and quasi-quantum-mechanical speculations any further. My intention was only to show that mentalistic ontologies have resources for a naturalistic theory of mind that are promising and capable of controlled scientific investigation, but are mainly unexplored; and therefore Putnam's declaration of the impossibility of a naturalistic theory of mind is premature. My subjective assessment of prospects is a combination of optimism and pessimism. On the one hand, I find it hard to believe that the remarkable psychophysical correlations that are familiar at the level of common sense and sharpened at the level of current physiological psychology are not susceptible of deep explanations. On the other hand, the ideas required for deep explanations are likely to be very different from those that are fruitful in natural science so far. New ideas may be needed, for example, about the structure of the space of states appropriate for psychophysical systems, about a principle of composition for building a space of states for a composite system out of the spaces associated with the components, and about the fundamental laws governing the temporal evolution of a psychophysical system.

That the mind–matter problem constitutes a formidable obstacle to a naturalistic account of the knowing subject should not eclipse the previously noted richness of our knowledge of psychophysical correlations and the partial accomplishments of the program of closing the circle. I shall add two observations. (i) The standard procedures of working scientists are repositories of much sophistication concerning the interaction of the knowing subject with the objects of investigation, the instruments employed, and the environment in which the investigations take place. Experimental data, both in the sense of physical registrations (on photographic plates, computer tape, etc.) and in the sense of the perceptions of the experimenter, are effects eventuating from intricate causal networks. Much of scientific method consists in systematizing the untwining of causal networks so as to reveal significant connections. Distortions due to the perceptual apparatus of the subject are particularly dangerous when this apparatus is the primary receptor of physical signals, as in direct telescopic observation, pattern recognition in biomedical investigations, and (in the early days of atomic physics) scintillation counting by eye; in fact, the astronomer Bessel developed an important part of error theory in order to correct for "personal" contributions to scientific data. Moreover, implicit in the invention and use of instruments that supplement the perceptual apparatus is some understanding of the process of perception: the existence of thresholds of perception and limits upon fine discrimination, differences of response of different observers to similar stimuli, and differences of response of the same observer under varying circumstances, and so forth. (ii) The second observation is that a naturalistic treatment of mind has thrown much light upon perceptual illusions (e.g., Gregory 1973, Ratliff 1965). Illusions can be identified, corrective measures can be taken to prevent us from being misled by them in practical activities, and reasonable scenarios can be constructed for the biological utility of having perceptual equipment that is subject to illusions under unusual conditions. The successful analysis of perceptual illusions by biologically oriented psychologists provides a good answer to the skeptical argument that "the trail of the human serpent is over everything" (quoted from James in Putnam 1990). The human serpent has the wisdom to acknowledge its own trail and to take countermeasures against it!

IV. IMPLICATIONS FOR REALISM

After long preparation I turn to the main purpose of this paper: to show how the program of closing the circle provides a fresh perspective on the problem of realism. The program envisages the identification of the knowing subject (or more generally, the experiencing subject) with a natural system that interacts with other natural systems. In other words, the program regards the first person and an appropriate third person as the same entity. From the subjective standpoint the knowing subject is at the center of the cognitive universe, and from the objective standpoint it is an unimportant system in a corner of the universe.

The relevance of closing the circle to realism can be partially expressed by a negative thesis: If either the subjective or the objective aspect of the knowing subject is played down, or if the substantial identity of these two aspects is neglected, then the problem of realism is flattened, or – to use quasimathematical language – it is projected into a subspace of smaller dimensionality than it deserves. The typical arguments among realists and antirealists are moves within this subspace. Among the arguments of realists are the following: that only the realist ascription of full ontological status to the putative entities of scientific theories can prevent the predictive success of the natural sciences from being a miracle; that this ascription is the best explanation of predictive success; that only realism can prevent the incommensurability of successive theories in the historical development of the sciences, because only it guarantees a common reference; that realism with respect to theoretical scientific terms is the natural extension of realism with respect to the locutions of ordinary language; that only the realist can explain the instrumental reliability of the methods of the sciences; that any program of explicating theoretical terms solely via their linguistic linkage to observational terms is fatally undercut by the absence of a sharp observational–theoretical dichotomy. Antirealists of various kinds (instrumentalists, constructive empiricists, conventionalists, etc.) typically deploy arguments like the following: best explanation arguments are dubious in all circumstances, and in particular their use for supporting realism is circular; the attempt to apply hypothetico-deductive and inductive procedures of inference to propositions about entities lying beyond human experience is flawed, because the legitimacy of these methods in ordinary scientific contexts is contingent upon remaining within the domain of possible experience; Bayesian considerations show that no body of evidence can endow the statement that a scientific theory is literally or approximately true with a larger probability than it gives to the statement that the theory is empirically adequate.

These and cognate arguments are presented in detail, often with great acumen, in the collection *Scientific Realism,* edited by Leplin (1984). The debate between realists and antirealists is framed in terms of the character of scientific language, of semantical considerations about reference and truth, of rules and guidelines for inductive inference, of pragmatic considerations about the use of theories, and of reflections on the historical development of the sciences. The space of the debate is evidently many-dimensional, but my point is that the omission of considerations of closing the circle diminishes the debate not only in breadth but in weight.

Some exceptions to this last statement deserve comment. As noted in Sections I and II, Putnam acknowledges the bearing upon realism of what he calls a "totalistic view," which is his pejorative term for closing the circle; but because he denies the possibility of a totalistic view, his conclusion concerning realism differs from mine. Boyd broadens the usual frame of the debate by saying that "a realist conception can be integrated into, and can serve to justify, a broader naturalistic conception of epistemology and of philosophy itself" (1984, p. 80) – a statement for which I have much sympathy, with the qualification that the case for realism itself depends upon far-ranging philosophical considerations. And Hooker recognizes the relevance of the program of closing the circle to realism in many passages, such as the following:

Scientific Realism then is to play an integral part in the construction of a naturalistic, evolutionary epistemology, evolving in dynamic interaction with science itself, and in the construction of a naturalistic, evolutionary account of mind, of the use of concepts, of language and thought, both linking smoothly into the scientific picture to form the unbroken web of our worldview. (1987, p. 8)

Hooker's proposals and those of the present paper are similar in general conception, but different in details and presentation. Little purpose would be served by frequently noting the similarities and differences, and therefore it is left to the reader to make comparisons, which will be illuminating to both texts.

Returning to the usual frame of the debate, I find that one symptom of its flattening of the problem of realism is the persistent inconclusiveness of the debate as a whole. Stein has made a strong case that both realism and instrumentalism are capable of refinements and retrenchments that protect each against standard refutations, and that the residual issue between them is negligible:

The issue between realism and instrumentalism seems to me not to be clearly posed; and what I really believe is that between a cogent and enlightened "realism" and a sophisticated "instrumentalism" there is no significant difference – no difference that *makes* a difference. (1989, p. 61)

Stein's conclusion seems to me to be correct concerning the debate within its usual frame; but broadening the frame by serious attention to the program of closing the circle has the consequence of identifying a real problem in the neighborhood of a *Scheinproblem*.

My thesis can be illustrated by examining some aspects of van Fraassen's constructive empiricist philosophy of science. He contrasts his position with scientific realism, which he characterizes by the following minimum commitment: "Science aims to give us, in its theories, a literally true story of what the world is like; and acceptance of a scientific theory involves the belief that it is true" (1980, p. 8). Constructive empiricism holds instead that "Science aims to give us theories which are empirically adequate; and acceptance of a theory involves as belief only that it is empirically adequate" (*ibid.*, p. 12). The nontechnical explication of calling a theory "empirically adequate" is that "what it says about the observable things and events in this world, is true – . . . it 'saves the phenomena'" (*ibid.*). Van Fraassen does not deny the meaningfulness of speaking about processes and structures which are not directly accessible to observation, and he allows that statements purporting to refer to them have definite truth values, but constructive empiricism is steadily agnostic about specifying these truth values. His agnosticism (carefully distinguished from a denial) ensures that characterizing a theory as empirically adequate is weaker than saying that it is literally true, with the consequence that on any evidence the probability of the latter cannot be higher than the probability of the former – an elementary consequence of Bayesian probability theory (van Fraassen 1985, p. 247). It should not be concluded that this argument commits van Fraassen to the weakest possible theory allowed by the empirical data (1980, p. 68), for a theory is of no interest without sufficient informational content; in fact, the claim that a theory is empirically adequate is quite risky, since what is observable far exceeds what has been observed (1985, p. 253). But whatever the belief in the literal truth of a theory adds to the belief in its empirical adequacy is "supererogatory" for all purposes of scientific investigation (1985, p. 268).

Scientific realists often claim that an interpretation of science cannot plausibly restrict its attention to what is observable, because empirical regularities would be coincidences or miracles if there were not unobservable microstructures underlying appearances. This argument is a special case of "inference to the best explanation," about which van Fraassen makes the following powerful criticism (in addition to some generic reservations about this mode of inference):

even if we were to grant the correctness (or worthiness) of the rule of inference to the best explanation, the realist needs some further premiss for his argument. For this rule is only one that dictates a choice when given a set of rival hypotheses. In other words, we need to be committed to belief in one of a range of hypotheses before the rule can be applied. Then, under favourable circumstances, it will tell us which of the hypotheses in that range to choose. The realist asks us to choose between different hypotheses that explain the regularities in certain ways, but his opponent always wishes to choose among hypotheses of the form 'theory T_i is empirically adequate'. So the realist will need his special extra premiss that every universal regularity in nature needs an explanation, before the rule will make realists of us all. (1980, p. 21)

Van Fraassen presents acute arguments about other aspects of philosophy of science which I shall not try to summarize, but shall only say that his claim that constructive empiricism does justice to scientific activity "without inflationary metaphysics" (1980, p. 73) is difficult to refute at the level of debate that he demarcates. That there may be more to say is admitted (though skeptically) as a possibility in the following passage:

Philosophy of science is not metaphysics – there may or may not be a deeper level of analysis on which that concept of the real world is subjected to scrutiny and found itself to be . . . what? I leave to others the question whether we can consistently and coherently go further with such a line of thought. Philosophy of science can surely stay nearer the ground. (*ibid.*, p. 82)

Inquiry about the closing of the circle, I believe, provides a natural entry into the "deeper level of analysis" that van Fraassen tentatively adumbrates.

The explicit use of first-person locutions in epistemological contexts is uncommon in van Fraassen's exposition, but it seems fair to say that a tacit reference to the first person is pervasive and inevitable. One can hardly characterize oneself as an empiricist without the tacit use of "I observe," even if the body of evidence relevant to the assessment of a scientific theory is drawn from the pooled experience of many persons. And one can hardly maintain that theory acceptance has a pragmatic dimension (*ibid.*, p. 88) concerning the three-termed relation among the theory, the world, and the person using the theory, unless the user is capable of understanding and asserting expressions like "my commitment is . . . ." It is nevertheless an important matter of philosophical principle for van Fraassen to conduct epistemological analyses of the knowing subject in terms of the third person, who is an object scrutinized by the sciences. A section entitled "The Hermeneutic Circle" (1980, pp. 56–59) deserves to be quoted *in toto,* but here are two crucial passages:

Not all philosophers who have discussed the observable/unobservable distinction, by any means, have done so in terms of vocabulary. But there has been a further assumption common also to critics of that distinction: that the distinction is a philosophical one. To draw it, they seem to assume, is in principle anyway the task of the philosophy of perception. To draw it, in principle anyway, philosophy must mobilize theories of sensing and perceiving, sense data and experiences, *Erlebnisse* and *Protokolsaetze.* If the distinction is a philosophical one, then it is to be drawn, if at all, by philosophical analysis, and to be attacked, if at all, by philosophical arguments.

This attitude needs a Grand Reversal. If there are limits to observation, these are a subject for empirical science, and not for philosophical analysis. (*ibid.*, pp. 56–57)

Science presents a picture of the world which is much richer in content than what the unaided eye discerns. But science itself teaches us also that it is richer than the unaided eye *can* discern. For science itself delineates, at least to some extent, the observable parts of the world it describes. Measurement interactions are a special subclass of physical interactions in general. The structures definable from measurement data are a subclass of the physical structures described. It is in this way that science itself distinguishes the observable which it postulates from the whole it postulates. The distinction, being in part a function of the limits science discloses on human observers, is an anthropocentric one. But since science places human observers among the physical systems it means to describe, it also gives itself the task of describing anthropocentric distinctions. It is in this way that even the scientific realist must observe a distinction between the phenomena and the trans-phenomenal in the scientific world-picture. (*ibid.*, p. 59)

Now it can be asked: is not van Fraassen's Grand Reversal precisely the program of closing the circle – the attempt to locate the knowing subject in the natural world, where its properties are studied with increasing precision by biology, physiology, and empirical psychology? Has he not "coopted" a program that prima facie favors realism (as Putnam recognized in his polemic against "totalistic explanation") and converted it into a strategy of the antirealist philosophy of constructive empiricism?

My answer is that the Grand Reversal must be regarded with reservations, because of a crucial dilemma confronting van Fraassen, whether he chooses to identify the first- and third-person aspects of the knowing subjects or not. Constructive empiricism regards a scientific theory as a story about what is observable, and if the theory is comprehensive (a unified science) then the story becomes a "large-scale *roman fleuve*" (1985, p. 268), constrained in its narrative by consistency and fidelity to the evidence. The third-person aspect of the knowing subject can be treated by a scientific theory which is consistent with neurophysiological and psychological evidence and comprehensive enough to explain cognitive processes; but constructive empiricism is agnostic about the ontological status of receptors, transducers, information processors, and any other putative entities of the theory. The knowing subject in the first person, by contrast, is not an element in the scientific story. It is the auditor of the story and the judge who has certain attitudes toward it, such as belief and commitment; and also the primary repository (or at least one of them) of the empirical evidence which the theory must fit. These radical differences between the third and the first persons generate a dilemma. If the first- and third-person aspects of the knowing subject are identified by van Fraassen, then his agnosticism about any claim for a scientific theory beyond empirical adequacy is breached, for an element in a theory would be identified with something that exists in an exemplary way; if constructive empiricism is taken seriously, this identification would be a solecism comparable to the identification of Kant's phenomenal and transcendental selves. The other option is to refrain from ontologically identifying the first- and third-person aspects of the knowing subject. But this choice risks losing the advantages that the Grand Reversal promises: to break down the theoretical–observational dichotomy, to remove the question of the limits of observation from armchair introspective psychology to scientific psychology, and to eliminate the misguided classical doctrine of an autonomous epistemology antecedent to natural science. Thus either option poses a serious difficulty for constructive empiricism.

I suggest the following strategy for seeking an alternative. First, accept the identification of the first-person and third-person aspects of the knowing subject, as the most natural accommodation to common sense and to much of empirical psychology – especially to the massive evidence that a normally maturing child, without any philosophical sophistication, constructs a rough picture of the world and places himself or herself within

it. Second, conservatively note the crucial features of each of these aspects. Finally, draw the implications of identifying the two conservatively
characterized aspects. The result is a modest but nontrivial version of
realism, with the virtue that it is designed to be extended, corrected, and
refined when introspective knowledge of the first person and scientific
knowledge of the third person increase (and, incidentally, when there is
communication between human beings, thereby engaging the *second-
person* aspect of the knowing subject).

The first step is an instance of inference to the best explanation, which
is generically challenged by antirealists, as noted previously. It will be
useful to delay until the end of this section the metaphilosophical question of the validity of this special inference.

The second step is formidable, because the nature, constitution, and
boundaries of the knowing subject *qua* subject, the first-person aspect,
are all problematic. It may indeed seem hardly to advance the clarification
of the problem of realism to link it to such an intricate tangle of psychological and philosophical questions. My answer is that, for the purpose
of the proposed strategy, only a minimal statement about the first-person
aspect of the knowing subject is required: that it unequivocally exists and
is an exemplary instance of concrete existence (as contrasted with the
formal existence of mathematical entities). The Cartesian *cogito* has, of
course, been repeatedly and often justifiably criticized, but in my opinion
it is the additional baggage of philosophical psychology loaded by Descartes upon his pristine argument that has caused difficulties – baggage
concerning the substantiality of the knowing subject, its temporal endur-
ance as a self, its alleged simplicity, its sharp distinction from the body,
and the alleged clarity and distinctness of our ideas about the self. Because
these are all questions that I do not need to be drawn into, I shall bracket
them by adapting a terminology that has been valuable in elementary
particle theory – of a "bare" and a "dressed" entity. I do not know how
much of the body, or even of the environment, ought to be included in
the "bare" subject (one source of uncertainty being the conjecture mentioned in Section III that mentality consists in an entangled state of neural
components); or how much and in what way the "bare" subject becomes
"dressed" by intimate causal interaction with the body, thus giving rise to
the phenomenological impression that the mind pervades the body. What
I do know is that the core of Descartes's argument can be salvaged, with
the conclusion that the subject does exist, whatever its actual constitution
may be. The vagueness of this fundamental existential claim is compensated by its reliability.

As to the other part of the second step, whatever there is to be said
about the knowing subject *qua* object is taken from those sciences that

directly concern cognition or indirectly bear upon it. Whatever fallibilist attitudes are appropriate to the natural sciences at any stage of their development will therefore adhere to a conservative characterization of the third-person aspect of the knowing subject. But the grand scientific picture – that the knowing subject is a system interacting with other systems in nature, detecting signals, transducing them into neural signals, undergoing internal changes of state in a rule-guided but stochastic manner, and so on – is most unlikely to be changed. Above all, for the purpose at hand, the sciences treat the knowing subject *qua* object as an entity with the same ontological status as those with which it interacts – existent in the same way they are, or a mere instrument for systematizing appearances if they are. This statement of the ontological parity of the various objects interacting with the knowing subject *qua* object is not subverted by the fallibility and revisability of scientific descriptions.

The third step in the proposed strategy is to combine the conservative characterizations of the two aspects of the knowing subject: *unequivocal existence is the contribution of the first person, ontological parity with other systems is the contribution of the third person.* The result is to establish a "beachhead" for a realistic philosophy. As in a military invasion, the beachhead is modest but indispensable, and it is meant to be extended. This modest realism provides an answer to the perennial challenge of antirealists to make sense of any attribution of "reality" to the putative entities of scientific theory beyond their value for ordering and anticipating experience: their reality is to have the same status as that entity whose reality is inescapable, the knowing subject itself. The modest realism also offers an answer to the challenge to make sense of the locution "outside the mind": for even if the extent of the "dressed self" is obscure, there are, according to our best scientific knowledge, sufficient distinctions between it and other systems with which it interacts to prevent a slide to solipsism or to Parmenidean monism. In addition, the modest realism offers a way progressively to remove the unclarity of the Cartesian starting point. The knowing subject is neither transparent to introspection and rich in content, as Descartes maintains; nor transparent and empty, as Hume thinks; nor opaque and rich, as Heidegger can be paraphrased to say; but translucent and rich. The interplay of the information provided by scientific investigations of the knower *qua* object with the introspective information about its subjective aspect may illuminate the intricacies of the constitution of the "dressed" self – its space of states, its internal dynamics, and its causal interaction with the rest of the world.

The spirit of the modest realism just proposed is elegantly expressed by Faber:

I submit that the essential spirit of science, the attitude that alone makes possible such an unprecedented relation between humankind and nature, is this: consciously and deliberately to approach inanimate things with the same humility, the same delicate hesitation to impose one's own interpretative scheme upon the other, that we accord (in our better moments) to other persons. (1986, p. 233)

In view of the centrality of first-person considerations in the foregoing argument, I would add (with some damage to Faber's prose) the phrase "and ourselves" to his sentence.

One important matter has been delayed. Was it legitimate in the first step of the proposed strategy to infer the identity of the first- and third-person aspects of the knowing subject from a body of common sense and psychological evidence? The identification does indeed appear to be an inference to the best explanation, but a milder one than the inference to a realistic ontology that van Fraassen criticized (1980, p. 21, quoted earlier). The conclusion of my inference is a proposition about an individual case, not a universal generalization; and the inference does not rest upon a strong auxiliary premiss like "every universal regularity in nature needs an explanation." Denying the identification of the first- and third-person aspects of the knowing subject is a statement of the same strength as asserting it. Those classical philosophers who denied the identification did not appeal to caution with regard to a mode of inference, but gave reasons based upon their fundamental metaphysical and epistemological position; Kant's distinction betwen the transcendental and the phenomenal selves is a corollary of his transcendental idealism, and Berkeley had his own definite reasons for distinguishing between a mind and any object of nature. Their explicitness makes it possible to engage them dialectically, as illustrated in Section II. By contrast, the pervasive strategy of van Fraassen's constructive empiricism is ontological agnosticism. It is the reticence of agnosticism that permits him to use a Bayesian argument (1985, p. 247, cited previously) favoring the constructive empiricist interpretation of a scientific theory over a realist interpretation. It would be consistent with his agnosticism to escape from the dilemma just posed, concerning the identification of the two aspects of the knowing subject, by simply refusing to take a stand on the ontological status of the knowing subject as one of the putative objects of scientific theory. Agnosticism is the plea of nolo contendere, which protects it against direct refutation.

But the great strength of agnosticism is its weakness – it is uninformative. There is no direct adjudication between a metaphilosophical preference for uninformative entrenchment and a preference for informative vulnerability, but I believe there is a good indirect adjudication: let us see which strategy is the more fruitful. I hold that engagement in a program of systematic philosophy – with all of its explorations from tentative starting points and its dialectical returns to reassess the starting points, with its reflections upon the assumptions and the results of the natural sciences, with its ambitious program of meshing epistemology and metaphysics, etc. – will be richer in heuristics and more inspiring as a guide to life than a carefully retrenched philosophy, contrary to van Fraassen's dictum that empiricism "leaves room for the freedom to be found in a life *without* a world picture" (1985, p. 301). The methods of systematic philosophy overlap with those of ordinary scientific practice, but cannot be the same in all respects, because the former is global and the latter is relatively local; it is appropriate to conduct a scientific investigation without questioning all of the overt and tacit assumptions of the inquiry, including the methods of scientific inference, but in the global enterprise of systematic philosophy all assumptions are subject to examination. Nowhere is dialectical examination needed more than in the assessment, generically and specifically, of inference to the best explanation, which plays a central role in the argumentation concerning realism. The assessment depends largely on the coherence of the world picture into which the conclusion of a particular inference of this kind is incorporated. Is there a danger that the dialectic exploration will lead to anarchy? Yes, there is, unless (as Plato believed) the human intelligence is such that the dialectic can discipline itself; and that, contrary to Plato, can only be judged a posteriori.

V. THE ONTOLOGY OF SCIENCE IN PROGRESS

The modest realism arrived at in the preceding section asserts the legitimacy of ontological inferences about entities causally connected with the knowing subject. But no answer has yet been proposed to questions about the nature of the entities whose existence is inferred, questions that traditionally were posed as "What are their essences?" or "What substances are they?"

If systematic philosophy were achieved and the program of closing the circle were completed, then presumably explicit answers would be given to these metaphysical questions. Suppose, for specificity, that the envisaged systematic philosophy is an objective idealism resembling Whitehead's philosophy of organism in its general theses. Then the ultimately existing things are occasions of experience, the overwhelming preponderance of which are nonhuman, so that the word "experience" must be extrapolated far from its familiar use. The characterizations of individual occasions, or much more commonly of societies of occasions, which are offered by fundamental physics are neither incorrect nor literally true, according to Whitehead, but are incomplete stories. For instance, "The notion of physical energy, which is at the base of physics, must then be conceived as an abstraction from the complex energy, emotional and purposeful, inherent in the subjective form of the final synthesis in which each occasion completes itself" (1933, p. 188). This proposal is a thoroughly metaphysical version of realism, but one that fails to conform to the conditions of Putnam's definition of "metaphysical realism" stated in Section III. The world would not consist of a fixed totality of objects, because actual occasions are evanescent, and those of the future do not even have the status of being concretely ingredient in the occasions of the fleeting present. There would be no "God's-eye view," in the sense of a compendium of all truths, whether there is a God or not, because a Whiteheadian God is subject to the metaphysical constraint of prehending only those occasions that have already come to be. Above all, even though Whitehead conceives of occasions as objective, they certainly cannot be characterized as "mind-independent," for they themselves *are* mental, or at least protomental. I have refrained from commenting on Putnam's arguments concerning reference (1981, pp. 1–48, 217–18; 1983, pp. 1–25), directed against what he understands by "metaphysical realism," because I do not see their relevance to the versions of realism which I envisage. Some good criticisms of these arguments are given by Hooker (1987, p. 438) and the authors whom he cites. There are resources of both metaphysics and semantics for replying to the following challenge posed by Putnam:

A metaphysical system will have to be rich enough to embrace what is indispensable to discourse, including talk of reference, talk of justification, talk of values in general; and it will have to be accompanied by some sketched-out story of how we have access to "metaphysical reality". (1990, pp. 39–40)

Meanwhile, what can be said about the character of natural entities in our state of incomplete knowledge, acknowledging the fallibility and corrigibility of the science we now possess, but still marvelling at its coherence, predictive power, explanatory depth, and progressive history? Following a little-known suggestion of Newton, which was elaborated by Stein, I believe that the most promising way to articulate the ontology of science in progress is to explicate questions about essence and substance in terms of causality (with due regard for the complexity of causality, as discussed in the latter part of Section II). The idea is that the involvement of an entity in a causal relation is ipso facto an ontological characterization; and when the causal relation exhibits a definite structure, as it must if it instantiates a law of nature, then this structure is an aspect of what the entity is; furthermore, it is an aspect that is accessible to human observers via the causal chains that link them to the entity in question.

The following is a passage in which Stein quotes, paraphrases, and comments upon Newton's remarkable proposal:

"And so let us feign" – "fingamus itaque" – says Newton, that God causes some previously empty region of space to become impervious to penetration by bodies; that, as it were . . . , he creates a *field of impenetrability.* "Feign" in the second place that this impenetrability is not always preserved in the same part of space, but rather is allowed "to be transferred hither and thither according to certain laws," but in such a way that the shape and size of the impenetrable region are conserved. If there are several such *mobile impenetrability fields,* then since *ex hypothesi* they cannot penetrate one another, the laws governing their migration must be such as to determine their behavior in the event of impingement upon one either of another, or – if we make such a distinction – of a body of the "ordinary" kind; in other words, those laws must include laws of impact.

Finally, according to Newton, a third supposition is necessary, if these newly created regions of impenetrability are to have all the essential attributes of the matter we know: they must be able to interact with minds – able, that is, to excite perceptions in the latter, and susceptible in turn of being moved by them. If all three conditions are met, . . . the mobile impenetrable spaces would be, to us, entirely indistinguishable from what we call "bodies".

In his further discussion of this analysis of the nature of body, Newton remarks that it replaces the obscure – he says "unintelligible" – notion of a substrate or "substance" endowed with a "substantial form" with the clear one of extension (of this, he has said, we have "an Idea the clearest of all") to which there have been imparted clearly specified "forms" (impenetrability, laws of motion, laws of interaction with minds; he adds, if there is any difficulty in this conception it is not with the form God imparts to space, but the manner by which he imparts it. But that is not to be taken for a difficulty since the same point occurs with regard to the manner by which we move our limbs, and nonetheless we believe ourselves able to move them.). (199–)

Stein then comments further that Newton anticipated James's and Whitehead's radical rejection of much of the philosophical tradition, in that "Newton . . . suggested not only that minds, but – and more emphatically – that *bodies* . . . are best conceived, not through the notion of 'stuff', but through that of 'functions'" (*ibid.*).

Had the passage cited by Stein from Newton's unpublished papers been widely known earlier, it would surely have enriched the documentation of those historians of ideas who regard Newton as a spiritual ancestor of positivism, supplementing their citation of "hypotheses non fingo" (e.g., Burtt 1932, pp. 207 and 226). Newton does indeed share with the positivists the desire to purge scientific and philosophical discourse of obscure language. But the traits of anthropocentrism and of retrenchment from making assertions about the world itself are not to be found in the passage cited from Newton. The bodies that he analyzes functionally are involved causally with each other, in a manner entirely independent of their being perceived by human beings. The mode of their interaction is governed by objective laws of nature, and even though the only laws referred to in the passage are laws of inertial motion and impact (as expected in an early manuscript, prior to Newton's formulation of the general concept of a force between two bodies), his idea is clearly applicable to arbitrary interactions between spatially separated bodies. Finally, even though Newton's third supposition does refer to perceiving minds, there is no suggestion of the ontological primacy of the latter, as in Berkeley's immaterialism and in versions of idealism descended from him; Newton insists rather upon an interaction between bodies and minds, placing them on the same level ontologically and unequivocally locating minds within nature, even if the laws governing the mind–body interaction are unknown.

Newton's functional treatment of bodies extends with ease to fields. As Stein points out in the passage just quoted, the concept of field is implicit in Newton's characterization of bodies, which can be paraphrased as "fields of impenetrability." Furthermore, the "accelerative force" in Newton's *Principia,* which is characterized as "a certain efficacy, spread from the center to the individual places around it, to move any bodies that are in them," clearly refers to what in modern locution is called "the gravitational field due to the source" (Stein 1970, pp. 265–66). The ontological status of the field, as something coordinate with bodies, consists in the fact that it is referred to a place whether that place is occupied by a body or not; and the ontological characterization is causal or functional, since the definition is given in terms of potential action upon any body that might occupy the place. It is interesting to cite a contemporary nonphilosophical physicist, who also relates the ontological status of a field to causation. The introduction to Purcell's textbook on electromagnetism makes a pedagogical plea for impressing upon students the idea of the reality of the electric field as something that "crackles" (1963, p. x). Is Purcell paralleling Newton's supposition that bodies excite perceptions in the mind, or is he merely construing a crackle physicalistically as an acoustical excitation resulting from ionization of the air? He does not say, but – however he is interpreted – his characterization of the field is causal and conforms to the archetype in the passage cited from Newton.

Newton's proposal is relevant to a frequently stated antirealist argument: that the history of science exhibits no convergence to a final theory, and a fortiori not to the ultimate truth, because even after a science has reached "maturity", so that a successor theory approximately agrees with its predecessor in experimental predictions, the transition typically shows a radical discontinuity in ontology (e.g., Laudan 1984). This argument loses much of its force if ontological distinctions are interpreted in terms of causal laws. As noted in points (ii) through (iv) at the end of Section II, a unitary view of nature is consistent with a hierarchy of causal laws at different levels of description, and the character of the laws at a coarse level may be different from that of the laws at a finer level. A scientific revolution (in a mature science) typically preserves previously established laws at a coarse level while introducing innovations at a finer level. Consequently, if ontological questions are construed causally, then the partial conservatism of a scientific revolution implies a partial continuity of ontology. This thesis can be illustrated by two episodes from the history of optics.

In the seventeenth century the main body of known optical phenomena was contained in what is now called "geometrical optics," which regards light as travelling in rays that are straight in a homogeneous medium but generally curved in a medium of varying constitution. Newton preferred a particle ontology of light to a wave ontology for two reasons, given in Query 28 of his *Optics* (Newton 1952). (i) He argued that a wave theory could not explain the casting of sharp shadows by opaque obstacles, the nonpropagation of light through crooked passages, and other manifestations of ray propagation of light. (ii) He could see no way for a wave theory of light to explain the manifestation of "sides" (i.e., polarization behavior) in the doubly refracted light emerging from Iceland spar. Both of these objections were effectively answered by Young, Fresnel, and their followers in the early nineteenth century, the first by a careful analysis of diffraction and the second by the theory that the vibrations of light are transverse to the direction of propagation, rather than along that direction (as is the case with sound waves) (Whittaker 1951, pp. 107–22). They established that the wave theory of light as a fine-level description is consistent with geometrical optics at a coarse level, and of course the wave theory is much more successful than the particle theory in accounting for the phenomena of diffraction and interference. Undoubtedly the prevailing opinion about the nature of light at a fine level of description (i.e., at a spatial scale of the order of a wavelength of the light) was radically changed by the work of Young, Fresnel, Arago, Foucault, Kirchoff et al. But the revolution was conservative in the sense of preserving the previously well-established laws of optics. If ontology is construed causally, then the ontology associated with the causal structure of geometrical optics is sketchy and that associated with wave optics is richer; but there is no inconsistency between them, only a fulfillment and enrichment of the former by the latter.

The second episode is the experimental disconfirmation of the neoclassical theory of the radiation field (Crisp and Jaynes 1969, Stroud and Jaynes 1970, Jaynes 1970) and the confirmation of the quantum theory of the radiation field. Both theories assume that the atoms with which the radiation field interacts are governed by quantum mechanics, but the former supposes that the radiation field itself is unquantized. It is surprising that the neoclassical theory can account perfectly well for the photo-electric effect, and cannot even be disproved without supplementary assumptions (see Clauser 1974) by experiments that seem to set limits upon the splitting of light beams (Ádám, Jánossy, and Varga 1955). The question was decisively settled by correlation experiments of Clauser (1974) and of Kimble, Dagenais, and Mandel (1977). Clauser studied pairs of photons emitted in an atomic cascade, the first of which is split by a beam-splitter feeding two detectors 1A and 1B, and the second by a beam-splitter feeding detectors 2A and 2B. The neoclassical theory predicts that the product of coincidence counts $C_{1A-1B}C_{2A-2B}$ will be greater than or equal to the product $C_{1A-2B}C_{1B-2A}$, while the quantum theory of the radiation field predicts a violation of this inequality; Clauser's experiment confirmed the latter prediction. Kimble et al. measured joint detections of light scattered from a single atom (resonance fluorescence) at two times separated by a small interval and found that for a sufficiently small time separation the joint counting rate is smaller than what is allowed by the semiclassical theory. As in the episode discussed in the preceding paragraph there is a hierarchy of phenomena, but here the hierarchy is constituted by the multiplicity of correlated interactions of radiation with matter, single events constituting a coarser level than two-event correlations, and so forth. Two theories of radiation which agree with regard to single interactions of radiation and matter deviate in their predictions regarding joint interactions.

What is the relevance of this episode in the history of optics to the question of ontological discontinuity as science advances? First of all, one can say that something of great significance would have been preserved whichever account of the radiation field had triumphed: specifically, the granularity of energy exchanges between light and matter, and the stochasticity of these exchanges. (Of course, these are two radical features uncovered by the quantum-mechanical revolution, thereby raising the question of ontological discontinuity at another locus. But that question can be answered by discriminating yet another level of description, that of high quantum numbers, where the correspondence principle ensures agreement between quantum and classical physics; and also by recognizing certain deep-lying structural continuities – see the following paragraphs.) Second, the crucial difference between the two rival theories of the radiation field lies less in the visual imagery accompanying them (continuity versus discontinuity) than in the conflicting causal (even though statistical) laws governing multi-event correlations and in other discrepancies that inevitably follow – for example, in energy levels and in scattering behavior. It is important, finally, to reiterate what was said above in rejecting the assimilation of Newton to the positivists: that the point is not that the difference between two rival theories is constituted only by their different predictions in actual or possible experiments, but rather that what a physical system is substantially, quite apart from our knowledge of it, is best expressed in terms of the causal laws that govern its behavior.

Extrapolating from these two episodes we can say (setting aside the possibility of human error in observation and inference) that any causal structure truly exhibited at a given level of description of a system is a solidly established aspect of what that system is, regardless of surprises and refinements at finer and deeper levels. From the point of view of the ultimate true theory regarding the system it must be the case (if error is bracketed) that at the specified level of description the specified causal structure holds. Part of what the system *is* is to have the specified structure. For this reason, the enterprise of science is not doomed to be a web of Penelope, with perpetual doing and undoing, just because of the existence of an indefinite number of levels of description.

This argument for continuity can be amplified by considering the deep comparison of utterly dissimilar theories that is afforded by the conceptual machinery of modern mathematics. For example: on the surface, the wave mechanics of Schrödinger is utterly different from classical particle mechanics. But in the Hamilton–Jacobi formulation of classical mechanics the characteristic function S is introduced, with the property that its gradient at a point in configuration space determines the direction of motion of the system. Hamilton recognized that this function is analogous to the phase function of physical optics, the gradient of which determines the direction of an optical ray; and Schrödinger deepened Hamilton's observation by showing that his wave mechanics bears the same relation to particle mechanics as wave optics does to geometrical optics (Schrödinger 1926). Heisenberg's and Dirac's formulations of quantum mechanics exhibit structural continuities with classical mechanics in other respects (see Primas 1981, pp. 54–55, for a valuable analysis). Stein makes the following philosophical comment on the prevalence of structural continuity in the evolution of modern physics:

this process of development, from the less to the more fundamental (*Tieferlegung der Fundamente*, as, according to Hermann Weyl, Hilbert liked to call it) has very little to do with the "referential" relations of theories or with the relationships of their "ontologies" in the sense of Quine. To borrow from the ancient philosophical tradition, what I believe the history of science has shown is that on a certain very deep question Aristotle was entirely wrong, and Plato – at least on one reading, the one I prefer – remarkably right: namely, our science comes closest to comprehending "the real", not in its account of "substances" and their kinds, but in its account of the "Forms" which phenomena imitate (for "Forms" read "theoretical structures", for "imitate", "are represented by") . . . in this structural deepening what tends to persist – to remain, as it were, quasi-invariant through the transformation of theories – is on the whole (and especially in what we think of as the "deepest" – or the most "revolutionary" – transformations) not the features most conspicuous in referential semantics: the substances or "entities" and their own "basic" properties and relations, but the more abstract mathematical forms. (1990, pp. 57–58)

VI. A DARK CLOUD

At several junctures in this paper I have adduced the astonishing progress of the natural sciences in order to remain optimistic regarding the program of closing the circle. Without withdrawing what I have said so far I wish to acknowledge a development in twentieth-century physics which darkens the prospect of the program. It is not an insuperable obstacle, in my opinion, but it suffices to show that any possible realization of the program will have to be quite intricate.

The dark cloud is a peculiar implication of quantum mechanics for causality, which is the central element in Newton's surrogate for substance. The difficulty is not the breakdown of determinism, which is often thought to be the most radical innovation of quantum mechanics; as sketched in Section III, the conception of causality as a constraint on temporal processes is compatible with a stochastic dynamics. The trouble is that the dynamical law of quantum mechanics – the time-dependent Schrödinger equation – is actually a deterministic law for a closed system, but what it determines is the temporal evolution of the quantum state, which is best understood not as a catalog of actual values of all the relevant dynamical variables but as a network of potentialities (see, e.g., Bell 1987, pp. 117–38). In a closed system (of which the universe as a whole is certainly an example) the actualization of potentialities – for instance, the occurrence or non-occurrence of a click of a Geiger counter, or the location of a definite spot on a photographic plate – is precluded by the conjunction of the time-dependent Schrödinger equation and certain mathematical properties of quantum potentialities. But how can Newton's causal analysis of physical objects be adapted to quantum mechanics if in crucial processes there are no definite events? Furthermore, even apart from a version of realism based upon Newton's proposal, quantum mechanics obviously implies an obstacle to the entire program of closing the circle. The evidence upon which quantum mechanics is founded is obtained from the functioning of measuring instruments, like particle detectors. If the world view of quantum mechanics precludes the definiteness of events involving particle detectors, then the evidential underpinnings of the theory are removed. This obstacle to closing the circle is independent of the "bifurcation of nature" into mental and physical aspects. The indefiniteness of detection concerns the physical registration of a process of interaction and amplification, not the registration of the event upon a psyche.

Without attempting to survey systematically the various proposals for solving the problem of the actualization of potentialities, I shall mention three major classes of proposals, with very different implications for the program of closing the circle.

(1) There is a search for modifications of quantum mechanics in a way that preserves the wonderfully fruitful ideas of the present theory and nevertheless provides a *physical* explanation of the actualization of potentialities. The most promising of these is the stochastic modification of the Schrödinger equation, so that sporadically a spontaneous wave-packet reduction would occur. A survey of such proposals by Károlyházy, Ghirardi, Rimini and Weber, Piron, Gisin, Pearle, Bell, Diósi, and others is given by Shimony (1991).

(2) There are attempts to solve the problem of actualizing potentialities by invoking the mind–body interaction in a crucial way (e.g., Wigner 1962, Faber 1986, Malin 1988).

(3) The actualization of a potentiality, and accordingly, the definite occurrence of a measurement result, is interpreted epistemically rather than as a physical process (as in proposal 1) or a psychophysical one (proposal 2). Essentially this is Bohr's proposal, and Putnam has endorsed it as part of his general program of internal realism.

I do not know how the third proposal could be proved definitively, because it is unclear what analytic and experimental discoveries would imply the impossibility of an account of actualization of a potentiality as a physical or psychophysical process. To be sure, after lengthy and unsuccessful searches for such an account, weariness may set in, and an epistemological solution to the problem would become attractive; but, of course, this capitulation might be premature. The plausibility of this conclusion would be enhanced, however, by a more systematic exposition of a philosophy renouncing "totalistic explanations" than either Bohr or Putnam has provided.

The success of the second proposal would be an unprecedented unification of theories of matter and mind, for not only would a problem within quantum mechanics be solved, but also new light would be cast upon the mind–body problem. My own subjective estimation is that the mind–body problem is several orders of magnitude more difficult than the quantum-mechanical reduction problem, and that the two problems will not be solved in tandem. I need hardly add that I would be delighted by discoveries contrary to my subjective expectation.

The first proposal seems to me the most promising, and in fact I think that it may be successfully carried out in our generation. If so, then there would be a partial achievement of the program of closing the circle, for the detection and registration of events that constitute the physical data of scientific theories would be theoretically understood. The program of closing the circle would still not be completely achieved until a naturalistic account is given of the registration of these data upon the consciousness of human observers.

Acknowledgments. This work was supported by the National Endowment for the Humanities. I am grateful to Dr. Jeremy Butterfield for his helpful comments.

REFERENCES

Ádám, A., Jánossy, L., and Varga, P. (1955), *Annalen der Physik* 16: 408.

Bacon, F. (1937), *Essays, Advancement of Learning, New Atlantis, and other Pieces.* New York: Odyssey Press.

Bell, John S. (1987), *Speakable and Unspeakable in Quantum Mechanics.* Cambridge: Cambridge University Press.

Bohm, D. (1980), *Wholeness and the Implicate Order.* London: Routledge & Kegan Paul.

Bohr, N. (1958), *Atomic Physics and Human Knowledge.* New York: Wiley.

Boyd, R. (1984), "The Current Status of Scientific Realism." In J. Leplin (ed.), *Scientific Realism,* pp. 41–82. Berkeley: University of California Press.

Boyd, R. (1985), "Lex Orandi est Lex Credendi." In P. M. Churchland and C. A. Hooker (eds.), *Images of Science,* pp. 3–34. Chicago: University of Chicago Press.

Burtt, E. A. (1932), *The Metaphysical Foundations of Modern Physical Science.* Garden City, NY: Doubleday.

Changeaux, J.-P. (1985), *Neuronal Man.* New York: Pantheon.

Clauser, J. F. (1974), "Experimental Distinction between the Quantum and Classical Field-theoretic Predictions for the Photoelectric Effect." *Physical Review D* 9: 853.

Crisp, M. D., and Jaynes, E. T. (1969), "Radiative Effects in Semiclassical Theory," *Physical Review* 179: 1253.

Cushing, J., and McMullin, E. (1989), *Philosophical Lessons from Quantum Theory.* Notre Dame, IN: Notre Dame University Press.

Dirac, P. A. M. (1938), *Proceedings of the Royal Society A* 165: 199.

Faber, R. (1986), *Clockwork Garden: On the Mechanistic Reduction of Living Things.* Amherst: University of Massachusetts Press.

Gregory, R. L. (1973), "The Confounded Eye." In R. L. Gregory and E. H. Gombrich (eds.), *Illusion in Nature and Art,* pp. 49–95. New York: Scribner's.

Hooker, C. A. (1987), *A Realistic Theory of Science.* Albany: State University of New York Press.

Jaynes, E. T. (1970), "Reply to Leiter's Comment." *Physical Review A* 2: 260.

Kant, I. (1929), *Critique of Pure Reason,* transl. N. Kemp Smith. New York: Humanities Press.

Kant, I. (1947), *Prolegomena to Any Future Metaphysics,* transl. P. Carus. La Salle, IL: Open Court.

Kimble, H. J., Dagenais, M., and Mandel, L. (1977), "Photon Antibunching in Resonance Fluorescence." *Physical Review Letters* 39: 691.

Laudan, L. (1984), "A Confutation of Convergent Realism." In J. Leplin (ed.), *Scientific Realism,* pp. 218–49. Berkeley: University of California Press.

Leplin, J. (ed.) (1984), *Scientific Realism.* Berkeley: University of California Press.

Lockwood, M. (1989), *Mind, Brain and the Quantum.* Oxford: Basil Blackwood.

Lovejoy, A. O. (1960), *The Revolt against Dualism.* La Salle, IL: Open Court.

Malin, S. (1988), "A Whiteheadian Approach to Bell's Correlations." *Foundations of Physics* 18: 1035.

Nahm, M. (1945), *Selections from Early Greek Philosophy.* New York: F. S. Crofts.

Newton, I. (1952), *Optics.* New York: Dover.

Peirce, C. S. (1935), *Collected Papers,* vol. 6, ed. C. Hartshorne and P. Weiss. Cambridge, MA: Harvard University Press.

Penrose, R. (1989), *The Emperor's New Mind: Concerning Computers, Minds, and the Laws of Physics.* Oxford: Oxford University Press.

Primas, H. (1981), *Chemistry, Quantum Mechanics and Reductionism.* Berlin: Springer.

Purcell, E. M. (1963), *Electricity and Electromagnetism* (Berkeley Physics Course, vol. 2). New York: McGraw-Hill.

Putnam, H. (1975), *Mind, Language, and Reality* (Philosophical Papers, vol. 2). Cambridge: Cambridge University Press.

Putnam, H. (1981), *Reason, Truth, and History.* Cambridge: Cambridge University Press.

Putnam, H. (1983), *Realism and Reason.* Cambridge: Cambridge University Press.

Putnam, H. (1987), *The Many Faces of Realism.* La Salle, IL: Open Court.

Putnam, H. (1990), *Realism with a Human Face.* Cambridge, MA: Harvard University Press.

Ratliff, F. (1965), *Mach Bands: Quantitative Studies on Neural Networks in the Retina.* San Francisco: Holden-Day.

Schrödinger, E. (1926), "Quantisierung als Eigenwertproblem," part 2. *Annalen der Physik* 79: 489.

Shimony, A. (1985), "Some Proposals Concerning Parts and Wholes." In P. Sällström (ed.), *An Inventory of Present Thinking on Parts and Wholes,* vol. 1, pp. 115–22. Stockholm: Swedish Council for Planning and Coordination of Research.

Shimony, A. (1987), "Methodology of Synthesis: Parts and Wholes in Low Energy Physics." In R. Kargon and P. Achinstein (eds.), *Kelvin's Baltimore Lectures and Modern Theoretical Physics,* pp. 399–423. Cambridge, MA: MIT Press.

Shimony, A. (1988), "Physical and Philosophical Issues in the Bohr–Einstein Debate." In H. Feshbach et al. (eds.), *Niels Bohr: Physics and the World,* pp. 285–303. Chur: Harwood.

Shimony, A. (1991), "Desiderata for Modifications of Quantum Dynamics." In A. Fine et al. (eds.), pp. 59–69. East Lansing, MI: Philosophy of Science Association.

Squires, E. (1990), *Conscious Mind in the Physical World.* Bristol: Adam-Hilger.

Stapp, H. P. (1990), "Quantum Measurement and the Mind–Brain Connection." In P. Lahti and P. Mittelstaedt (eds.), *Symposium on the Foundations of Modern Physics.* Singapore: World Scientific.

Stein, H. (1970), "On the Notion of Field in Newton, Maxwell, and Beyond." In R. Steuwer (ed.), *Historical and Philosophical Perspectives of Science* (Minnesota Studies in the Philosophy of Science, vol. 5), pp. 264–87. Minneapolis: University of Minnesota Press.

Stein, H. (1989), "Yes, but . . . – Some Skeptical Remarks on Realism and Antirealism." *Dialectica* 43: 47.

Stein, H. (1990), "On Locke, 'the Great Huygenius, and the Incomparable Mr. Newton'." In P. Bricker and R. I. G. Hughes (eds.), *Philosophical Perspectives on Newtonian Science.* Cambridge, MA: MIT Press.

Stein, H. (199–), "On Metaphysics and Method in Newton." Unpublished manuscript, Philosophy Department, University of Chicago.

Stroud, C. R., and Jaynes, E. T. (1970), "Long-term Solutions in Semiclassical Radiation Theory." *Physical Review A* 1: 106.

Tisza, L. (1963), "The Conceptual Structure of Nature." *Reviews of Modern Physics* 35: 151. Reprinted in L. Tisza (1966), *Generalized Thermodynamics,* pp. 343–77. Cambridge, MA: MIT Press.

van Fraassen, B. (1980), *The Scientific Image.* Oxford: Clarendon Press.

van Fraassen, B. (1985), "Empiricism in the Philosophy of Science." In P. M. Churchland and C. A. Hooker (eds.), *Images of Science,* pp. 245–308. Chicago: University of Chicago Press.

Walsh, W. H. (1975), *Kant's Criticism of Metaphysics.* Chicago: University of Chicago Press.

Weisskopf, V. (1979), *Knowledge and Wonder,* 2nd ed. Cambridge, MA: MIT Press.

Wheeler, J. A. (1983), "On Recognizing Law without Law." *American Journal of Physics* 51: 398.

Wheeler, J. A. (1985), "Bohr's 'Phenomenon' and 'Law without Law'." In G. Casati (ed.), *Chaotic Behavior in Quantum Systems,* pp. 363–78. New York: Plenum.

Whitehead, A. N. (1929), *Process and Reality.* New York: Macmillan.

Whitehead, A. N. (1933), *Adventures of Ideas.* New York: Macmillan.

Whittaker, E. (1951), *A History of Theories of Aether and Electricity,* vol. 1. London: Thomas Nelson.

Wigner, E. P. (1962), "Remarks on the Mind–Body Question." In I. J. Good (ed.), *The Scientist Speculates,* pp. 284–302. London: Heinemann. Reprinted in E. P. Wigner (1967), *Symmetries and Reflections,* pp. 171–84. Bloomington: University of Indiana Press.

3

Search for a worldview which can accommodate our knowledge of microphysics

I. CONTRIBUTION OF THE NATURAL SCIENCES TO A WORLDVIEW

In a sense every human being with rudimentary intelligence has a worldview, that is, a set of attitudes on a wide range of fundamental matters. A philosophical worldview must be more than this, however. It must be articulate, systematic, and coherent. A philosophical worldview may be naturally subdivided into at least the following parts: (1) a metaphysics, which identifies the types of entities constituting the universe (possibly organizing them into a hierarchy, with some fundamental and others derivative), and which in addition asserts fundamental principles, such as those of cause and chance, that govern these entities; (2) an epistemology, concerned with the assessment of human claims to knowledge or justified belief; (3) a theory of language; and (4) a theory of ethical and aesthetic values. For reasons of professional training and preoccupation, and also of relevance to this collection of essays, I shall confine my comments almost entirely to metaphysics and epistemology.

A necessary condition for the coherence of a philosophical worldview is the meshing of its metaphysics and its epistemology. The metaphysics should be capable of understanding the existence and the status of subjects like ourselves who are capable of deploying the normative procedures of epistemology. And the epistemology should suffice to account for the capability of human beings to achieve something like a good approximation to knowledge of metaphysical principles, in spite of the spatial and temporal limitations of our experience and the flaws and distortions of our sensory and cognitive apparatus. This meshing can be briefly called "the closing of the circle."

In my opinion the twentieth century is one of the golden ages of metaphysics, probably surpassed by the fourth century B.C. in conceptual

Article is taken from *Philosophical Consequences of Quantum Theory: Reflections on Bell's Theorem,* edited by James T. Cushing and Ernan McMullin. © 1989 by the University of Notre Dame Press. Reprinted by permission.

innovations, but probably surpassing all previous ages in the control and precision of the best metaphysical thinking. At least four important factors can be cited to account for the flowering of twentieth-century metaphysics. (1) There has been an unparalleled sharpening of logical, mathematical, and semantical analysis, which has provided powerful tools for philosophical criticism. The positivists used some of these tools to challenge the legitimacy of the enterprise of metaphysics, but this challenge has been a stimulus to careful constructive work.[1] (We learned from Carnap that in every neighborhood of a real problem there is a *Scheinproblem,* arising from abuse of language; but we also learned – from whom? maybe Plato – that in the neighborhood of every *Scheinproblem* there is a real problem awaiting intelligent extraction.) (2) The methodology of metaphysics has been greatly extended by the use of the hypothetico-deductive method – often tacitly, but in some cases (like Whitehead's *Process and Reality*) self-consciously and explicitly. There are, of course, classical instances – in the work of Newton, for example, and even earlier in the Greek atomists – of renouncing the assumption that a fundamental principle concerning the world has to be known with greater certainty than derivative principles. Aristotle already made the famous distinction between "things which are more obscure by nature but clearer to us" and "those which are clearer and more knowable by nature,"[2] but the full exploitation of his profound perception required the replacement of his own method of intuitive induction by the indirect procedures of the hypothetico-deductive method. (3) Parts of the natural sciences have been deepened to the point where evidence can be brought to bear in a controlled way upon problems traditionally classified as metaphysical. The results of relevant investigations in the natural sciences make it possible to use the hypothetico-deductive method specifically and fruitfully. (4) Phenomenological investigations have provided refined reports of ordinary experience, which are relevant to metaphysics. Although I am skeptical that phenomenology provides direct access to metaphysical principles, I believe that phenomenological reports are valuable supplements to typical scientific evidence in metaphysical investigations using the hypothetico-deductive method.[3]

Of the foregoing, I am most preoccupied with (3) and can cite several spectacular examples. General relativity has shown that the space–time field interacts dynamically with matter, thereby behaving more like

1. A fine example is Hao Wang, *Beyond Analytic Philosophy* (Cambridge, Mass.: MIT Press, 1986). See especially his remark about metaphysics on p. 131.
2. Aristotle, *Physics* 184a 20–1, translated in *The Basic Works of Aristotle,* ed. R. McKeon (New York: Random House, 1941).
3. See, for example, A. N. Whitehead, *Adventures of Ideas* (New York: Macmillan, 1933), chap. 15.

a substance than did the fixed space–times of Newtonian kinematics or of special relativity. Molecular biology has provided, in some detail, an explanation in terms of chemical cybernetics of the crucial organic properties of teleological behavior, spontaneous morphogenesis, and reproduction, thereby eliminating the need for postulating nonphysical vitalistic elements in primitive biology. We can discern that elementary particle physics and cosmology are in the process of making contributions to metaphysics, but it is probably premature to say exactly what these are. The metaphysical implications of quantum mechanics will be discussed in some detail below.

It is legitimate, in view of all these rich results, to speak of the enterprise of *experimental metaphysics*. A warning is needed, however, against possible misunderstanding of this term. One should not anticipate straightforward and decisive resolution of metaphysical disputes by the outcomes of experiments. We know, in fact, that laboratory tests alone do not settle without careful conceptual analysis even those problems which are commonly classified as scientific, and *a fortiori* such analysis is indispensable in coming to grips with metaphysical problems.

2. METAPHYSICAL IMPLICATIONS OF QUANTUM MECHANICS

Whether quantum mechanics has definite metaphysical implications depends upon two conditions: (a) the legitimacy of interpreting the quantum state of a physical system "realistically," as something independent of the knowledge of an observer, or indeed of all observers; and (b) the completeness of the quantum-mechanical description of a physical system. If these two conditions are satisfied, then we are obliged to accept a literal understanding of the peculiarities of the quantum formalism. The fact that in any pure quantum state there are physical quantities that are not assigned sharp values will then mean that there is *objective indefiniteness* of these quantities. The fact that the outcome of a measurement of such an indefinite quantity is not determined by the quantum state of the object, even in conjunction with the quantum state of the measuring apparatus, means that there is *objective chance*. The fact that the quantum state does determine (with due consideration for statistical fluctuations) the frequencies in an ensemble of measurements means that there is *objective nonepistemic probability* (sometimes called "propensity"). The fact that there exist quantum states of two-body systems which cannot be factorized into products of one-body quantum states means that there is *objective entanglement* of the two bodies, and hence a kind of *holism*. And the fact that spatially well-separated bodies can be entangled means that there is a peculiar kind of *quantum nonlocality* in nature, which is manifested in correlations of the outcomes of measurements performed upon the separated bodies.

The challenge to condition (b) is the program of "hidden variables," which are putative properties of a physical system not specified by the system's quantum state. The simplest type of hidden-variables theory is one in which a complete state of a system assigns to every proposition (or "eventuality") of the system a definite truth value (so that the quantum-mechanical indefiniteness of a quantity is interpreted epistemically as due to ignorance or to practical limitations upon measurement). It was proved by Gleason and others[4] that in almost all interesting cases such a simple hidden-variables theory is mathematically excluded by the algebraic structure of the quantum formalism. Bell (1966) noted, however, that Gleason's theorem does not preclude a more complex type of hidden-variables theory, called "contextual," according to which a complete state assigns a definite truth value to a proposition only relative to a specified context. The success of a contextual hidden-variables program would permit all of the peculiarities of the quantum-mechanical formalism to be interpreted epistemically, thereby avoiding the metaphysical innovations implied by a realistic interpretation.

The encouragement which Bell provided to advocates of a more conservative metaphysics by exhibiting the family of contextual hidden-variables theories was undermined by a great theorem which he proved under restrictive conditions in 1964 but much more generally in 1971: Any hidden-variables theory which obeys a certain "locality condition" will imply that the correlations of certain quantities of two spatially separated systems must satisfy a certain inequality, and that under specified conditions the predictions of quantum mechanics violate that inequality. Bell's theorem precludes the interpretation of quantum mechanics by a contextual hidden-variables theory satisfying the locality condition. The discrepancy exhibited in Bell's theorem has the further consequence that in principle an experimental test can be made between quantum mechanics and the entire family of physical theories satisfying the locality condition. At least thirteen tests have been performed, of which all but two confirm the predictions of quantum mechanics and violate Bell's inequality. And there are good reasons for suspecting systematic errors in the two anomalous cases.[5] The most spectacular of these experiments is that of Aspect and his collaborators, in which the choice of analyzers used to examine the polarization

4. Gleason (1957); Bell (1966); Kochen and Specker (1967); Belinfante (1973).
5. A detailed survey of experiments up to 1978, and also of variant proofs of Bell's theorem up to that date, is given by Clauser and Shimony (1978). A tabulation of experiments up to 1987 is given by Redhead (1987b); the experiment with time-varying analyzers is reported by Aspect, Dalibard, and Roger (1982).

of each of a pair of photons in a quantum-mechanically entangled state is made too rapidly for information to be conveyed subluminally from the apparatus used for one of the photons to the apparatus used for the other; consequently, the violation of Bell's locality condition by the correlations exhibited in this experiment cannot be explained (except by rather contrived and desperate assumptions) as instances of information transfer in accordance with relativity theory.

The foregoing summary of Bell's theorem and the experiments which it inspired is extremely condensed, but details may justifiably be omitted since I have given detailed expositions elsewhere (1984a; 1984b; 1986) and many other papers are devoted to detailed analysis. The following remarks will suffice for my purposes here.

1. Jarrett (1984) made the valuable discovery that Bell's locality condition is equivalent to the conjunction of two conditions, which he calls "locality" and "completeness" (for which I have suggested the names "parameter independence" and "outcome independence"). The former holds if the probability of an outcome of an observation on one particle is independent (given the complete state of the particle pair at the moment of emission) of the parameter chosen for the analyzer of the other particle; the latter holds if the probability of an outcome of an observation on one particle is independent (again given the complete state of the pair) of the outcome of the observation of the other particle. The violation of Bell's inequality by the experimental correlations implies by *modus tollens* that Bell's locality condition is violated in the experimental situation, and therefore one or the other of Jarrett's locality and completeness conditions must be false. But which one?

Since the experimental results not only violate Bell's inequality but confirm with accuracy the quantitative predictions of quantum mechanics, it is reasonable to answer this question by consulting quantum mechanics. It is easily seen that the quantum predictions of an entangled state can violate completeness (outcome independence), but analysis also shows[6] that the quantum predictions do not violate locality (parameter independence). Jarrett demonstrated that the violation of locality would in principle permit a message to be sent between the two pieces of apparatus with superluminal velocity. By contrast, even though a message can be sent by exploiting a violation of completeness, it would be subluminal (Jarrett 1984; Shimony 1986). Consequently, even though there is evidently some tension between relativity theory and the violation of completeness by quantum mechanics (and, by the argument just given, by the experimental results), there is nevertheless a kind of "peaceful coexistence" between quantum mechanics and relativity theory.

6. Eberhard (1978); Page (1982); Ghirardi, Rimini and Weber (1980).

2. Because of peaceful coexistence, the tension between quantum mechanics and relativity theory is not a crisis, resolvable only by the retrenchment of one or both. Nevertheless, a deeper understanding of the relation between these two fundamental parts of physical theory is desirable. One possibility is that success in quantizing space–time structure (perhaps more radically than envisaged in current programs of quantizing general relativity) will throw some unexpected light upon peaceful coexistence. Specifically, modification of the topology of space–time in the small may yield a new interpretation of nonlocality. Another possibility is that illumination will come more from conceptual analysis than from new physics. The failure of completeness (outcome independence) indicates that the classical concept of a localized event needs to be broadened and that the classical concept of causality must be modified. Investigations in both of these directions are reported in the papers by Henry Stapp, Don Howard, and Paul Teller in this volume [J. T. Cushing and E. McMullin (eds.), *Philosophical Consequences of Quantum Theory* (Notre Dame, IN: University of Notre Dame Press, 1989)].

3. Someone might propose to bypass Bell's theorem and the experiments which it inspired by accepting a hidden-variables theory that violates Bell's locality condition. To be sure, in an experimental situation like Aspect's such a strategy would have been unpalatable to Einstein, but from the standpoint of a conservative metaphysician it might be attractive: one could exorcize the quantum-mechanical innovations of objective indefiniteness, objective chance, objective probability, and entanglement by paying the price of accepting nonlocality. In my opinion, the crucial weakness of this proposal is the unlikelihood that a nonlocal hidden-variables theory can be devised which achieves the peaceful coexistence with relativity theory that quantum mechanics enjoys.[7] Unless revolutionary experiments overthrow the relativistic prohibition of superluminal signals, such peaceful coexistence remains a desideratum for any physical theory.

4. In the foregoing reasoning about the significance of Aspect's experiment several loopholes have been pointed out, the importance of which should be assessed.

---

7. The hidden-variables model of Bohm (1952) is so designed that it replicates all the predictions of standard quantum mechanics, though details of a relativistic version are not worked out. It would seem, therefore, that if standard quantum mechanics can achieve peaceful coexistence with relativity theory, so can this model of Bohm. However, Bohm himself was not content with a completely phenomenological postulation of a "quantum potential," and in later work he spoke of objectively real fields which are the locus of the quantum potential (e.g., 1957, 112–114). The question of peaceful coexistence of his models with relativity theory is discussed in the much more recent work by Bohm and Hiley (1984) and Bell (1987).

(a) In Aspect's experiment, the choice of analyzers for examining the polarization of each photon of a pair is made by switches which are periodically switched off and on. Even though relativity precludes the direct transfer of information from one switch at a given moment in the laboratory frame of reference to the other switch at the same moment, it does not preclude sufficiently clever hidden variables located at one of the switches from using inductive logic to infer the contemporaneous state of the other switch. This criticism can be partially answered by pointing out that two switches are driven independently and have independent phase slippage from time to time, and furthermore that there is no obvious mechanism for the hypothetical exercise in inductive logic. Nevertheless, a revised experiment in which the switches are operated stochastically rather than periodically is desirable.

(b) Most of the pairs which pass through the analyzers on the two sides of the experiment are not detected in experiments performed to date. There is a hidden-variables model which reproduces the quantum-mechanical predictions of observed correlations because the hidden variables not only determine passage or non-passage of a particle through an analyzer but also detection or non-detection; and furthermore, all the operations postulated by this model are local.[8] The model requires, however, a limit upon the efficiency of the detectors, and hence this loophole can be blocked by performing a correlation experiment in which both the analyzers and the

---

8. Clauser and Horne (1974). This model yields quantum-mechanical counting rates (provided the detectors are sufficiently inefficient) for any choice by the experimenter of analyzer orientations. The experimenter's choice does not affect the rules of operation of the model, but only specifies initial and boundary conditions, as in any physical experiment. Fine (1982d) has presented two models of a genre which he calls "prism models," in which there is response or nonresponse by the two parts of the apparatus, subsequent to the specification of the orientations of the analyzers only for a subspace of the space of states. His two models are much simpler than that of Clauser and Horne, but each has the limitation of being tailored to a specific choice of two alternative orientations of one analyzer and two alternative orientations of the other analyzer. For instance, in the "maximal" model on p. 287, the response function $A_i(\lambda)$ when the first analyzer is put in the ith orientation ($i = 1$ or 2 only), depends upon θ and θ', which in turn depend on the two alternative orientations chosen for the second analyzer. Although Fine says that there exist many more prism models, he has not explicitly exhibited one in which the response function $A_i(\lambda)$ for all the infinitely many values of i are defined in a way that is independent of the choice of two orientations of the second analyzer to be used in an actual experiment; and likewise for the response functions $B_j(\lambda)$ for all the infinitely many values of j of the second analyzer. If a different prism model has to be provided for each choice of the two pairs of orientations in an actual experiment, this would be tantamount to endowing the experimenter not only with the power to choose initial and boundary conditions but also with the ability to influence the laws of physics. Consequently, Fine has not justified his claim: "There is, therefore, no question of the consistency of prism models, at least in principle, with the quantum theory" (1986, 52).

detectors are sufficiently efficient (Lo and Shimony 1981; Mermin 1986; Fry 1992).

(c) Although the actions of the two switches for one pair of particles in Aspect's experiment are events with spacelike separation, and furthermore with spacelike separation from the event of particle emission, there nevertheless exists a region of overlap of the three backward light cones extending from the switching and emission events; and it is logically possible that events in this overlap region can control the emission and analysis of each pair in a manner fully consistent with relativity theory, but in such a way that the quantum-mechanical correlations are achieved. There is, however, no known or even proposed mechanism for achieving this result. Consequently, general considerations of scientific methodology suggest that this loophole not be taken seriously. After all, any experimental result which seems to disconfirm a favored hypothesis could be attributed to a conspiracy with an unknown mechanism, but at the price of generic skepticism about experimental investigations of nature.

In spite of the reservations which have been indicated, the case is strong for rejecting hidden-variable theories and accepting the completeness of the quantum-mechanical description of physical systems. Before taking the next step and accepting the metaphysical implications of a realistic interpretation of quantum mechanics, a discussion is required of the realistic interpretation, labeled above in Section 2 as condition (a). There is an important philosophical tradition, extending from Berkeley to Mach to some of the positivists, according to which any scientific theory is nothing more than an instrument for organizing and predicting human experience. The usual arguments for an instrumentalist interpretation of scientific theories are independent of the content of the theories and apply just as well, or just as poorly, to Newtonian particle mechanics as to quantum mechanics. There are semantical arguments, to the effect that terms putatively referring to unobserved entities are meaningless; epistemological arguments, advocating maximum caution in performing inductions; and methodological arguments, espousing Ockham's principle.[9] But how good is the case for instrumentalism? The *a priori* arguments of Berkeley and his successors against physical realism rest upon unjustifiably restrictive and arbitrary semantical, epistemological, and methodological theses, as a vast body of critical literature has attempted to demonstrate.[10] If the *a priori* arguments fail, then instrumentalism must be

9. I discovered, with great delight, Machs' [sic] General Store in Pawlet, Vermont; it reveals the Yankee storekeeper within the instrumentalist: "What we have is on the shelves; what you don't see, don't exist; and we don't give credit."

10. Three papers from this literature which I have found particularly illuminating are C. S. Peirce's review of Fraser's edition of the works of George Berkeley, *Collected Papers,*

judged *a posteriori* in competition with various versions of physical realism, preference being given to the point of view with the greatest explanatory power. It is particularly difficult to incorporate instrumentalism into a coherent philosophy which "closes the circle" between metaphysics and epistemology, if the metaphysics takes cognizance of the evidence that human beings evolved only very recently in the history of the universe.[11]

3. REDUCTION OF THE WAVEPACKET AND STOCHASTIC VARIANTS OF QUANTUM MECHANICS

Although the content of a scientific theory is irrelevant to the usual arguments for interpreting it instrumentally, there is a peculiarity in quantum theory that supplies an entirely novel argument for instrumentalism. The peculiarity is the problem of measurement (also known as "the problem of the reduction of the wavepacket" and as "the problem of the actualization of potentialities"), and it arises from the linearity of quantum dynamics. Here is a highly idealized formulation of the problem, which suffices for present purposes. Suppose that u_1 and u_2 are normalized vectors representing states of a microscopic object in which a physical quantity A has distinct values a_1 and a_2; that v_0 is a normalized vector representing the initial state of a measuring apparatus; that v_1 and v_2 are normalized vectors representing states of the measuring apparatus in which a macroscopic quantity B (such as the position of a pointer needle) has distinct values b_1 and b_2 respectively; and finally that the interaction between the microscopic object and the measuring apparatus is such that

$$u_i \otimes v_0 \to u_i \otimes v_i \quad (i = 1 \ or \ 2),$$

where the arrow stands for temporal evolution during an interval of specified duration t. Under these circumstances, if it is known that initially the microscopic object was either in the state represented by u_1 or the state represented by u_2, but it was not known which, then the missing information can be obtained simply by examining the quantity B of the apparatus at time t after the interaction commenced and ascertaining whether the value of this macroscopic quantity is b_1 or b_2. Furthermore,

vol. 8, ed. A. W. Burks (Cambridge, Mass.: Harvard University Press, 1958), 9–38; J. F. Thomson's article on Berkeley in *A Critical History of Western Philosophy*, ed. D. J. O'Connor (New York: Free Press, 1964), 236–252; and H. Stein (1972), 267–438, especially pp. 370–373 and pp. 405–409.

11. This argument is often presented in works on evolutionary epistemology. One version may be found in Shimony (1971). Mach himself was an ardent advocate of the theory of evolution; there was evident tension between this part of his thought and his dictum that "the world consists only of our sensations," *The Analysis of Sensations* (New York: Dover, 1959), 12.

the observation of B will permit the inference of the value of A both at the beginning and at the termination of the measurement. Thus far there is no conceptual difficulty.

Suppose, however, that initially the microscopic object is prepared in the state represented by the superposition $c_1 u_1 + c_2 u_2$, where the sum of the absolute squares of c_1 and c_2 is 1, and neither c_1 nor c_2 is zero. States of this kind are physically possible, according to the formalism of quantum mechanics, and indeed it is often experimentally feasible to prepare such states. The linear dynamics of quantum mechanics implies then that

$$(c_1 u_1 + c_2 u_2) \otimes v_0 \rightarrow c_1 u_1 \otimes v_1 + c_2 u_2 \otimes v_2.$$

In the state represented by the sum on the right hand side of this process, the macroscopic quantity B does not have a definite value. This fact in itself is peculiar, because our ordinary experience indicates that macroscopic physical quantities always have definite values. Furthermore, there is a conceptual difficulty in understanding the quantum formalism. The standard interpretation of the superposition $c_1 u_1 + c_2 u_2$ is that the quantity A has an indefinite value, but in the event that A is actualized, there is a probability $|c_1|^2$ that the result will be a_1 and a probability $|c_2|^2$ that it will be a_2. Now if the quantum dynamics precludes a definite measurement result, what sense does it make to speak of the probabilities of various outcomes? A literal and realistic interpretation of the quantum dynamics undermines the literal and realistic interpretation of the quantum state! This problem evidently is dissolved, however, if the quantum formalism is interpreted instrumentally. For then the quantum state is interpreted as a shorthand for a procedure which permits the reliable anticipation of a statistical distribution in observations of a certain kind. The superposition $c_1 u_1 + c_2 u_2$ can be used effectively for this purpose, and one should avoid the *Scheinproblem* of demanding the values of quantities A or B apart from their role in human experience.

My opinion is that a realistic interpretation of the quantum formalism, and thereby the metaphysical implications of quantum mechanics summarized earlier, can be salvaged by a change of physics: *viz.,* the abandonment of linear dynamics. This proposal – admittedly conjectural – is the antithesis of my attempt to draw philosophical consequences from scientific results, for it indicates rather a reliance on philosophical considerations to supply the heuristics for a scientific investigation. Working physicists are happy with standard quantum dynamics (the time-dependent Schrödinger equation and its relativistic surrogates), and there are no puzzling phenomena at present which point to a modification of this dynamics. The motivation for this suggestion is that of saving physical realism. The title of my paper – "Search for a worldview which can accommodate our knowledge of microphysics" – in a way is not quite appropriate, since I am proposing to modify the microphysics in order to achieve a coherent worldview.

One way to modify linearity is to maintain a differential equation governing dynamic evolution of a state, but to replace the linear differential equation of Schrödinger by a nonlinear one. The hope would be that the nonlinear equation would agree very closely with the Schrödinger equation for systems with few degrees of freedom, where the latter has been very well confirmed, but would deviate sharply from it for systems with many degrees of freedom. Specifically, a superposition of states in which a macroscopic quantity has distinct values should spontaneously and rapidly evolve into a state in which that quantity has a sharp value. The requirement "rapidly" is imposed to agree with the phenomenology of laboratory measurements, which are often completed in time intervals of the order of microseconds or less. No proposals have been made concretely which satisfy all of these *desiderata,* and it may be mathematically impossible to satisfy them simultaneously.

More promising is the family of stochastic modifications of the Schrödinger equation, in which the initial state does not determine the final state unequivocally, and hence a role of chance in the outcome is explicitly postulated – and such modifications have been proposed.[12]

The theory of Ghirardi, Rimini, and Weber (1986) seems to me the most promising to date, and indeed they have sketched a very impressive unification of microdynamics and macrodynamics. They postulate that during all but a discrete set of instants the Schrödinger equation governs the evolution of the quantum state of a physical system; however, at instants which themselves are selected stochastically, there is a spontaneous reduction of the spread of the quantum state in configuration space. The rate of such stochastic reductions of spread is slow for systems with few degrees of freedom and very fast (automatically from their formalism, with no additional postulation) for systems with many degrees of freedom. One anticipates, therefore, that even if a measurement should result in a state which is a superposition of different positions of a pointer needle, which is a macroscopic system with a number of the order of 10^{23} of degrees of freedom, there would be a spontaneous transition within a microsecond or so into a well-localized state of the needle.

A number of objections may be raised against the theory of Ghirardi, Rimini, and Weber. It implies that energy and momentum are not conserved, though the amount of nonconservation is small. The theory is nonrelativistic. It is very much tied to the position representation, even

---

12. Pearle (1976; 1979); Gisin (1981; 1984); Gisin and Piron (1981); and Ghirardi, Rimini, and Weber (1986).

though experimental considerations may indicate more desirable choices of the representation in which reduction takes place. It may not suffice to explain the rapidity of results in most measurement situations, for as Albert and Vaidman pointed out,[13] the excitation of a few molecules, with few degrees of freedom, by an incident particle is the crucial precipitating step of a typical measurement. My strongest objection, however, is that the theory permits the formation for a short time of monstrous states of macroscopic objects – states in which the pointer needle has an indefinite position, or Schrödinger's cat is neither dead nor alive – and then it rapidly aborts the monstrosity. It would be much better to have a stochastic theory which provides contraception against the formation of such a monstrosity, by destroying unwanted superpositions in the earliest stages of the interaction of the microscopic object with the macroscopic apparatus. I have no suggestions at present for the details of such a theory, but it is my preoccupation and the subject of my current research.

A stochastic modification of quantum mechanics will not be acceptable unless it has clear experimental consequences which are confirmed. At present I have no way of saying what those consequences might be. My hope is that some of them will concern macromolecules, which occupy a strategic position between microscopic and macroscopic bodies. The quantum mechanics of macromolecules has provided essential ingredients in the triumphs of molecular biology of the last three decades: the stability of DNA, the folding of proteins to form characteristic globular shapes, and the stereospecific recognition of a substrate by an enzyme. I am not convinced, however, that current quantum dynamics is entirely satisfactory for biological purposes. It is biologically important that an enzyme behave like a switch, which is definitely off or definitely on with regard to mediating a chemical reaction. It would be troublesome to have macromolecular analogues of Schrödinger's cat – enzymes which are in a superposition of switching on and switching off. Pragmatically, molecular biology pays no attention to the possibility of such superpositions, which seems to indicate that they seldom if ever occur. It is not clear to me, however, that the non-occurrence of such superpositions is well understood within the framework of standard linear quantum dynamics, whereas a stochastic modification of this dynamics might provide a natural explanation (see Pattee 1967).

4. THE MIND–BODY PROBLEM

The conjectured stochastic modification of quantum dynamics would contribute greatly to "the closing of the circle," which is one of the chief

13. D. Albert (1990).

requirements for a coherent worldview, for it would explain in principle how there are definite outcomes of experiments – e.g., definite registrations on photographic plates and definite firings of Geiger counters – and these outcomes provide the evidence upon which the immense superstructure of physical theory is based. But unless a physicalistic explanation of mental events is correct – which seems incredible to me, for reasons that are naive but strong – the definite physical outcomes of experiments constitute only a necessary but not a sufficient condition for human experience and inference. The greatest obstacle to "closing the circle" is the ancient one which haunted Descartes and Locke – the mind–body problem. Does contemporary microphysics have any implications concerning this problem?

One very interesting implication is negative. Whitehead proposed to solve the mind–body problem by a modern version of Leibniz's monadology, according to which the fundamental constituents of natural things are endowed with protomental characteristics. (Whitehead's actual occasions differ from Leibniz's monads by being short-lived and possessing "windows.") In ordinary matter the protomental characteristics of neighboring occasions are "incoherent," and therefore the descriptions of matter by physical theory are statistically very accurate; but a sketch of an explanation is also provided for the possibility of emergence of high level mentality, when the protomental characteristics of a society of occasions achieve "coherence."[14] It is, however, very difficult to reconcile Whitehead's theory with the great body of evidence favoring the intrinsic identity of electrons, for his postulated protomental features at the subphysical level would endow each electron with unequivocal individuality.

On the other hand, the metaphysical innovations of quantum mechanics – objective indefiniteness, objective chance, objective probability, and entanglement – have obvious analogies to some phenomenological features of mentality. It is premature to judge the significance of these analogies. Nevertheless, I conjecture that a world in which these metaphysical principles hold is somehow more hospitable to a dualism of mind and body than a world governed by the metaphysical principles associated with classical physics. I certainly do not want to suggest that quantum mechanics by itself provides a resolution to what Whitehead calls "the bifurcation of nature," but it may provide a framework within which a resolution can be sought.[15] In making this conjecture I run the risk that my position may be conflated with that of Maharishi Mahesh Yogi, who

14. A. N. Whitehead, *Adventures of Ideas,* 266–267.
15. See R. Faber, *Clockwork Garden* (Amherst, Mass.: University of Massachusetts Press, 1986), chaps. 10–11. This book is reviewed by Shimony, *Foundations of Physics* 17 (1987): 1041–1043.

expounds the thesis that macroscopic quantum coherence is exhibited by a meditating cohort (the "Maharishi effect"). I suspect, however, that Maharishi is not as cognizant as I of the need to determine with precision the mathematical structure of the space of mental states or to design careful and informative experiments.

REFERENCES

Albert, D. (1990), "On the Collapse of the Wave Function." In A. Miller (ed.), *Sixty-two Years of Uncertainty,* pp. 153–65. New York: Plenum.

Aspect, A., Dalibard, J., and Roger, G. (1982), "Experimental Tests of Bell's Inequalities Using Time-varying Analyzers." *Physical Review Letters* 49: 1804–7.

Belinfante, F. J. (1973), *A Survey of Hidden-Variable Theories.* Oxford: Pergamon.

Bell, J. S. (1964), "On the Einstein–Podolsky–Rosen Paradox." *Physics* 1: 195–200. Reprinted in Bell (1987).

Bell, J. S. (1966), "On the Problem of Hidden Variables in Quantum Mechanics." *Reviews of Modern Physics* 38: 447–52. Reprinted in Bell (1987).

Bell, J. S. (1987), *Speakable and Unspeakable in Quantum Mechanics.* Cambridge: Cambridge University Press.

Bohm, D. (1952), "A Suggested Interpretation of the Quantum Theory in Terms of 'Hidden' Variables, I." *Physical Review* 85: 166–79.

Bohm, D. (1957), *Causality and Chance in Modern Physics.* New York: Harper.

Bohm, D., and Hiley, B. (1984), "Quantum Potential Model for the Quantum Theory." In S. Kamefuchi et al. (eds.), *Foundations of Quantum Mechanics in the Light of New Technology.* Tokyo: Physical Society of Japan.

Clauser, J. F., and Horne, M. A. (1974), "Experimental Consequences of Objective Local Theories." *Physical Review D* 10: 526–35.

Clauser, J. F., and Shimony, A. (1978), "Bell's Theorem: Experimental Tests and Implications." *Reports on Progress in Physics* 41: 1881–927.

Eberhard, P. (1978), "Bell's Theorem and the Different Concepts of Locality." *Nuovo Cimento* 46B: 392–419.

Faber, R. (1986), *Clockwork Garden.* Amherst, MA: University of Massachusetts Press.

Fine, A. (1982), "Some Local Models for Correlation Experiments." *Synthèse* 50: 279–94.

Fine, A. (1986), *The Shaky Game: Einstein, Realism and the Quantum Theory.* Chicago: University of Chicago Press.

Fry, E. (1992), "An Experimental Test of the Strong Bell Inequalities." Unpublished manuscript, Physics Department, Texas A&M University.

Ghirardi, G.-C., Rimini, A., and Weber, T. (1980), "A General Argument against Superluminal Transmission through the Quantum-mechanical Measurement." *Lettere al Nuovo Cimento* 27: 293–98.

Ghirardi, G.-C., Rimini, A., and Weber, T. (1986), "Unified Dynamics for Microscopic and Macroscopic Systems." *Physical Review D* 34: 470–91.

Gisin, N. (1981), "A Simple Non-linear Quantum Evolution Equation." *Journal of Physics A* 14: 2259–67.
Gisin, N. (1984), "Quantum Measurements and Stochastic Processes." *Physical Review Letters* 52: 1657–60.
Gisin, N., and Piron, C. (1981), "Collapse of the Wave-function without Mixture." *Letters in Mathematical Physics* 5: 379–85.
Gleason, A. M. (1957), "Measures on the Closed Sub-spaces of Hilbert Spaces." *Journal of Mathematics and Mechanics* 6: 885–93.
Jarrett, J. (1984), "On the Physical Significance of the Locality Conditions in the Bell Arguments." *Noûs* 18: 569–89.
Kochen, S., and Specker, E. P. (1967), "The Problem of Hidden Variables in Quantum Mechanics." *Journal of Mathematics and Mechanics* 17: 59–87.
Lo, T. K., and Shimony, A. (1981), "Proposed Molecular Test of Local Hidden-variable Theories." *Physical Review* 23A: 3003–12.
Mach, E. (1959), *The Analysis of Sensations.* New York: Dover.
Mermin, N. D. (1986), "The EPR Experiment: Thoughts about the 'Loophole'." In D. M. Greenberger (ed.), *New Techniques and Ideas in Quantum Measurement Theory. Annals of the New York Academy of Sciences,* vol. 480, New York: New York Academy of Sciences.
Page, D. (1982), "The Einstein–Podolsky–Rosen Physical Reality Is Completely Described by Quantum Mechanics." *Physics Letters* 91A: 57–60.
Pattee, H. H. (1967), "Quantum Mechanics, Heredity, and the Origin of Life." *Journal of Theoretical Biology* 17: 410–20.
Pearle, P. (1976), "Reduction of the Wave-packet by a Non-linear Schrödinger Equation." *Physical Review D* 13: 857–68.
Pearle, P. (1979), "Toward Explaining Why Events Occur." *International Journal of Theoretical Physics* 18: 489–518.
Redhead, M. (1987), *Incompleteness, Nonlocality, and Realism: A Prolegomenon to the Philosophy of Quantum Mechanics.* Oxford: Clarendon.
Shimony, A. (1971), "Perception from an Evolutionary Point of View." *Journal of Philosophy* 68: 571–83.
Shimony, A. (1984a), "Controllable and Uncontrollable Nonlocality." In S. Kamafuchi et al. (eds.), *Foundations of Quantum Mechanics in the Light of New Technology.* Tokyo: Physical Society of Japan.
Shimony, A. (1984b), "Contextual Hidden Variables and Bell's Inequalities." *British Journal for the Philosophy of Science* 35: 25–45.
Shimony, A. (1986), "Events and Processes in the Quantum World." In R. Penrose and C. Isham (eds.), *Quantum Concepts of Space and Time,* pp. 182–203. Oxford: Oxford University Press.
Stein, H. (1972), "On the Conceptual Structure of Quantum Mechanics." In R. G. Colodny (ed.), *Paradigms and Paradoxes: The Philosophical Challenge of the Quantum Domain,* pp. 367–438. Pittsburgh: Pittsburgh University Press.
Thomson, J. F. (1964), "Berkeley." In D. J. O'Connor (ed.), *A Critical History of Western Philosophy,* pp. 236–52. New York: Free Press.
Whitehead, A. N. (1933), *Adventures of Ideas.* New York: Macmillan.

PART B

Perception and conception

4

Perception from an evolutionary point of view*

Recently there have been several extremely interesting expositions[1] of a thesis which is implicit in the evolutionary view of man, but which had (at least after the early waves of enthusiasm for Darwinian ideas[2]) been rather neglected: that human perceptual powers are as much a result of natural selection as any feature of organisms, with selection generally favoring improved recognition of objective features of the environment in which our pre-human ancestors lived. That these studies of perception from an evolutionary point of view have not influenced most writers on epistemology is one of many symptoms of the continuing dissociation of philosophy from the natural sciences. To be sure, at least one reason for abstaining from evolutionary considerations in epistemological inquiry is obvious and weighty: namely, to avoid begging questions, since the evolutionary view of human perceptual powers seems to be antecedently committed to the propositions of a specific epistemological position, the causal

This work originally appeared in *Journal of Philosophy* 68 (1971), pp. 571–83. Reprinted by permission of the publisher.

*Presented in an APA symposium on Evolution and the Causal Theory of Perception, December 28, 1971. Fred I. Dretske and Floyd Ratliff were co-symposiasts.

I am grateful to Howard Stein for discussions over many years of the ideas in this paper, as well as for suggestions on the final draft.

1. K. Lorenz, "Gestalt Perception as Fundamental to Scientific Knowledge," pp. 37–56 of L. v. Bertalanffy and A. Rapoport, eds. *General Systems,* **7** (Ann Arbor: Society for General Systems Research, 1962), translated from a paper in *Zeitschrift für experimentelle und angewandte Psychologie,* **6** (1959): 118–165; D. T. Campbell, "Pattern Matching as Essential in Distal Knowing," pp. 81–106 of K. R. Hammond, ed., *The Psychology of Egon Brunswik* (New York: Holt, Rinehart & Winston, 1966), and "Evolutionary Epistemology," to appear in P. A. Schilpp, ed., *The Philosophy of Karl R. Popper;* H. Yilmaz, "On Color Vision and a New Approach to General Perception," pp. 126–141 of E. E. Bernard and M. R. Kare, eds., *Biological Prototypes and Synthetic Systems* (New York: Plenum, 1962); F. Ratliff, *Mach Bands: Quantitative Studies on Neural Networks in the Retina* (San Francisco: Holden-Day, 1965); D. Bohm, *The Special Theory of Relativity* (New York: Benjamin, 1965), Appendix: "Physics and Perception."

2. See Campbell, "Evolutionary Epistemology" (note 1), and M. Čapek, "Ernst Mach's Biological Theory of Knowledge," *Synthese,* **18** (1968): 171–191, for many references to early formulations of evolutionary epistemology.

theory of perception. This reason will appear decisive, however, only if the mission of epistemology is conceived to be the establishment of an irrefutable foundation upon which the entire edifice of knowledge is to be erected. If, instead, epistemology is conceived to have a dialectical structure,[3] in which the starting point of an investigation is subject to subsequent correction and refinement, then one can legitimately ask whether evolutionary considerations, despite their presuppositions, can strengthen the case for the causal theory. The main purpose of this paper is to outline a positive answer to this question, by summarizing some of the insights that have been achieved by an evolutionary view of perception and by suggesting some points in the dialectic structure of epistemology at which these insights may be properly applied.

I

Since the philosophy of John Locke, who gave the classical formulation of the causal theory of perception, was in large part an extrapolation from mechanics,[4] it may be useful to state the general theses of the causal theory in language borrowed from classical physics. The causal theory regards the perceiver as one of many interacting systems in a common space–time. At any moment each system is in a definite state, which completely specifies its intrinsic properties, independently of the states of all other systems; but, because of dynamical interactions, the temporal development of the state of any one system does depend upon the states of others. (Note that this way of speaking temporarily evades the question of whether the perceiver should be regarded as a composite system, with mind and body as components, each in a definite state.) What may be called "the cognitive state" of the perceiver, consisting of the entire network of sensations, perceptions, memories, dispositions to organize information, inferences, etc., constitutes a partial specification of the complete state. Apart from all considerations of acuity, sanity, etc. of the perceiver, his cognitive state is, in principle, an index of the existence and properties of other systems only because of the dynamical interactions among systems. The entire state of the perceiver, and *a fortiori* his cognitive state, is kinematically independent of the states of all other systems

3. See, for example, W. V. Quine, *Ontological Relativity and Other Essays* (New York: Columbia, 1969), chs. 3 and 5 (with references to Campbell and Yilmaz on p. 90), and my "Scientific Inference," pp. 79–172 of R. G. Colodny, ed., *The Nature and Function of Scientific Theories: Essays in Contemporary Science and Philosophy* (Pittsburgh: University Press, 1971).
4. M. Mandelbaum, *Philosophy, Science, and Sense Perception* (Baltimore: Johns Hopkins Press, 1965), ch. 1.

in the world; but, because of the dynamical interactions, the cognitive state is in fact usually a quite good index of the existence of some other systems and of some of their properties. The numerous variants of the causal theory of perception (types of representative realism, critical realism, etc.) are distinguished from one another by their treatments of three questions: (1) what is the justification for this general picture? (2) what are the components and the structure of the cognitive state? and (3) in what respects is the cognitive state a good index of the properties of the environment?

It has often been objected that the causal theory of perception is incoherent. For the cognitive state of the perceiver reveals nothing directly about other systems; and knowledge of causal laws whereby something could be inferred about them is debarred by the inaccessibility to the perceiver of the external systems in the alleged causal relations. ("The mind has never anything present to it but the perceptions, and cannot possibly reach any experience of their connection with objects."[5]) The standard reply to this objection – and a partial answer to question 1 – is that the causal theory of perception is a hypothesis, and its confirmation, like that of any well-established scientific hypothesis, consists in its power to predict and explain facts of our experience with greater accuracy, comprehensiveness, and simplicity than any rival hypothesis.[6] However, it has been claimed that this reply begs the question, for the hypothetico-deductive method is nothing more than a systematic means for deciding among alternative schemes for ordering experience, and hence it tacitly and unavoidably operates within the cognitive state of the perceiver.[7] According to this view of the hypothetico-deductive method, the entities postulated by the causal theory of perception or by any scientific hypothesis have the ontological status of convenient fictions, and cannot be considered to be on the same footing as the elements of experience themselves.

The foregoing argument shows in a very condensed way how the question of the correctness of the causal theory of perception can be converted into a *Scheinproblem*. There seems to be a confrontation of frameworks, each giving on its own terms a different answer to a commonly formulated question. Carnap, more than any one else, has reasoned that the attempt to establish one or another framework as the correct one is futile, since each framework provides its own criterion for the reality of entities. He

5. D. Hume, *An Inquiry concerning Human Understanding,* C. Hendel, ed. (New York: Liberal Arts Press, 1955), sec. XII, part I.
6. E.g., A. O. Lovejoy, *The Revolt against Dualism* (Chicago: Open Court, 1929), ch. 8, and D. Williams, "The Inductive Argument for Realism," *Monist,* **45** (1934): 186–209.
7. A. J. Ayer, *The Origins of Pragmatism* (San Francisco: Freeman, Cooper, 1968), pp. 288–324.

recommends that the construction of alternative frameworks be viewed with tolerance and that the choice among them be regarded only as a practical matter "judged as being more or less expedient, fruitful, conducive to the aim for which the language is intended."[8]

It seems to me possible to present considerations that reduce the arbitrariness of choice among frameworks, without relinquishing Carnap's fundamental wisdom that we should not attempt to prove what is intrinsically incapable of proof and refute that which is hedged against refutation. One consideration is that there can be great differences of a methodological character among frameworks. Carnap's selection of examples makes it appear that each framework is somehow the linguistic counterpart of some philosophical dogma. There is a framework of things and a framework of sense data, which are, respectively, the counterparts of the old-fashioned "positions" of realism and phenomenalism; and similarly there are frameworks that are the counterparts of Platonism and nominalism. The principle of tolerance is therefore applied by him on a metaphilosophical level, toward frameworks which internally are manifestations of dogmatism. But might it not be possible to construct a framework that internalizes the principle of tolerance by a deliberate ontological openness? And is not just such a framework being employed informally when one appeals to the hypothetico-deductive method in judging the causal theory of perception against phenomenalism, Berkeleyan idealism, Kantian idealism, etc., all these alternatives being admitted initially as candidates without prejudgment?[9]

Another consideration is that the appeal to practicality in choosing a framework cannot in fact be accomplished without theory, as Carnap himself recognizes.[10] It is possible, for example, to explain in terms of biological and psychological theory why a thing language is efficient in coping with the ordinary problems of life, whereas a phenomenalistic language would be highly inefficient (or even psychosis-inducing). In this way results from the natural sciences can play a guiding role, though admittedly without conclusiveness, in the choice of a framework.

A historical comment may help, finally, to achieve perspective on the question of whether the correctness of the causal theory of perception is

8. R. Carnap, "Empiricism, Semantics, and Ontology," *Revue Internationale de Philosophie*, No. 11 (January 1950): 20–40.
9. Hilbert's program can be interpreted suggestively (though not with textual accuracy) as recommending openness toward the domain of mathematics and as criticizing Cantor on the one hand and Brouwer on the other for prejudging foundational questions prior to investigation; cf. Judson Webb, "Mentalism, Mechanism, and Meta-mathematics," Ph.D. thesis, Case-Western Reserve University, 1971.
10. Carnap, *op. cit.,* pp. 23 and 40.

a *Scheinproblem*. Berkeley's primary argument for immaterialism was evidently logical or conceptual, centering upon an analysis of what we can possibly mean by "existence."[11] And it is possible to construe his analysis (though counter to his intentions), as Carnap would do, as a proposal for a framework more parsimonious than Locke's but still adequate for ordinary life. What may be unfair to Berkeley, however, is an imputation that he intended by choosing a framework to entrench his immaterialism against all considerations. An indication of openness may be found in some subsidiary arguments which he directed against Locke, to supplement his primary one:

> . . . all phenomena are, to speak truly, appearances in the soul or mind; and it hath never been explained, nor can it be explained, how external bodies, figures, and motions, should produce an appearance in the mind. These principles, therefore, do not solve, if by solving it meant assigning the real, either efficient or final cause of appearances, but only reduce them to general rules.[12]

Berkeley suggests, on the other hand, that true causes are adduced in the immaterialist philosophy: efficient causes, of which volition is the prototype, and final causes, which are the designs of a "wise and good agent." It is reasonable to suggest, then, that success in providing explanations in some detail within a framework different from his immaterialism, contrasted with vagueness in teleological explanations within his own framework (because the designs of God are mysterious to us[13]), should be a weighty consideration by Berkeley's own criteria.

II

The evolutionary point of view supports the causal theory of perception by providing quasi-teleological explanations for many features of the perceptual powers of human beings and other animals, thereby answering one of the challenges posed by Berkeley.

The evolutionary explanation of an organic feature typically proceeds in the following way. An environment, biotic as well as physical, is assumed to be given, even though it is not constant through time. The population of interest is recognized to have a general life-strategy, which itself is an evolutionary achievement and therefore in flux, but at a much slower rate than specific features. There is variability in any breeding population, ultimately traceable to mutations, and in successive generations

---

11. G. Berkeley, *Essay, Principles, Dialogues,* M. W. Calkins, ed. (New York: Scribner's, 1929), pp. 124–137 (secs. 1–24 of *Principles of Human Knowledge*).
12. *Ibid.,* p. 413 (sec. 251 of *Siris*).
13. *Ibid.,* pp. 157–158 (sec. 61 of *Principles of Human Knowledge*).

a statistical selection among variants occurs. If one organic feature exhibits a net superiority over alternative accessible features in performing functions required by the life-strategy, then it is statistically favored and in sufficiently many generations becomes a modal feature for the population. In this manner, the relative optimum in a certain respect serves as a kind of final cause for the evolutionary development of the population, and a quasi-teleological explanation is given for a modal feature which is close to the optimum. The prefix 'quasi' is appropriate, because current evolutionary theory dispenses with the hypothesis that there is any conscious or unconscious aim in a phylogenetic sequence toward optimality, and takes statistical selection as the surrogate for such an aim. It is clearly out of place here to discuss the intricate ways in which the foregoing simplified pattern of evolutionary explanation is qualified by modern population genetics, but we have it upon high authority that these qualifications do not change the fundamental Darwinian thesis that evolutionary changes are mainly adaptive.[14]

The life-strategy of almost every mature higher animal with respect to its environment can be characterized in very general terms as one of exercising mobility for such purposes as seeking out food and mates, avoiding predators, and finding suitable conditions of temperature. The environment is almost completely independent of the animal, but the details of the life-strategy of the animal determine which aspects of the environment are important for it. The function of the perceptual faculties of the animal is to recognize these important aspects with sufficient accuracy and speed that it can behave effectively with respect to them. In practice, the evolutionary explanation of a feature of an animal's perceptual faculties consists in reasonably exhibiting its efficiency in performing the requisite function of recognition. It is impossible at the present state of scientific development and perhaps, in view of the complexity of the situation, forever, to measure how close a feature is to optimality. However, this kind of limitation of fineness is characteristic of evolutionary explanation generally, not just of the evolutionary explanation of perceptual features. Furthermore, in spite of the nonexistence of a quantitative measure of closeness to optimality, one is repeatedly struck by the appearance of great precision of adaptation: e.g., the sensitivity of butterflies to extremely minute quantities of aromatic chemicals characteristic of potential mates; the extreme acuity of vision of high-flying hunting birds; the specialization of a layer of the retina of the frog to respond to the visual

14. G. G. Simpson, *The Major Features of Evolution* (New York: Columbia, 1953), chs. 5 and 6; H. J. Muller, "Evidence of the Precision of Genetic Adaptation," *The Harvey Lectures 1947–48* (Springfield: Charles C. Thomas, 1950), pp. 165–229.

stimulus of a moving insect;[15] the visibility of ultraviolet light (which is polarized more strongly upon scattering than is lower-frequency radiation) to bees, who utilize the polarization of scattered sunlight for orientation.[16] It must be acknowledged that in practice the evolutionary explanation of perceptual features is usually elliptical, since very seldom is there direct evidence regarding the phylogenetic sequence eventuating in a feature. Nevertheless, the tacit premise that precise adaptation is the end-product of a long sequence of evolutionary modification is so well established in the case of anatomical features, for which the paleontological record is available, and it fits the evidence of comparative zoology so well, that it can hardly be questioned in the case of features of the perceptual apparatus.

It is largely a matter of convention whether the term 'perception' is to be applied to those interactions of an animal with its environment which are linked to appropriate behavioral responses without intervening conscious episodes. Whatever the convention, an essay attempting to answer Berkeley's challenge about the final causes of "appearances in the mind" must be concerned primarily with those cognitive states (of human beings, and presumably also of other animals having large repertoires of possible responses for a single stimulus) which are associated with consciousness. Here, then is a list of some of the most striking features of human perception, together with some comments on the adaptive character of each.

1. *Very many of the physical stimuli impinging upon the sense organs are censored, so that they give rise to negligible conscious experience, and the differentiation of those which are censored from those which are admitted is systematic.* A major function performed by this systematic selection is to play down the visual and auditory "noise" arising from the body in favor of "signals" from the environment.[17] For example, the insensitivity of the auditory system to low frequencies prevents the pulse from overwhelming information-bearing sounds. Similarly, the low-frequency cutoff of the visible spectrum prevents us from seeing the thermal radiation produced by the eye itself. And one of the consequences of the remarkable rapid fading of stabilized images upon the retina is the invisibility,

15. J. Y. Lettvin, H. R. Maturana, W. S. McCulloch, and W. H. Pitts, "What the Frog's Eye Tells the Frog's Brain," *Proceedings of the Institute of Radio Engineers,* **47** (1959): 1940–1951.

16. The visibility of ultra-violet light to bees and their sensitivity to polarization were discovered by K. v. Frisch, *Bees, Their Vision, Chemical Sense, and Language* (Ithaca, N.Y.: Cornell, 1950), but the evolutionary explanation of the correlation of these two traits was suggested in unpublished work by Yilmaz.

17. G. v. Békésy, *Sensory Inhibition* (Princeton, N.J.: University Press, 1967), ch. 1.

under normal circumstances, of the network of capillaries overlaying the photoreceptors in the retina – which otherwise would intrude into every visual scene.[18]

2. *The subjective impression felt at a point in the visual field is not a function only of the intensity of the physical stimulation of the corresponding point on the retina, but depends strongly upon the intensity of stimulation of neighboring regions.* The phenomenon of "Mach bands" is an outstanding manifestation of this characteristic of visual perception (which also has auditory and tactile analogues). Mach, who first observed this phenomenon, not only correctly conjectured that it is physiologically based upon some mechanism of neural inhibition, but pointed out that it served the biologically valuable function of accentuating the boundaries of objects in the visual field.[19]

3. *There are many types of constancy phenomena, whereby the appearance of an object alters comparatively little when quite radical changes in viewing conditions occur.* Thus, when a cloud passes between the perceiver and the sun, changing the modal wave length of the illuminant from yellow to blue, the colors of objects hardly appear to change.[20] And tilted circles appear to most perceivers (as one judges from their sketches) to be ellipses of smaller eccentricity than one would expect from noting the geometrical projections of the circles upon the retina.[21] These constancy phenomena evidently expedite the tracking of moving objects, which pass from one illuminant to another and change their orientation relative to the eye.

4. *Perhaps the most striking characteristic of human perception is its propensity toward organization into wholes and patterns, by filling in broken contours, integrating heterogeneous cues, and generally playing up or down the contributions from different stimuli so as to achieve a "good Gestalt."* The extent to which this propensity is innate – which has been one of the central problems in the psychology of perception – is relatively unimportant for our purposes, since nativistic or empirical mechanisms can equally well be understood as evolutionary achievements if it is biologically advantageous to form Gestalten. The biological advantage is evident, if one considers that higher animals are largely concerned with objects in the environment at some distance from themselves and, furthermore, that the behavior of these objects is characterized by

18. Ratliff, "On the Objective Study of Subjective Phenomena: The Purkyne Tree," pp. 77–92 of V. Kruta, ed., *J. E. Purkyně 1787–1869 Centenary Symposium. Prague 8–10 Sept. 1969* (Brno: 1971).
19. Ratliff, *Mach Bands* (note 1).
20. See, for instance, Yilmaz, "On Color Vision" (note 1).
21. R. H. Thouless, "Phenomenal Regression to the Real Object," *British Journal of Psychology,* **21** (1931): 339–359, and **22** (1931): 1–30.

more regularity than is found in the proximal stimuli to which they give rise, since the latter depend upon such contingencies as orientation, interception, and illumination. As Brunswik has said,

. . . in the psychological applications of communication theory it is usually the organism that appears as the source of noise; in reality, however, the limited ecological validities of proximal cues relative to distal object variables furnish a perfect environmental counterpart to this internal noise.[22]

Organization into Gestalten has the effect of shifting the focus of perception from the erratic proximal stimuli to the distal objects, which are more regular in their behavior and, partly for that reason, more important from the standpoint of effective behavior. Of course there is a chance of error in the process of going beyond the cues, but the success of any procedure for separating a signal from noise is a matter of probability. However, when we review human perceptual representations of practically relevant parts of the environment, we must admit that statistically the extraction of signal from noise in Gestalt perception is amazingly accurate.[23]

5. *The propensity toward perceptual organization is to some extent qualified by an ability to shift attention to cues in their individuality.* This feature of human perceptual powers is described and conceptualized very differently by different psychologists. Thus, Gibson speaks of the "visual field," which he characterizes in terms similar to those of sense-data theorists, in contrast to the "visual world."[24] And Bruner refers to "close looking" at cues when a perceiver's expectations are not fulfilled.[25] One can conjecture that the ability to switch from a primary mode of perception, focusing upon objects, to a secondary mode, focusing upon cues, is of great biological advantage. The first is essential for swift action directed toward distant objects; the second for checking impetuous bad judgments, recognizing illusions, and applying past experience and thought to the grading of cues. The ability to switch, voluntarily or not, from one to another mode is surely bound up with the richness of the range of responses available to human beings on the basis of stimulation, and also with the interplay of perception and conception. Here is a reason, quite distinct from the anatomical considerations that originally evoked the remark, for saying that "the human eye is definitely an eye for all purposes."[26]

22. E. Brunswik, "Scope and Aspects of the Cognitive Problem," pp. 5–31 of *Contemporary Approaches to Cognition* (Cambridge, Mass.: Harvard, 1957), p. 29.
23. J. Platt, "Two Faces of Perception," pp. 63–116 of B. Rothblatt, ed., *Changing Perspectives on Man* (Chicago: University Press, 1968).
24. J. J. Gibson, *The Perception of the Visual World* (Boston: Houghton Mifflin, 1950).
25. J. Bruner, "On Perceptual Readiness," *Psychological Review,* **64** (1957): 123–152.
26. E. N. Willmer, "Colour Vision and Its Evolution in the Vertebrates," pp. 306–325 of J. Huxley, A. C. Hardy, and E. B. Ford, eds., *Evolution as a Process* (New York: Collier, 1963).

The foregoing summary of respects in which human perception is adaptive is greatly in need of refinement. Moreover, the list could be augmented by discussing depth perception, perception of causality, etc. Enough has been said, however, to show that the evolutionary point of view permits quasi-teleological explanations of perceptual features to be incorporated in an entirely natural way into the causal theory of perception. If there is no action at a distance in nature, then the causal theory of perception must maintain that the stimuli to which an animal is sensitive are proximal; but evolution has eventuated in animals which transform their sensitive reactions so that their resulting cognitive states are quite accurate indices of crucial distal characteristics of the environment. It is a matter of physical law that, under the boundary conditions provided by a more or less fixed geological milieu, statistical correlations exist between sets of proximal and sets of distal variables, and the development of the perceptual faculties of higher animals by natural selection has capitalized upon these correlations.[27] Evolutionary considerations also enhance the explanatory power of the causal theory of perception in another respect: perceptual illusions can be understood as resulting, under abnormal conditions, from the very mechanisms that provide reliable information under normal conditions of perception.[28] It is hard to imagine how the type of teleological explanation of appearances in the mind which Berkeley envisaged could so naturally treat misleading and veridical perceptions in a unitary manner.

Finally, I wish to note without elaboration that, even though we have been preoccupied with the first of the three questions mentioned in the first paragraph of section I, that of the justification of the causal theory of perception, the appeal to evolution has yielded fragmentary answers to the second and third, and potentially can yield much more.

III

An essential component in the dialectical establishment of a philosophical thesis is to show that rival theses possess motivations which are understandable and plausible, yet fully satisfiable by the thesis advocated. Evolutionary considerations are quite fertile in providing dialectical niches for rivals to the causal theory of perception. A classical example is the suggestion by Baldwin, Lorenz, and others,[29] that a priori principles need

27. Campbell, "Pattern Matching" (note 1), p. 86.
28. Lorenz, "Gestalt Perception" (note 1), pp. 45–47.
29. J. M. Baldwin, *Thought and Things,* 3 vols. (New York: Macmillan, 1906, 8, 11); Lorenz, "Kant's Doctrine of the A Priori in the Light of Contemporary Biology," pp. 23–35 of Bertalanffy and Rapoport (note 1), translated from a paper in *Blätter für Deutsche Philosophie,* **15** (1941): 94–125. A history of this suggestion is given in Campbell, "Evolutionary Epistemology" (note 1).

not be considered transcendental and hence beyond the limits of theoretical explanation, but can be interpreted as evolutionary achievements drawing upon the experience of our pre-human ancestors.

As another example, the phenomenology of Merleau-Ponty[30] is susceptible of a comparable re-interpretation. One can respect Merleau-Ponty's insistence that "being-in-the-world" characterizes our perceptions taken in their integrity, so that there is something prima facie perverse about trying to prove the existence of the external world. And we can agree that the central objects of perception are not merely façades, but things that have other sides to them. However, as fundamental as these characteristics of perception are in practical life and in pre-reflective cognition, they may have a derivative status in a biological view of human knowledge. As indicated in the above discussion of Gestalt perception, the focusing upon objects rather than cues involves a kind of unconscious statistical inference. Such inferences are sometimes fallacious, for instance, when we are deceived by a *trompe-l'oeil* painting, and there are possible universes in which they would mostly be fallacious. However, the process of natural selection that resulted in the propensity to make this kind of unconscious inference occurred in our universe just because they are on the whole successful. The modality of "being-in-the-world" which characterizes perception is also comprehensible from an evolutionary point of view, since it enhances the emotional intensity and the sense of urgency in reacting to putative perceptual objects. The claim here, in other words, is that the insights of Merleau-Ponty on the phenomenology of perception can be embedded into a naturalistic epistemology with a complex dialectical structure, which explains the cognitive processes of ordinary life in terms of principles of the natural sciences and which closes the circle by showing how the objective scientific view of the world emerges from these ordinary cognitive processes.

IV

My advocacy of the causal theory of perception must, in conclusion, be qualified by an admission of the force of the second of Berkeley's criticisms in the passage quoted above: that the efficient causation of appearances in the mind by external bodies remains a mystery. I construe this criticism to refer to our ignorance, at the present stage of scientific development, of a set of fundamental laws governing mind–body interactions which are comparable in scope and precision to the laws governing the interactions of fields and particles in physics. (Berkeley, who took volition to be

30. M. Merleau-Ponty, *The Phenomenology of Perception,* C. Smith, trans. (London: Routledge & Kegan Paul, 1962); *The Primacy of Perception and Other Essays,* J. M. Edie, trans. (Evanston, Ill.: Northwestern Univ. Press, 1964).

the prototype of efficient causation, undoubtedly construed his criticism differently, but I am setting aside historical concern at this point.)

It is hardly news to assert that the mind–body problem is still unsolved and that, as a consequence, the theory of knowledge is incomplete. However, I should like to use some of the terminology of section I to point out something that has not been much noticed about the unsolved problem. In physics the space of possible states of a physical system can be completely specified without knowing the law of temporal development of the state of the system; one may know the kinematics of the system but not its dynamics, though the converse is not true. What I wish to insist regarding systems composed of both bodies and minds is that we are profoundly ignorant even about the preliminary kinematical question. Although we have a great deal of phenomenological information about minds and although some of it can be cast in mathematical form, we do not know at all the geometry of the space of states of a system endowed with mentality.

We cannot even say whether certain important general kinematical propositions are true or not. It is particularly important for epistemology to know whether the state of each of a set of interacting systems can be in principle specified without reference to the states of the others. This is the case in classical physics, and it was assumed in the formulation of the causal theory of perception in section I, in the statement "the entire state of the perceiver . . . is kinematically independent of the states of all other systems in the world." However, generalizing a kinematical proposition of classical physics to a set of interacting minds and bodies is an extravagant extrapolation. It is all the more extravagant in view of the fact that this proposition happens to have been discarded by modern physics; for, according to quantum mechanics, the composite system $A + B + C + \cdots$ can be in a definite pure state while the individual systems $A, B, C, \ldots$ are not in definite pure states.[31] A few very speculative writers have even conjectured that this peculiarity of the kinematics of quantum mechanics is a clue to the mind–body problem, for it shows that holistic descriptions, which previously were reserved for mental phenomena, must be applied to physical phenomena as well.[32] Possibly we have here a confirmation within physics of Whitehead's warning against the "fallacy of simple location."[33]

31. See, for instance, B. d'Espagnat, *Conceptual Foundations of Quantum Mechanics* (Reading, Mass.: Benjamin, 1971).
32. See, especially, H. Stapp, "S-Matrix Interpretation of Quantum Theory," *Physical Review,* **3** (1971): 1303–1320.
33. For a critical examination of this possibility, see my "Quantum Physics and the Philosophy of Whitehead," pp. 307–330 of R. S. Cohen and M. W. Wartofsky, eds., *Boston Studies in the Philosophy of Science,* vol. II (New York: Humanities, 1965), and J. S. Burgers, "Comments on Shimony's Paper," pp. 331–342, *ibid.*

We are now in very deep waters, for there are mysteries in the foundations of quantum mechanics as well as in the mind–body relationship. The scientific world view that will eventually emerge from investigating these mysteries may require a very different formulation of the causal theory of perception from what was given in the past. Nevertheless, I find it hard to believe that the picture of an environment interacting causally with biologically well-adapted perceivers will not be a good macroscopic approximation to the more refined pictures that will be presented by the science of the future.

5

Is observation theory-laden? A problem in naturalistic epistemology

> Erect your schemes with as much method and skill as you please; yet
> if the materials be . . . spun out of your own entrails . . . the edifice
> will conclude at last in a cobweb. . . . As for us the ancients, we are
> content with the bee to pretend to nothing of our own, beyond our
> wings and our voice, that is to say, our flights and our language. For
> the rest, whatever we have got, has been by infinite labour and search,
> and ranging through every corner of nature.
>
> Jonathan Swift, *The Battle of the Books*

I. INTRODUCTION

The orthodox logical empiricist treatment of the relation between scientific
theories and observations (as exemplified in the work of Rudolf Carnap,
Ernest Nagel, Carl Hempel, and R. B. Braithwaite) abstained as a matter
of principle from considerations of empirical psychology. Since psychology is the least developed of the natural sciences, an appeal to it was
supposed to subvert the epistemological program of establishing a firm
foundation for all the sciences. Furthermore, epistemology was not considered to be in need of the answers to the questions typically investigated
by empirical psychology. It is also likely that the abstention of the orthodox logical empiricists from empirical psychology in their treatment of
the relation between theories and observations was in large part due to
their acceptance of Gottlob Frege's polemic against psychologism in logic,
for they regarded the central problem concerning this relation to be one
of logic.

This work originally appeared as "Is Observation Theory-Laden? A Problem for Naturalistic Epistemology" in *Logic, Laws, and Life,* Robert G. Colodny, editor. Published in 1977
by the University of Pittsburgh Press. Used by permission of the publisher.

This essay is an expansion of a lecture entitled "Theory, Observation, and Common
Sense" delivered in 1969 at the Center for Philosophy of Science of the University of Pittsburgh. That lecture, in turn, was an expansion of part of a talk, "Proposals for a Naturalistic Epistemology," given at a workshop of the Center in 1965. Two other publications (see
additional references) based upon the workshop talk present theses relevant to the present
essay. I am deeply grateful to my friends Howard Stein and Richard Burian for their constructive criticisms of earlier drafts.

One of the most influential criticisms of orthodox logical empiricism has been N. R. Hanson's argument that perception is intertwined with concepts, so that observation is usually, or in crucial circumstances, "theory-laden." Hanson cites a large number of experiments in empirical psychology in developing his argument, and these citations appear to be indispensable to his case.[1] The primary purpose of this essay is to evaluate critically Hanson's argument.[2] It is important to do so partly because of widespread acceptance of his views or similar ones,[3] but even more because of their bearing upon the general program of naturalistic epistemology. In this program, scientific results concerning the place of human beings in nature have important implications for epistemological problems. Accordingly, the abstention of the orthodox logical empiricists from considerations of empirical psychology deprived them of valuable information for epistemological investigations. However, even though I strongly concur with Hanson's view that psychology is relevant to epistemology, I disagree with some of the citations from empirical psychology which he deploys, and even more with the mode of deployment. If the disagreement only concerned the former matter, then I would be in effect asserting that epistemological problems can be settled by an appeal to psychology. Rather, I believe that there should be a dialectic interplay between psychology and epistemology, and a secondary purpose of this essay is to illustrate this interplay, even if it is difficult to articulate the process.

As regards the thesis that observation is theory-laden, I shall maintain that it is a great oversimplification of the functioning of our cognitive

1. N. R. Hanson, *Patterns of Discovery* (Cambridge: Cambridge University Press, 1958), and *Perception and Discovery: An Introduction to Scientific Inquiry* (San Francisco: Freeman, Cooper, 1969). His claim that considerations of empirical psychology are only ancillary to a logical analysis of sentences expressing perceptual judgments will be examined in section 4.
2. Kordig's analysis ("The Theory-ladenness of Observation," *Review of Metaphysics,* 24 [1971]: 448–84) is penetrating on questions of linguistic usage and scientific methodology, but it is little concerned with psychological questions. There have also been several good discussions of particular aspects of Hanson's position, for example, F. Dretske, *Seeing and Knowing* (Chicago: University of Chicago Press, 1969); I. Scheffler, *Science and Subjectivity* (Indianapolis: Bobbs-Merrill, 1967), p. 14; and R. A. Putnam, "Seeing and Observing," *Mind,* 78 (1969): 493–500.
3. For example, T. Kuhn, *The Structure of Scientific Revolutions* (Chicago: University of Chicago Press, 1962); S. Toulmin, review of Hanson's *Patterns of Discovery,* in *British Journal for the Philosophy of Science,* 10 (1959–1960): 346–49; G. Gale and E. Walter, "Kordig and the Theory-Ladenness of Observation," *Philosophy of Science,* 40 (1973): 415–32; F. Suppe, *The Structure of Scientific Theories* (Urbana: University of Illinois Press, 1974); and M. Polanyi, *Personal Knowledge* (New York: Harper and Row, 1964), p. 101.

faculties. It seems to me that what emerges from the psychological literature on perception is a complex picture of various kinds and degrees of integration between percepts and concepts together with mechanisms for switching from close integration to relative autonomy. There are strategies of perception, which can appropriately be called "integrative," in which sensory clues seem to be suppressed from consciousness, and the resulting perceptual judgments extrapolate far beyond any input that can reasonably be characterized as "given." The existence and importance of integrative strategies give some support to the thesis that observation is theory-laden. However, there are also strategies which can appropriately be called "analytic," in which beliefs and theories are to some extent held in abeyance and clues are brought to the surface of consciousness. The neglect of these strategies is Hanson's fundamental error. I shall maintain that the ability of human beings to switch among strategies of different kinds is essential to the reliability of our cognitive processes in general, and particularly of the controlled processes of scientific investigation.

The logical empiricist thesis that there is a sharp distinction between theory and observation has been challenged from other points of view than Hanson's. Thus, P. Achinstein and F. Dretske point out that the techniques of observation, and hence the extension of the term "observation," are changed by the development of scientific theories. H. Putnam argues that there is no clear dichotomy between observational and theoretical terms, such that observation reports employ only the former.[4] These challenges are important for correcting an artificially schematized picture of scientific knowledge presented by the logical empiricists. Nevertheless, there is a core of good sense in the assumption of the logical empiricists that "publicly observable" things, properties, and events can be acknowledged by the advocates of conflicting theories. Psychological theory is able to supply a vindication (with some reservations) of their assumption, thereby supporting to some extent a position developed in deliberate abstention from psychological considerations.[5]

2. HANSON'S ARGUMENTS

In this section I shall summarize Hanson's arguments that observations are theory-laden, with particular attention to his reliance upon the results

4. P. Achinstein, *Concepts of Science* (Baltimore: Johns Hopkins Press, 1968); Dretske, *Seeing and Knowing;* and H. Putnam, "What Theories Are Not," in *Logic, Methodology and Philosophy of Science,* ed. E. Nagel, P. Suppes, and A. Tarski (Stanford: Stanford University Press, 1962), pp. 240–51.

5. In the last section of this essay, the qualifications of this sentence will be explained and some criticisms of logical empiricism will be briefly stated.

of empirical psychology. The discussion will be almost entirely concerned with visual observation, since nearly all of Hanson's examples are visual, though occasional remarks suggest that he takes essentially the same position with regard to all the senses.[6]

Although Hanson acknowledges the existence of genuine cases of phenomenal seeing, in which there is little or no conceptual element, he considers them to be atypical.[7] The typical case of seeing is *seeing that . . .* , where the ellipsis indicates a sentential clause, a clause that could stand separately as a complete declarative sentence.[8] *Seeing that* takes place even when the report of the seeing has the grammatical form "*N* sees *X*," where *X* is a common noun rather than a sentential clause, for Hanson makes the following exegesis of such reports: "What is it to see boxes, staircases, birds, antelopes, bears, goblets, X-ray tubes? . . . It is to see that, were certain things done to objects before our eyes, other things would result." *Seeing that* is the sense of seeing which Hanson considers to be crucial in research in the natural sciences, for

'Seeing that' threads knowledge into our seeing; it saves us from re-identifying everything that meets our eye; it allows physicists to observe new data as physicists, and not as cameras. . . . Observation in physics is not an encounter with unfamiliar and unconnected flashes, sounds and bumps, but rather a calculated meeting with these as flashes, sounds and bumps of a particular kind.[9]

Finally, it appears from Hanson's examples that the implicit sentential clauses in *seeing that* can be supplied by any part of the subject's system of knowledge and beliefs. Thus, "Tycho and Simplicius see that the universe is geocentric; Kepler and Galileo see that it is heliocentric."[10]

The argument for these theses commences with a consideration of reversible perspective drawings and ambiguous figures, which are familiar from textbooks of Gestalt psychology. In seeing one of these figures in two different ways – as bird or antelope, as convex or as concave – he says that it is natural to say that "we see different things." But the difference is neither in the way the retina is affected nor in the sensations, since the same lines are involved in the two instances of seeing.[11] The difference cannot be due to differing interpretations, because, as Ludwig Wittgenstein says, if seeing an ambiguous figure as a box means "I am having a particular visual experience which I always have when I interpret the figure as a box, or when I look at a box," then "if I meant this, I ought to

6. For example, see his *Perception and Discovery,* pp. 73, 79.
7. *Patterns of Discovery,* p. 20.
8. *Ibid.,* pp. 24–25.
9. *Ibid.,* pp. 20–21, 22, 24.
10. *Perception and Discovery,* p. 102.
11. *Patterns of Discovery,* pp. 11–12.

know it. I ought to be able to refer to the experience directly and not only indirectly."[12] A second reason given by Hanson does not occur in Wittgenstein's text, namely, that in ordinary usage when we use the term "interpretation" we refer to a process which takes time, and this is not the case in seeing an ambiguous figure in one way rather than another; but instantaneous interpretation is inadmissible, for it "hails from the Limbo that produced unsensed sensibilia, unconscious inference, incorrigible statements, negative facts and Objektive." What is different in seeing the ambiguous figure as a bird or as antelope is the *organization,* which "is not itself seen as are the lines and colours of a drawing. . . . Yet it gives the lines and shapes a pattern. Were this lacking we would be left with nothing but an unintelligible configuration of lines." Immediately after this answer, Hanson asks a question, which sounds like a request for a causal explanation: "How do visual experiences become organized?" He answers that "the context gives us the clue" and then explains that "such a context, however, need not be set out explicitly. Often it is 'built into' thinking, imagining and picturing. We are set to appreciate the visual aspect of things in certain ways."[13] Hanson then makes reference to a large body of literature in experimental psychology on "set" and "Aufgabe."[14] He comments on this reference that "philosophy has no concern with fact, only with conceptual matters (cf. Wittgenstein, *Tractatus,* 4.111); but discussions of perception could not but be improved by the reading of these twenty papers";[15] and "one can talk philosophy by way of a factual discipline without thereby allowing his case to stand or fall on the success of that discipline."[16]

In both of his general books on philosophy of science, Hanson shifts to a second stage of argument in which the reliance upon evidence from empirical psychology is minimized, and at the same time his conclusions are strengthened in the sense that "an alternative account would be not merely false, but absurd."[17]

The argument proceeds as follows: (1) Scientific knowledge has an inescapable linguistic element, for it is expressible in sentences and presumably only in this way.[18] (2) A sentence is not a picture of a state of

---

12. Wittgenstein, *Philosophical Investigations* (Oxford: Basil Blackwell, 1953), p. 194, quoted in Hanson, *Patterns of Discovery,* p. 10.
13. *Patterns of Discovery,* pp. 10, 13, 15.
14. *Ibid.,* note to p. 15, on pp. 180–81; for more detail, see *Perception and Discovery,* pp. 161–66.
15. *Patterns of Discovery,* p. 181.
16. *Perception and Discovery,* p. 131.
17. *Patterns of Discovery,* p. 24; *Perception and Discovery,* p. 131.
18. *Patterns of Discovery,* p. 25; *Perception and Discovery,* pp. 126–27.

affairs, for assertion has a character – as Wittgenstein emphasized in his later work, when he rejected the views of the *Tractatus* – which is distinct from representation.[19] (3) The sense of "seeing" which is important for natural science is determined by the requirement that it be significant and relevant for scientific knowledge. But,

> if seeing were a purely visual phenomenon untainted by any of the effects of language, then nothing that we saw with our eyes would ever be relevant to what we know about the world, and nothing that we know about the world could ever have significance with respect to what we say we see. . . . Our vision would be without understanding, our understanding without light. If this is absurd it will show again how different is seeing from the mere formation of retinal and mental pictures, and how central to the connection between vision and knowledge is the concept of *seeing that*.[20]

Hanson concludes that "it is a matter of logic, not merely a matter of fact, that seeing as and seeing that are indispensable to what is called in science, *seeing* or *observing*."[21] Since for Hanson, who clearly follows Wittgenstein on this matter, logic is concerned with the rules of language,[22] it is important to identify the language to which he is implicitly referring when he draws his conclusion. Although his texts are not entirely clear on this point, one can reasonably say that the language he has in mind is that of ordinary discourse, with perhaps some enrichment (of which ordinary language is susceptible) with technical terminology. The evidence for this interpretation is his insistence that the second stage of his argument does not rely upon psychology or neurophysiology,[23] which surely implies that the language to which he is referring is not any theoretical language of one of these disciplines. Furthermore, he asserts that what "ordinary people mean . . . is of maximum importance when we are concerned to understand a general philosophical concept like seeing."[24] The component of ordinary-language philosophy in Hanson's thought must be kept in mind even in an examination of the first stage of his argument, which explicitly draws upon some results of empirical psychology.[25]

19. *Patterns of Discovery*, pp. 26–29.
20. *Ibid.*, p. 132; see also *Perception and Discovery*, p. 86.
21. *Perception and Discovery*, p. 147.
22. For example, *ibid.*, pp. 184–85.
23. *Ibid.*, pp. 130, 132.
24. *Ibid.*, p. 79.
25. An important departure of Hanson from Wittgenstein is his extrapolation of the analysis of Gestalt switches to situations in which the subjects hold different sets of beliefs concerning matters of theoretical physics. Since Wittgenstein believed that theoretical science is a different "language game" from ordinary discourse, it seems reasonable to suppose that he would object to this extrapolation.

3. THE CONFLATION OF SENSES OF "SEEING"

Hanson recognizes (with evident pleasure) the extraordinary variety of syntactical and semantic constructions of the verb "to see."[26] Two of these senses he separates from the others: seeing in the intransitive sense "indeterminately or in general, as do infants and lunatics";[27] and phenomenal seeing, which occurs in artificial situations such as an oculist's tests, or when one is confused by unfamiliar phenomena. His dominant tendency, however, is to conflate rather than to discriminate senses of the verb. Although Hanson repeatedly cites Wittgenstein on matters of philosophical method and on specific questions of analyzing locutions of seeing, he does not heed the warning of *Philosophical Investigations* (p. 66) against seeking for one essential use underlying the varied uses of a word. Indeed, at one point, Hanson explicitly takes exception to the warning. In typically exploratory and tentative manner Wittgenstein says, "Seeing as . . . is not part of perception. And for that reason it is like seeing and again not like" (p. 197). Hanson comments:

Something of the concept of seeing can be discerned from tracing uses of 'seeing as . . .'. Wittgenstein is reluctant to concede this, but his reasons are not clear to me. On the contrary, the logic of 'seeing as' seems to illuminate the general perceptual case.[28]

In some cases the conflation of senses of "to see" is crucial to Hanson's epistemological position, but in order to avoid becoming enmeshed in peripheral issues it will be helpful to point out one case of conflation which is both unconvincing and inessential to his central thesis. In a passage quoted at the beginning of section 2, Hanson construes "*N* sees *X*" – where *X* denotes an object – as implicitly meaning "*N* sees that . . ." with the blank filled by an appropriate sentential clause. However, great confusion often results unless the context following the verb in "*N* sees *X*" is extensional. For example:

The boy points to an X-ray tube and asks, what's that? The father answers 'that is an X-ray tube'. The answer is not only true but relevant. Yet how could it be relevant if the son did not already see an X-ray tube?[29]

For similar reasons, a language which lacked the extensional usage of "to see" would require circumlocutions in order to express the fact that two

26. A fine collection of various constructions of "to see" is given by G. Warnock, "Seeing," *Proceedings of the Aristotelian Society,* 55 (1954–1955): 201–18.
27. *Perception and Discovery,* p. 107.
28. *Patterns of Discovery,* p. 19.
29. R. A. Putnam, "Seeing and Observing," p. 494. See also Dretske, *Seeing and Knowing,* p. 37; and A. Collins, "The Epistemological Status of the Concept of Perception," *Philosophical Review,* 76 (1967): 436–59.

observers disciminate visually a common object which they differently describe or identify or theorize about. Except for the diminution of the rhetoric in certain passages, Hanson's exposition of his main thesis would not suffer from the recognition of an extensional sense of "to see."

The circumstances in which Hanson's epistemology requires the conflation of "seeing X," "seeing X as . . . ," and "seeing that . . ." are rather special, but in them the conflation has a *prima facie* plausibility. They are circumstances in which the locutions are either used in the first-person singular,[30] or else in the second or third person but with a sympathetic modality, as if to represent that person's point of view. First-person singular usage excludes the extensional sense of "to see" discussed in the preceding paragraph, for my naming the object will involve recognizing it, even erroneously, in terms of some identifying characteristic. Hence the thing I see is seen *as* an object having these characteristics, and I see *that* it has these characteristics and whatever characteristics I believe to be associated with them (or at least I can say this if I disgregard the possibility of error). Exactly the same justification for the conflation of senses applies to second- or third-person usage with a sympathetic modality, for then there is not only an assertion of a visual discrimination but tacit reference to the mental state associated with that discrimination. Since Hanson is evidently committed to writing the history of science from the points of view of the historic figures, it is understandable that his third-person locutions should be in a sympathetic modality (and, conversely, it is understandable why his conflation of senses of "to see" has appealed to some historians of science).

A historical example, however, will undermine the plausibility of the foregoing defense of conflation. Because Hanson maintains that observation is theory-laden (the conflation of "seeing X" and "seeing that . . ."), he declares that "Tycho and Simplicius see a mobile sun, Kepler and Galileo a static sun."[31] If so, what did Newton see when he looked at the sun? Now Newton tells us what his dynamical theory implies about the motion of the sun: "The sun is agitated by a perpetual motion, but never recedes far from the common center of gravity of all the planets" (Bk. III, Prop. XII). But it would surely be ironical to attribute to Newton the opinion that he *saw* a sun agitated by perpetual motion, for Newton insisted upon a strong principle of visual relativity, that only relative motion among bodies could be discerned by sight; the criterion for absolute motion, which is what his theory is ascribing to the sun, is a dynamical one, applicable to phenomena only by means of the systematic combination of

30. See Warnock, "Seeing," p. 204, on the consequences of restricting attention to the first-person singular in analyzing "to see."
31. *Patterns of Discovery,* p. 17.

observation, theory, and analysis presented in the *Principia*.[32] To generalize, if a scientist's theoretical view of the world is sufficiently comprehensive to include a theory of vision, and if that theory sharply distinguishes visual evidence from cognition, then Hanson's view that seeing is theory-laden will generate a paradox, and perhaps an outright contradiction. The example suffices to show that something is amiss in Hanson's claim, but further analysis is required to diagnose the difficulty more precisely.

4. HANSON'S "LOGICAL" ARGUMENT AND ITS RELATION TO KANT

I shall first examine the later stage of Hanson's argument, which allegedly dispenses with empirical evidence and establishes a necessary connection between seeing and knowing. The argument is so loose in texture that it is hard to reconstruct it in a compelling manner. There is certainly something correct in the insistence that an integration of perception with thought occurs in our experience, at least in verbalized experience. Otherwise, he says, in a phrase that deliberately echoes Kant, "our vision would be without understanding, our understanding without light."[33] The trouble is that this dictum is a commonplace, meaning different things to different philosophers – a possibility of which Hanson should be particularly aware, in view of his general doctrine of relativity of meanings to conceptual framework. When Kant says in the "Transcendental Logic" that "intuitions without concepts are blind," he states which are the concepts that are given *a priori* and are indispensable for nonblind experience, and he provides an explicit method (strained though it may seem to us) for exhibiting these concepts systematically. Furthermore, he complicates the account by admitting judgments of perception which, in contrast to judgments of experience, do not involve the subsumption of intuitions under the *a priori* concepts, and nevertheless are apparently conscious. Hanson, of course, does not subscribe to the details of Kant's analytic, but the point is that, in the absence of a comparably detailed treatment, one cannot conclude from his argument whether there are any differentiations among the elements of a conceptual system with regard to their integration with seeing. But information on this question is indispensable to Hanson's program. If only the concepts in a universal core of common sense, or in P. F. Strawson's universal "descriptive metaphysics," or in the innate equipment which enables any normal infant to acquire a langauge, are integrated in a necessary manner with seeing, then the thesis

32. Bk. I, Scholium to the definitions; see also H. Stein, "Newtonian Space-Time," *Texas Quarterly* (1967): 174–200.
33. *Patterns of Discovery*, p. 26.

of the theory-ladenness of observation loses the strong sense which Hanson evidently intends. Not only is the range of concepts integrated with seeing undetermined by Hanson's argument, but the mode of integration is entirely obscure. How tight is the integration; how much under control; how much affected by physical, emotional, and social variables? One example will suffice to show how Hanson slips from the commonplace thesis of integration to a logically unwarranted specification of it. He illustrates "bridging" by saying that in seeing objects there is "seeing that if x were done to them y would follow."[34] But visual reports can be expressed in sentential form – for example, "That appears cowlike" – without any commitments regarding the dispositions of visual objects. It may, of course, be argued that discourse in terms of appearances is derivative from discourse concerning things which have dispositions, but such an argument introduces considerations beyond the meager ones upon which Hanson has chosen to rest his case.

The foregoing remarks suggest that Hanson's position can be best defended by disregarding his claims concerning the force of the second stage of his argument, and by taking his deployment of psychological evidence as indispensable to his argument as a whole. Throughout the second stage of his argument, I suggest, he tacitly relied upon the evidence concerning Gestalt switches, illusions, and "sets" to justify his strong conclusions concerning the scope and tightness of the integration of a conceptual framework with seeing. It is necessary, therefore, to appraise his use of the evidence of empirical psychology.

It would be unwise, however, to attempt to examine the relevance of empirical psychology for all the issues Hanson has raised concerning the relation between perception and thought. Two of the issues are raised in any philosophy which is Kantian in inspiration: whether experience is possible without some minimal conceptual ordering, and whether this conceptual ordering is in some sense *a priori*. Unfortunately, the implications of empirical psychology for these issues is obscured by deep disagreements among psychologists – for instance among J. J. Gibson, the Gestalt psychologists, and Egon Brunswik on the conceptual elements of experience, and among the behaviorists, Jean Piaget, and Noam Chomsky on the question of innateness. The generically Kantian issues are so profound that they will not be illuminated by the results of empirical psychology unless the latter are organized into precise and comprehensive theories. By contrast, the special issues which Hanson raises over and above the generic Kantian ones are less profound. There are interesting psychological findings, acknowledged by adherents of different schools,

34. *Ibid.*, p. 30.

that bear upon Hanson's epistemological theses. As far as possible, I shall rely upon these results in appraising Hanson's theses, though occasionally (especially on the question of unconscious inference) controversial matters will not be avoidable.

5. PSYCHOLOGICAL CONSIDERATIONS: GESTALT SWITCHES AND ILLUSIONS

This section will be devoted to collecting and commenting on some findings about ambiguous figures and perceptual illusions that are relevant to Hanson's thesis that observation is theory-laden.

Hanson contends that the characteristics of the perception of ambiguous figures reveal something important about scientific observation ("the two astronomers are to the sun as you and I might be to the duck-rabbit when you see only a duck and I only a rabbit").[35] The unnaturalness of the ambiguous figure is compensated for by the fact that it provides a kind of laboratory control: namely, one can be sure that the images on the retinas of the two observers are virtually identical, and hence that their different identifications cannot be attributed to differences in physical details, but rather to differences in organization. It is important, then, to determine empirically what factors influence perceptual organization of an ambiguous figure. One of the most striking of the psychological findings is the resistance of the Gestalt perception to conceptual suggestion. When R. Leeper's famous wife–mother-in-law figure is seen as the former by the subject, the verbal communication to him by the experimenter that the latter is also to be seen does not help him make a Gestalt switch, even though there is no question of doubting the experimenter's word. Verbal instructions about where in the drawing to look in order to pick out features of the mother-in-law's physiognomy help more than the simple conceptual suggestion; and much more helpful yet is a graduated series beginning with a drawing in which the features of the mother-in-law are unambiguous and concluding in the ambiguous drawing under observation.[36]

An even more striking example of the subject's resistance to seeing as his conceptual framework would suggest is the variant of Leeper's experiment by W. Epstein and I. Rock. The subject is shown an alternating sequence W-M-W-M-W-M-W-M of drawings which are unambiguously "wife" and "mother-in-law" respectively, and a "wife" is anticipated at the next turn; but when the ambiguous drawing is then shown, it is seen as the mother-in-law: "a clear victory for recency over expectation," and

35. *Perception and Discovery,* p. 107.
36. U. Neisser, *Cognitive Psychology* (New York: Appleton-Century-Crofts, 1967), p. 61.

one could add that it is also a clear victory of a nonconceptual component in Gestalt organization over a conceptual component.[37]

Hanson analyzes one perceptual illusion in detail, namely the distorted room of A. Ames, in which objects in the room appear to be abnormally large or small against a background of apparently normal rectangular walls, whereas in fact the objects are of standard size and the walls are trapezoidal and tilted. He cites with apparent approbation Ames's explanation "that we wish to keep our environment relatively stable if possible, so that we can move about in it with confidence and surety" and "that we are always construing the visual stimulus pattern in terms of past experience and knowledge as well as present and future needs."[38] One further empirical fact about the experiment of Ames, however, throws great doubt on Hanson's interpretation of it: The illusion persists even after the subject has examined the construction of the room and is fully aware of the trick that is being played upon his visual system, and in fact even when the subject is a sophisticated psychologist with a theory about the mechanism of the illusion.[39] Ames's experiment allows the possibility that some elements of knowledge influence perceptual illusions, but shows that some, which *a priori* might be expected to be relevant, have no influence.

A body of data that can reasonably be interpreted to support Hanson's position concerns illusions of length and angle in plane figures, such as the arrow illusion of Müller-Lyer, the railway lines illusion of Ponzo, and the fan illusion of Hering.[40] The occurrence of these illusions can be understood as a by-product of the propensity of the apparent linear dimensions of an object to increase roughly in proportion to its apparent distance, if the retinal image is kept unchanged in size as the distance cues vary. This propensity, known as Emmert's law,[41] is evidently important for maintaining the apparent constancy of size of objects which are in motion relative to the subject. In order to apply this law to the illusions in question, it must be supposed that the figures are viewed as perspective drawings, so that, for example, converging straight lines are viewed as parallels receding into the distance.[42] The closeness of corresponding

37. *Ibid.,* pp. 141–42.
38. *Perception and Discovery,* pp. 157, 158.
39. See J. Bruner, "On Perceptual Readiness," *Psychological Review,* 64 (1957): 123–52.
40. See, for example, R. Gregory, *Eye and Brain* (New York: McGraw-Hill, 1966), pp. 136ff.
41. Dr. John Heffner has pointed out to me that Emmert's own statement of his law concerned only the apparent size of after-images, but I am following Gregory's wise application of the term "Emmert's law."
42. Gregory (*Eye and Brain*) summarizes much evidence in favor of this hypothesized connection between constancy scaling and the geometrical illusions. For example, he reports a case of a man blind since infancy whose sight was restored by corneal grafts and who acquired the ability to recognize by sight objects which he knew by touch, but whose

points on the converging lines then serves as a distance cue for an object lying between those points, and Emmert's law then implies that the apparent size of the object increases with closeness of the two points.

If the foregoing interpretation is accepted, then the geometrical illusions support Hanson's contention that there is "a propensity to 'alter' the details of one's visual field in order to get things sorted out intellectually,"[43] and it does so whether the achievement of constancy scaling is considered to be learned, as the authors cited maintain, or partially innate. Even more favorable to Hanson's thesis that observation is theory-laden is one further finding: that the susceptibility to the geometrical illusions decreases with age, beginning at age four or age six, according to two different studies,[44] which permits the interpretation that the acquisition of knowledge regarding spatial relations affects visual perception. If this interpretation is correct, then (in contrast to the situation of Ames's experiment) there is integration of knowledge with perception at two levels – the rather primitive level at which constancy scaling is acquired, and a more sophisticated level at which compensation is made for some of the errors caused by constancy scaling.

A quite different interpretation, however, can be given to the diminution of the geometrical illusions with age: namely, that maturation brings greater power to analyze a visual scene in a systematic manner, so that the visual impression is not controlled by what Piaget calls "centration."[45] Piaget himself suggests that decentration is intellectual rather than perceptual in character, even though it has consequences in controlling the subjectivity of perception. I do not wish to rely upon Piaget's complex and controversial theory at this point, and only mention it in order to indicate the existence of an interpretation of the empirical data quite different from that which supports Hanson.

For the purpose of this essay it suffices to consider another visual illusion which, like the geometrical illusions, is probably the by-product of constancy scaling, but which clearly resists relevant sophisticated knowledge. This is the illusion that the moon is larger when it stands near the horizon than when it is higher in the sky. Distance cues, such as occlusion

visual depth perception and ability to identify two-dimensional representations of objects remained very faulty. This man exhibited little or no susceptibility to the geometrical illusions involving two-dimensional figures, though he was normally susceptible to the Ames illusion, in which the experimental arrangement is three-dimensional.

43. *Perception and Discovery*, p. 154.

44. M. Segall, D. Campbell, and M. Herskovits, *The Influence of Culture on Visual Perception* (Indianapolis: Bobbs-Merrill, 1966), pp. 196–97.

45. J. Piaget, *The Mechanisms of Perception*, trans. G. Seagrim (New York: Basic Books, 1969), pp. 70–71 and *passim*.

and shift of angle consequent upon movements of the subject, are operative relative to terrestrial landmarks only when the moon is near the
horizon,[46] and these cues combined with Emmert's law yield the illusion.
Unlike the case of the geometrical illusions, however, the variation in
apparent size of the moon does not seem to be affected by maturation or
by theoretical knowledge. The moral is that discrimination and qualifications are necessary for a correct theory of the integration of a conceptual framework with perception. It is worthwhile to quote Rock at some
length for his articulation of this moral, whether the details of his theory
of perception are correct or not:

> Knowledge as such does not seem to affect perception. Yet it must be acknowl-
> edged that perception seems to move in the direction of veridicality, reflecting
> what has been learned about the environment. This is a problem that should not
> be glossed over. Still there is a difference between perceptual change produced
> by certain specific experiences and perceptual change produced by mere knowl-
> edge. . . . The core of the answer I am suggesting then is that when a concrete trace
> containing *visual* information accrues to a specific stimulus, it can affect the way
> it appears. This is not the same as "knowledge about the situation."[47]

6. PSYCHOLOGICAL CONSIDERATIONS: INTEGRATIVE AND ANALYTIC STRATEGIES OF PERCEPTION

The empirical data which are most significant for evaluating Hanson's
epistemological thesis concern the perception of ordinary objects, and for
the most part they are not very esoteric data. An evident characteristic of
the stimuli which the environment presents to our senses under ordinary
circumstances is their immense richness, both in actual presentation and
in potentiality for further presentations to the active observer. The stimuli
presented by ambiguous drawings and typical situations of visual illusions are very meager, whereas by contrast, the abundance of detail in
ordinary circumstancs makes it imperative for the organism to filter the
stimulus information and make discriminations. One source of richness
is the simultaneous involvement of several senses. Another is the array
of "higher order variables of stimulus,"[48] such as spatial and temporal
gradients, which are capable of conveying decisive information. Finally,
in ordinary situations there are usually opportunities for exploration, by

46. L. Kaufman and I. Rock, "The Moon Illusion," *Scientific American,* 207 (July 1962): 120–30.
47. I. Rock, *The Nature of Perceptual Adaptation* (New York: Basic Books, 1966), pp. 264–65.
48. J. Gibson, *The Perception of the Visual World* (Cambridge, Mass.: Riverside Press, 1950), p. 3.

motion of the organism as a whole or by movements of the eyes, hands, and head, for the purpose of bringing small clues into prominence and achieving new perspectives. The actual and potential richness of the field of sensory stimuli provides an opportunity for a diversity of strategies of perception and helps to explain why it is possible, and biologically desirable, that perception be a wonderfully flexible activity.

Two major classes of perceptual strategies can be recognized, without precluding a finer taxonomy and intermediate types: one which may be called "integrative" and the other "analytic." This dichotomy seems to be recognized, with various names and descriptions, by psychologists holding very different theories. For example, for Bruner, the integrative strategy involves a complex decision process in which clues are sought, a tentative categorization is made and then checked by examining other clues, and finally an identification is performed which terminates search – and the details of this process are normally subconscious. The analytic strategy, by contrast, is a "constant close look," with conscious attention to the clues and their compatibility with available categories, and with the possible option of introducing new categories.[49] J. Gibson distinguishes between "perception," which is the gathering of information via the senses about objects and events in the world, and "sensation," which is not a preliminary process to such information-gathering, but is rather the direction of the subject's attention upon his own sensory reactions.[50] Without minimizing the deep theoretical differences between Gibson and Bruner, one can say that they agree in ascribing functional and biological primacy[51] to the integrative strategies; and I think that it is fair to make the same ascription to most other psychologists, even to the classical empiricists who maintain that sensations are constitutive of and temporally antecedent to integrated perceptions.

The thesis of the theory-ladenness of perception is exemplified very well in certain types of integrative perception, but it does not do justice to the detailed taxonomy of the entire class. Consider first the perceptual identification by means of visual clues of objects which have been experienced by various senses, for instance, different kinds of fruit. The perceptual judgment that an object is an orange is normally immediate (not preceded by deliberate conscious scrutiny of the color and visual texture prior to the judgment), and it is integrated with an expectation of taste which could never be extracted from the visual clues alone. By stretching

49. Bruner, "On Perceptual Readiness."
50. *Perception of the Visual World,* and *The Senses Considered as Perceptual Systems* (Boston: Houghton Mifflin, 1966).
51. I give an argument from an evolutionary point of view in favor of this primacy in "Perception from an Evolutionary Point of View," *Journal of Philosophy,* 68 (1971): 571–83.

the term "theory" to apply to knowledge of any general proposition, even of a low level of sophistication, one has here a good example of theory-laden perception. Now consider, by contrast, the perceptual discrimination of male from female chicks, whose genital eminences look very similar to a novice.[52] To make the discrimination accurately (as high as 99.5 percent) learning is essential, and the perceptual judgment of the expert is integrative in the sense that clues are not antecedently consciously discriminated. There is, however, a difference between this case and the preceding one, which is at least as important as the common denominator of requiring a background of learning. In the latter case the information for distinguishing between male and female chicks (though, of course, not all the biological potentialities associated with the distinction) is already present in the total visual stimulus, but it is there subtly as a "higher-order variable," namely a pattern. The existence of such higher-order variables is an instance of the previously noted characteristic of perception under ordinary circumstances, that the stimulus is immensely rich in details. Perceptual learning in this case consists of acquiring the skill to discriminate the pattern. In the case of visually perceiving the orange as a fruit with an expected flavor, the visual stimulus does not contain the flavor as a higher-order variable.[53] The difference between these two cases of integrative perception is blurred when Hanson draws the following parallel: "The microscopist sees coelenterate mesoglea, his new student sees only a gooey, formless stuff. Tycho and Simplicius see a mobile sun, Kepler and Galileo see a static sun."[54] He has conflated pattern recognition, in which higher-order variables in the visual stimulus are discerned as a result of training, with a type of cognitive act in which the elements integrated with visual clues are not themselves visual variables.[55] The terms "perception" and "observation" can be used broadly so as to apply to this type of cognitive act, and the advantage of doing so in describing the activity of scientists is great:

Once the theory has achieved a status where it is no longer in the process of confirmation . . . then it is free to augment the range of observation. It gets swallowed up, so to speak, in proto-knowledge and new background beliefs and a new plateau of observation is achieved. . . . What we know is dependent, to a greater or

52. E. Gibson, *Principles of Perceptual Learning and Development* (New York: Appleton-Century-Crofts, 1969), p. 6.
53. A subtle intermediate case, which is not fully understood, involves intermodal transfer between senses. See Gibson, *Principles of Perceptual Learning,* pp. 215ff.
54. *Patterns of Discovery,* p. 17.
55. Hanson also conflates pattern recognition in ordinary circumstances with Gestalt switches, but it is an essential feature of the latter that the artificially drawn ambiguous figures lack appropriate higher-order variables for resolving the ambiguity (see J. Gibson, *Perception of the Visual World,* p. 211).

less degree, on what we have seen to be the case; but what we are capable of seeing, epistemically, is expanded by the accumulation of more information. The greater our experience, the more inclined we are to take certain things for granted; the more inclined we are to take things for granted (providing always, of course, that they are true) the more pregnant with information become the things we can see to be the case.[56]

However, such breadth of usage is harmful if it is associated with a coarsening of the phenomenological description of perception, which can blur epistemologically important distinctions and hamper our understanding of perceptual mechanisms.

The consideration from empirical psychology which is most damaging to the thesis that observation is theory-laden is the existence of analytic strategies of perception. These strategies are frequently adopted when observers find discordance between sensory clues and the expectations suggested by their beliefs and theories. It is indeed fair to say that the availability of analytic strategies to an observer is an index of his capability to relax theoretical preconceptions.

One striking example will show how Hanson overlooks the role of analytic strategies. He cites the experiment of Bruner and Postman in which subjects are shown in tachistoscopic exposure a nonstandard playing card, such as a red six of spades, and often report seeing a black six of spades or a red six of hearts, thereby integratively perceiving the card in accordance with a normal category. Hanson fails to mention, however, some of the further results of the experiment. When the exposure time was increased, more than half of the subjects frequently experienced perceptual disruption, that is, "a gross failure of the subject to organize the perceptual field at a level of efficiency usually associated with a given viewing condition," and made comments like "I don't know what color it is now or whether it's a spade or a heart. I'm not even sure now what a spade looks like!" And upon even longer exposure most of the subjects (89.7 percent at one second, the longest exposure used) finally recognized the characteristics of the nonstandard card. A "close look," or adoption of an analytic strategy of perception, succeeded finally in disentangling features which are conjoined in the usual classification of playing cards, and which therefore are inseparable when this classification has been internalized and an integrative strategy of perception is followed.[57]

An answer on Hanson's behalf to the charge that he has neglected the role of analytic strategies of perception might be given by pointing to

56. Dretske, *Seeing and Knowing,* pp. 256–57.
57. J. Bruner and L. Postman, "On the Perception of Incongruity: A Paradigm," *Journal of Personality,* 18 (1949–1950): 208–23.

those passages in which he acknowledges the occurrence of phenomenal seeing, particularly in situations of confusion and conceptual muddle:

In microscopy one often reports sensations in a phenomenal, lustreless way: 'it is green in this light; darkened areas mark the broad end. . . .' So too the physicist may say: 'the needle oscillates, and there is a faint streak near the neon parabola. Scintillations appear on the periphery of the cathodescope.'[58]

This answer, however, is inadequate for two reasons.

First, it misses the importance of an analytic strategy of perception when adherents of conflicting theories examine the results of an important experiment. Neither party may be in a state of confusion (at least subjectively), and therefore Hanson's account provides no reason for either to forgo an integrative strategy of perception in making his observations. There would then be the danger of the hedging of theories against adverse empirical data which I. Scheffler and C. Kordig have recognized in Hanson's epistemology.[59] What opposing theorists *should do,* in accordance with C. S. Peirce's admonition not to block the way of inquiry, and what they very often *in fact do,* because they are interested in finding out the truth about nature, is to adopt an analytic strategy of perception. Thus, bubble chamber photographs can be scanned cooperatively, with detailed attention to such isolable features as the thickness, length, and curvature of a track, and the number of tracks emerging from a vertex, in abstention from integratively perceiving the pictures as characteristic of certain elementary particle processes. To be sure, a communion of analytic scrutiny of experimental results is far from a sufficient condition for adjudication of theoretical disagreements. The questioning of auxiliary hypotheses provides immense resources for defending a favored theory against unfavorable data, as P. Duhem especially emphasized, and these resources can be expanded by exploiting statistical uncertainties. Stubbornness need not take the crude form of refusing to see what is plainly before one's eyes.

The second reason is more complex and has to do with the potentialities for switching to an analytic strategy of perception which are implicit in most instances of integrative strategy. A perceptual judgment resulting from an integrative strategy is potentially subject to being wondered about, doubted, censored, and reconstructed in memory as a consequence of subsequent experience and thought of the subject. He therefore has dispositions for a diversity of perceptual responses to any recurrence of the physical stimuli which resulted in the initial perceptual judgment, and among them are dispositions for adopting an analytic strategy, especially

---

58. *Patterns of Discovery,* p. 20.
59. Scheffler, *Science and Subjectivity;* Kordig, "The Theory-ladenness of Observation."

if the intervening experience should throw doubt upon the tacit premises of the initial integration. For example, an intense auditory stimulus may result, because of an integrative strategy, in a perceptual judgment that a firecracker has exploded; but given appropriate experience or beliefs, the occurrence of an almost identical stimulus at a later time could result in a perceptual judgment of a grenade attack, or in the analytic strategy of hearing the noise and holding its interpretation in abeyance. The potentiality for such a switch to an analytic strategy or perception retrospectively throws light upon the initial perceptual judgment. It shows something about the elements of that judgment and the way in which they have been integrated, in other words, about its *internal structure.* (An analogy to a realistic treatment of the dispositions of physical objects is obvious. The disposition of a crystal to melt when heated is understood in terms of its internal structure, and conversely, the internal structure of the crystal can be probed by subjecting it, or replicas of it, to a variety of external conditions, so that various dispositions are actualized.) The possibilities for switching to an analytic strategy of perception differ greatly in the face of different types of integration, and some, such as the temporal integration of the succession of presentations during the specious present, may be greatly resistant to an analytic strategy. However, the instances of integration which are Hanson's central concern, namely, of sensory clues with conceptual elements from theories of considerable scientific sophistication, are evidently associated with strong potentialities for switching to an analytic strategy. The potentialities are easy to discern, just because the content of sophisticated theories, the evidence for them, their logical organization, and the procedures for learning them are quite well articulated; whereas the "theories" (if the term is proper) of everyday life are so deeply engrained by heredity or infantile experience that their articulation requires great effort.

7. INTERPRETATION AND THE INTERNAL STRUCTURE OF PERCEPTION

The analysis of the preceding section skirted Hanson's contention that the integration of conceptual elements into perception could not be ascribed to interpretation. This topic deserves a separate treatment, since it involves questions of the mechanism of perception more crucially than topics discussed previously.

One may wonder why Hanson so insistently separates his epistemological position from the tradition represented by H. Helmholtz and more recently by Brunswik and Bruner, in which perceptual integration is considered to be a quasi-rational process, characterized variously as interpretation, unconscious inference, or probabilistic decision-making. There appear to be two philosophical reasons for Hanson's unwillingness to ally himself to this tradition. The first is his belief that the processes, postulated in this tradition, of unconscious inference and instantaneous interpretation (where "instantaneous" is applied broadly to any event with a temporal duration too brief to be consciously discerned) have very dubious ontological status.[60] The second reason is his suspicion that the tradition gives epistemological primacy to sense data or some other uninterpreted components of perception, thereby falsifying the actual phenomenology of experience and incurring the danger of excluding the subject from a public world.[61]

The first of these reasons would be expected from a philosopher of science like E. Mach, who thinks it worthless or harmful to attribute to natural things an internal structure which cannot be directly discerned; but it seems out of place in a philosopher who rejects a fictionalist understanding of the formalism of elementary particle theory.[62] A possible move at this point is to admit the meaningfulness of a microscopic description of physical, but not of psychological, entities, and in fact there is some textual evidence that Hanson does dichotomize in this way:

While what ordinary people usually mean when they speak of electrons, or waves, or genes, or functions is of little or no importance to the understanding of such purely scientific concepts, it is of maximum importance when we are concerned to understand a general philosophical concept like *seeing*.[63]

This passage suggests that Hanson subscribes to Wittgenstein's rejection of the comparability of psychology and physics.[64] However, to deny an internal structure to a well-integrated perceptual process is grossly incompatible with the psychological evidence. One could as well maintain in the face of biological evidence that blood is a homogeneous substance. That perceptual processes admit of a meaningful "microscopic" description is initially made plausible by the fact that they are inseparable from neurophysiological processes of great rapidity and fantastic complexity. This *prima facie* argument is reinforced by a large body of psychological findings on such matters as scanning under various types of instructions, search and reaction times, suppression of visual sensitivity during saccadic movements, rhythmic patterns in auditory recognition, and so forth.

---

60. *Patterns of Discovery,* p. 10.
61. *Perception and Discovery,* pp. 70–74.
62. N. R. Hanson, *The Concept of the Positron: A Philosophical Analysis* (Cambridge: Cambridge University Press, 1963), pp. 46–47.
63. *Perception and Discovery,* p. 79.
64. Wittgenstein, *Philosophical Investigations,* p. 232.

Two quotations will indicate the complexity of the evidence as well as reasonable interpretations:

Perception generally does seem to have the redundancy, wastefulness, and freedom from gross misrepresentation that characterize a parallel process. . . . Moreover, one might expect a parallel process to resist introspection, since so much unrelated activity is going on simultaneously.[65]

Further, the result of each process must be stored or temporarily maintained, so that the information-processing sequence can be subdivided into stages, stores, and processes – each with its own sequences, time constants, and interactions.[66]

Finally, without such considerations of internal structure it is hard to make sense of the fact that the perceptual system is "tunable" with great sensitivity in extracting information from the enormously rich array of sensory input.

An answer to the second of the reasons for Hanson's aloofness from Helmholtz and his followers is that their position does not essentially require the attribution of epistemological primacy to uninterpreted components of perception. They could recognize, as Bruner explicitly does, that under normal conditions it is the perceptual judgment issuing from an integrative strategy which is the basis for articulated reports about the environment and the guide to action. The elements which are integrated – whether they are sensory clues or less familiar elements – cannot by themselves serve these purposes. When one examines the efforts of epistemologists like B. Russell, H. H. Price, and A. J. Ayer to find an error-free foundation for empirical knowledge in uninterpreted sense data, the object of their quest seems elusive, because propositional formulation and articulation require some kind of cognitive integration, thereby reintroducing the possibility of error. The stages and subprocesses in the internal structure of perception, which were hypothesized in the preceding paragraph, are more remote from consciousness than sense data and even less capable of guiding action and serving as the basis for articulated reports. To demand that they do so would be as unreasonable as demanding that a complete calculation be performed by each distinct electronic process in a computer.

What is essential to the tradition of Helmholtz is first that the elements integrated in a perceptual judgment can be discerned if the subject switches to an analytic strategy, and second that a (quasi-)rational process, inferential or interpretative, can in principle be reconstructed in terms of these elements so as to yield the perceptual judgment as its conclusion. (The qualification "quasi" is needed in order to acknowledge the likelihood

65. Neisser, *Cognitive Psychology,* p. 74.
66. R. Haber and M. Hershenson, *The Psychology of Visual Perception* (New York: Holt, Rinehart, and Winston, 1973), p. 175.

that high critical standards of logic will not be satisfied by the untrained operations of human faculties.) The psychological evidence bearing on these two propositions is very complex and far from conclusive, as can be seen by considering a single area, pattern recognition. Those theories which stress feature analysis effectively incorporate the two essential propositions of Helmholtz's position, and indeed there is much evidence to "support the view that pattern recognition involves some kind of hierarchy of feature-analyzers."[67] On the other hand, much of the empirical data indicate that holistic and constructive operations occur which are akin to visual imagination,[68] and this thesis seems to be in conflict with those two propositions. The present question is bound up with the amenability of any instance of perceptual integration to an analytic treatment, and, as indicated in the preceding section, there are varying degrees of resistance to the adoption of an analytic strategy of perception. It was also pointed out there, however, that the integrations with which Hanson is concerned, involving concepts from sophisticated scientific theories, seem to be unusually amenable to an analytic strategy and also unusually rational in structure, and therefore for them the epistemological claims of Helmholtz and his school are quite reasonable.

The epistemological implications of the internal structure of perception extend beyond the question of unconscious interpretation, and also beyond the scope of this essay. A crude analogy may be suggestive. From a naturalistic point of view the human perceptual apparatus is an instrument for obtaining information about the environment, and knowledge of its principles of construction and operation, like comparable knowledge of any scientific instrument, permits us to know how reliable it is, to what errors it is subject, and what precautions may be taken to correct these errors.[69] As an illustration, the relation between integrative and analytic strategies of perception can be reconsidered. The information implicit in sensory stimuli is masked by various kinds of distortion and "noise," both external and internal to the organism. Integrative strategies of perception are particularly well adapted to overcoming certain defects inherent in the sense organs or distortions due to the position of the subject relative to objects of interest. For example, the temporal integration of successive presentations to the eyes, which are subject to spontaneous saccadic movements, counteracts the effects of the blind spots of the

---

67. Neisser, *Cognitive Psychology,* p. 85.
68. *Ibid.,* chap. 4, esp. p. 95.
69. By elaborating this remark, I believe, one can refute the contention of D. Hamlyn (*The Psychology of Perception* [London: Routledge and Kegan Paul, 1957]) and N. Malcolm ("The Myth of Cognitive Processes and Structures," in *Cognitive Development and Epistemology,* ed. T. Mischel [New York: Academic Press, 1973]) that empirical psychology has little to offer to epistemology.

retinae and of the angular limits of the field of vision. The integration of visual clues with knowledge of the appearance of common objects is the fundamental technique for deriving information about interesting things which are distant from the subject. But the integrative mechanism is itself a source of possible distortion, especially in the second of the two examples just mentioned, in which the "unconscious inference" from a visual clue to a perceptual identification can be erroneous. Hence, an analytic strategy is valuable for examining and correcting the result of an uncritical integrative strategy. It is fair to say that the general reliability of the human perceptual system, which is truly astonishing if one considers the multiplicity of sources of "noise," is largely due to a metastrategy of switching between integrative and analytic strategies.

8. REMARKS ON LOGICAL EMPIRICISM

In conclusion I shall consider to what extent the foregoing criticism of Hanson confirms the orthodox logical empiricists' position concerning observation. In view of the complexities of perception that are exhibited in even an incomplete and oversimplified survey, their neglect of empirical psychology appears to be unjustifiable. How many difficulties are disregarded in a characterization like the following?

The terms included in the observational vocabulary must refer to attributes (properties and relationships) that are directly and publicly observable – that is, whose presence or absence can be ascertained, under suitable conditions, by direct observation, and with good agreement among different observers.[70]

Surely empirical psychology supports H. Putnam's remark that "'Being an observable thing' is, in a sense, highly theoretical."[71] Nevertheless, I claim that the orthodox logical empiricists are partly vindicated by the empirical psychology which they neglected.

What psychology in its sophistication recovers is that the naive assumption of a "publicly observable" world is largely correct. The recovery is in no sense a scientific resurrection of the epistemological thesis of direct realism, for it is accomplished by the information-processing theory of perception summarized in the preceding section, which is closer to critical realism than to any other classical epistemological position. Under suitable circumstances, perception is usually veridical in the pragmatic sense of concurrence with subsequent impressions and reliability as a guide to action. If a group of observers, not under emotional stress and with sense organs free from major defects, is placed in circumstances of good illumination, adequate opportunity for inspection, and so forth, then the perceptions of each can be expected to be veridical in the pragmatic sense,

70. C. Hempel, *Aspects of Scientific Explanation* (New York: Free Press, 1965), p. 127.
71. "What Theories Are Not," p. 241.

and public observation is achieved. Indeed, rough concurrence with the responses of other observers is implicit in the phrase "reliable as a guide to action." Under these circumstances an integrative strategy of perception, yielding an identification of an object or event, is usually successful, so that the agreement among observers can be readily articulated in a physicalistic vocabulary. However, the perceptual mechanism permits flexibility on this point. If there are discrepancies among the reports of the observers, a switch to an analytic strategy of perception is possible, and disagreements can be adjudicated by reference to discriminated features. The ensuing dialectic among the observers may also bring out into the open previously unarticulated disagreements in their premises, and the result may be a retreat by one or more of the group to more cautious integrative strategies of perception. Flexibility of this sort in achieving public observation could probably be acknowledged by the orthodox logical empiricists without essentially changing their epistemological position. Had Carnap's suggestion that the observation language can be either physicalistic or phenomenalistic[72] been modified to indicate the virtues of combining both, the resulting recommendation would have agreed well with the above discussion of integrative and analytic strategies of perception.

This sketch of a partial vindication of the logical empiricists' position on observation is not intended as an endorsement of their epistemology as a whole. From the standpoint of a naturalistic philosophy, the attempt to understand theoretical terms exclusively by means of logical relations to observation terms is inadequate. What is needed is an analysis of the causal relations between those entities referred to in scientific theories and the discriminations by which human beings can become aware of these entities. Such an analysis evidently requires scientific knowledge of the place of human beings in the world and of the characteristics of their perceptual apparatus. But it also requires philosophical investigations which are not simply subsumable under the natural sciences: specifically, concerning the concept of causality itself, and concerning the applicability of the semantic notion of reference to terms in theories which are only approximately true. A number of philosophers of science are pursuing programs like the one envisaged here, and there have already been some excellent results.[73]

72. R. Carnap, "Empiricism, Semantics, and Ontology," *Revue Internationale de Philosophie*, 11 (1950): 20–40.
73. For example, W. Sellars, *Science, Perception and Reality* (London: Routledge and Kegan Paul, 1963); R. Burian, "Scientific Realism, Commensurability, and Conceptual Change: A Critique of Paul Feyerabend's Philosophy of Science" (Ph.D. dissertation, University of Pittsburgh, 1971); W. Rottschaefer, "Ordinary Knowledge in the Scientific Realism of Wilfred Sellars" (Ph.D. dissertation, Boston University, 1972); C. Hooker, "Systematic Realism" (ms.); and R. Boyd, "Realism and Scientific Epistemology" (ms.).

ADDITIONAL REFERENCES

Feyerabend, P. "Problems of Empiricism." In *Beyond the Edge of Certainty,* edited by R. Colodny, Englewood Cliffs, N.J.: Prentice-Hall, 1965.

Helmholtz, H. von. *Treatise on Physiological Optics.* Translated from the 3d German edition (1867) by J. Southall. New York: Dover, 1962.

Newton, I. *Mathematical Principles of Natural Philosophy.* Edited by F. Cajori. Berkeley: University of California Press, 1934.

Shimony, A. "Perception from an Evolutionary Point of View." *Journal of Philosophy,* 68 (1971): 571–83.

Shimony, A. "Scientific Inference." In *The Nature and Function of Scientific Theories,* edited by R. Colodny. Pittsburgh: University of Pittsburgh Press, 1970.

Stein, H. "On the Conceptual Structure of Quantum Mechanics." In *Paradigms and Paradoxes: the Philosophical Challenge of the Quantum Domain,* edited by R. Colodny. Pittsburgh: University of Pittsburgh Press, 1972.

PART C
Epistemic probability

6

Coherence and the axioms
of confirmation[1]

I. INTRODUCTION

It has been pointed out by Carnap[2] that 'probability' is an equivocal term, which is used currently in two senses: (i) the degree to which it is rational to believe a hypothesis h on specified evidence e, and (ii) the relative frequency (in an indefinitely long run) of one property of events or things with respect to another. This paper is concerned only with the first of these two senses, which will be referred to as 'the concept of confirmation,' in order to avoid equivocation.

We may distinguish a quantitative and a comparative concept of confirmation. The general form of statements involving the former is

The degree of confirmation of the proposition h, given the proposition e as evidence, is r (where r is a real number between 0 and 1).

In this paper the notation for a statement of this form is

$$C(h/e) = r.$$

The general form of statements involving the comparative concept is

The proposition h is equally or less confirmed on e than is the proposition h' on e',

which may be symbolized as

$$SC(h/e, h'/e').$$

This work originally appeared in *Journal of Symbolic Logic* 20 (1955), pp. 1–28. Copyright © 1955 by the Association for Symbolic Logic. All rights reserved. Reprinted by permission of Kluwer Academic Publishers.

1. From a dissertation in partial fulfillment of the requirements for the degree of Ph.D. in the Department of Philosophy of Yale University, and written in part while the author was a Sterling Fellow in 1951–52. I am grateful to Professors Rudolf Carnap, John Myhill, John Kemeny, Frederic Fitch, and Howard Stein for their very helpful suggestions on the material in this paper.

2. *Logical foundations of probability,* Chicago, 1950, pp. 23–26.

The following abbreviations will also be used in discussing comparative confirmation:

'$LC(h/e, h'/e')$' for '$SC(h/e, h'/e')$ & $\sim SC(h'/e', h/e)$',
'$EC(h/e, h'/e')$' for '$SC(h/e, h'/e')$ & $SC(h'/e', h/e)$'.

The former of these two abbreviations means intuitively

The proposition h is less confirmed on evidence e than h' is on e'.

The latter means

The proposition h is confirmed on evidence e to the same degree that h' is on e'.

It will be assumed throughout this paper that the evidential propositions e and e' in $C(h/e) = r$ and in $SC(h/e, h'/e')$ are logically possible. There is no loss of generality in making this assumption, since propositions which are not logically possible can never in fact serve as evidence.

It is very easy to define SC in terms of C,[3] and it is also possible to define C in terms of SC, if a sufficiently strong set of axioms is given for the comparative concept.[4] For the purposes of this paper, neither of these reductions is necessary, and both C and SC may be taken as primitive concepts. Epistemologically, however, SC is the prior concept, for in practice, and perhaps also in principle, it is often possible to compare the degrees of confirmation of two hypotheses on their respective evidences without being able to evaluate these degrees numerically.[5]

In common usage and in most of the systematic writing on confirmation, it is generally acknowledged that there are objective grounds for confirmation statements. The truth of statements such as

The degree of confirmation that the coin will turn up heads, on the evidence that the coin appears symmetrical, is 1/2

and

The proposition that it will rain continuously for forty days and forty nights during the coming year is less confirmed on our meteorological evidence that is the denial of this proposition on the same evidence

3. R. Carnap, *op. cit.*, p. 431.
4. B. O. Koopman, "The axioms and algebra of intuitive probability," *Annals of mathematics,* Series 2, vol. 41 (1940), pp. 269–92.
5. J. von Kries, *Die Prinzipien der Wahrscheinlichkeitsrechnung. Eine logische Untersuchung,* Freiburg i.B., 1886, pp. 29ff; J. M. Keynes, *A treatise on probability,* London and New York, 1921, ch. III.

seems to be independent of our private beliefs and psychological peculiarities, and to depend only upon the propositions mentioned in these statements and upon the concepts C and SC themselves. It is extremely difficult, however, to formulate an exhaustive and adequate set of rules for employing the concepts C and SC, and even to judge the correctness of many individual statements involving these concepts. This shows that our comprehension of the comparative and quantitative concepts of confirmation is quite vague. In brief, the task of the theory of confirmation is to remove this vagueness as much as possible.

The only precise rules for C on which there is fairly general agreement are the 'axioms' of confirmation. Axiom systems for this concept have been formulated by a number of logicians, notably Jeffreys,[6] Mazurkiewicz,[7] Hosiasson-Lindenbaum,[8] Von Wright,[9] Carnap,[10] and Kneale.[11] These differ from each other and from the following system of axioms in only relatively minor respects:[12]

(1) $0 \leq C(h/e) \leq 1$.
(2) If $e \dashv h$, then $C(h/e) = 1$.
(3) If $e \dashv \sim(h \& h')$, then $C(h \vee h'/e) = C(h/e) + C(h'/e)$.
(4) $C(h \& h'/e) = C(h/e) \cdot C(h'/h \& e)$ and
      $C(h \& h'/e) = C(h'/e) \cdot C(h/e \& h')$.

The great importance of this axiom system is that from it all of the classical mathematical theory of probability (though not all of modern probability theory, which depends in part upon the Principle of Complete Addi-

---

6. *Theory of probability* (second edition), Oxford, 1948, pp. 17–25.
7. "Zur Axiomatik der Wahrscheinlichkeitsrechnung," *Comptes rendus de la Société des Sciences de Varsovie,* Cl. III, xxv (1932), pp. 1–4.
8. "On confirmation," *Journal of symbolic logic,* vol. 5 (1940), pp. 133–48. "Induction et analogie: comparaison de leur fondement," *Mind,* vol. 50 (1941), pp. 351–65.
9. *The logical problem of induction,* Helsingfors, 1941, pp. 106–7.
10. *Op. cit.,* p. 285.
11. *Probability and induction,* Oxford, 1949, pp. 125–27.
12. For a comparison of some of these axiomatizations see section 62 of Carnap's *Logical foundations of probability.* It should be particularly noted that the arguments h, e of the function C in the above set of axioms are propositions rather than sentences. Consequently, the modal operator '$\dashv$' is used rather than the corresponding metalinguistic operator; likewise the modal operator '$\Diamond$' will be used in subsequent discussions. '$\Diamond p$' means 'p is possible', and '$p \dashv q$' means 'p logically implies q,' but for a detailed discussion of these operators see F. B. Fitch, *Symbolic logic: an introduction,* especially p. 66 and p. 71. The assumption made at the beginning of this paper that all evidential propositions are logically possible makes it superfluous to write explicitly the propositions '$\Diamond e$', '$\Diamond(h \& e)$' and '$\Diamond(h' \& e)$' as antecedents to the above axioms. Finally it should be noted that in subsequent discussions I shall consider a stronger set of axioms in which (2) is replaced by the following (2'): $C(h/e) = 1$ if and only if $e \dashv h$.

tivity[13]) can be derived. Indeed, the derivation of classical probability theory from (1)–(4) seems to be the teleological explanation of the general agreement on these axioms, for the justifications usually provided for these axioms are unsatisfactory. Those of the above-cited authors who attempt to justify the axioms for C follow one or the other of the two following patterns.

(a) The axioms (1)–(4) may be claimed to be analytically true in virtue of the meaning of 'C', and their truth to be intuitively evident.[14] This claim may be challenged, however. It is not self-evident that (1)–(4) are simultaneously satisfied by any adequate confirmation function, and indeed this has been doubted even by some logicians sympathetic to using the concept of confirmation as the central concept of inductive logic.[15] Since there is always a certain amount of conventionalism in specifying a metrical or quantitative concept, it is plausible that the function C could be so specified as to satisfy the conditions of range of values and of additivity of axioms (1), (2), (3); but it is not clear that when C is suitably transformed so as to satisfy (1), (2), and (3), it will also satisfy axiom (4).

(b) If the concept C is sufficiently restricted, it becomes easy to demonstrate (1)–(4). In particular, suppose '$C(h/e) = m/n$' is well-defined only when h and e satisfy the conditions of the Laplacian definition of probability, i.e., there is a set of alternatives $h_1, \ldots, h_n$ such that $e \dashv (h_1 \vee \cdots \vee h_n)$, $e \dashv \sim(h_i \,\&\, h_j)$ for $i \neq j$, and h_i and h_j are equally confirmed on e for all i and j between 1 and n, and finally, h is equivalent to the disjunction $h_{k_1} \vee \cdots \vee h_{k_m}$ of m of the alternatives $h_1, \ldots, h_n$. Then elementary algebra suffices to demonstrate (1)–(4).[16] This procedure is unsatisfactory, however, since we use the concept of confirmation and intuitively understand the meaning of this concept in situations where h and e cannot be analyzed in the Laplacian fashion. Jeffreys and Wrinch[17] modify defense (b) by first proving the axioms for a class of arguments satisfying the Laplacian conditions, and then claiming that it is natural and convenient to extrapolate the domain of these axioms to arguments not satisfying the Laplacian conditions. This seems to be an inductive inference, however, which cannot pretend to be decisive, and indeed seems to me quite unconvincing.

13. See section 5 of this paper.
14. J. Hosiasson-Lindenbaum, "Induction et analogie: comparaison de leur fondement," *Mind,* vol. 50 (1941), p. 353.
15. J. Kemeny, "Carnap on confirmation," *The review of metaphysics,* vol. 5 (1941), pp. 152–53.
16. H. Poincaré, *Calcul des probabilités* (second edition), Paris, 1912, pp. 37–39; W. Kneale, *Probability and induction,* pp. 125–27.
17. "On some aspects of the theory of probability," *Philosophical magazine,* 6th series, vol. 38 (1919), pp. 715–31, especially p. 722.

An axiomatization for *SC* was first given by Keynes,[18] and a more precise and stronger set of axioms was later formulated by Koopman.[19] Both Keynes and Koopman justify their axiomatizations by a claim of self-evidence. However, if it is meaningful to speak of degrees of self-evidence, many of Keynes' axioms and several of Koopman's (especially his axioms of composition and decomposition) are less self-evident than is desirable. Consequently, a more adequate justification of the axioms of *SC* as well as of *C* is needed.

To my knowledge, only two justifications of the axioms of *C* have been given which neither make dubious appeals to intuition nor too narrowly restrict the quantitative concept of confirmation.

One is the procedure of R. T. Cox,[20] who deduces the quantitative axioms from the very plausible assumptions that

$$C(h \,\&\, h'/e) = F(C(h/e), C(h'/h \,\&\, e)) \text{ and}$$
$$C(\sim h/e) = S(C(h/e))$$

for some moderately well-behaved functions *F* and *S* (i.e., functions possessing continuous second derivatives).

The second satisfactory justification is due independently to F. P. Ramsey and Bruno DeFinetti.[21] Unlike the majority of writers on confirmation, Ramsey and DeFinetti maintain a subjective theory of confirmation, i.e., that the degree of confirmation of *h* on *e* is not determined by the logical relations between these propositions, but is evaluated with equal validity by each person in accordance with his own habits of belief-formation. They do acknowledge, however, that the set of beliefs of a rational person must be internally consistent, and indeed require that these beliefs satisfy the condition of *coherence,* which is a natural extension of the condition of consistency. The notion of coherence used by Ramsey and DeFinetti is the following. When a person X has a set of beliefs precise enough to determine decisions in betting, there are two possibilities: (i) one may propose a set of stakes such that in betting for these stakes in accordance with his evaluations, X is bound to lose, or (ii) there exists no such set of stakes. In the latter case, X's beliefs are *coherent,* in the former they are *incoherent.* The remarkable result obtained by Ramsey and

18. *Op. cit.,* pp. 135–38.
19. *Op. cit.,* pp. 275–76.
20. "Probability, frequency, and reasonable expectation," *American journal of physics,* vol. 14 (1946), pp. 1–13.
21. F. P. Ramsey, "Truth and probability," *The foundations of mathematics and other logical essays,* London and New York, 1931; B. DeFinetti, "Sul significato soggetivo della probabilità," *Fundamenta mathematicae,* vol. 17 (1931), pp. 298–329, and "La prévision: ses lois logiques, ses sources subjectives," *Annales de l'Institut Henri Poincaré,* vol. 7 (1937), pp. 1–68.

DeFinetti is this: the confirmation evaluations made by an individual must satisfy the axioms of C, if these evaluations are to constitute a coherent set of beliefs. Thus the axioms of the quantitative concept of confirmation are justified, in that they are necessary conditions for the coherence, and hence for the rationality, of beliefs.

I shall not in this paper discuss the argument of Cox, though it seems valid and ingenious. My purpose here is to continue and to reinterpret the work of Ramsey and DeFinetti on the axioms of confirmation. I have adopted their method of justifying these axioms, with the following departures.

(a) The concept of coherence which I propose is a somewhat stronger and, I believe, also a more adequate explication of the intuitive notion of coherence than is Ramsey's and DeFinetti's.

(b) The principle by which Ramsey and DeFinetti justify the axioms of confirmation, which I call 'the Principle of Coherence', is formulated explicitly, so that its epistemological status may be examined. When thus formulated, the Principle of Coherence seems to be analytic.

(c) It is pointed out that the justification of both the comparative and the quantitative axioms presupposes that the ranges of C and SC satisfy certain minimum conditions. These conditions are summarized in what I call 'the Closure Rule', and the validity of this rule is defended.

(d) Since a strong concept of coherence is used, a somewhat stronger set of quantitative axioms than set (1)–(4), which Ramsey and DeFinetti are able to justify, is derived. The stronger set consists of the old axioms (1), (3), and (4), with axiom (2) replaced by the following axiom (2'): $C(h/e) = 1$ if and only if $e \dashv h$.

(e) The procedure of Ramsey and DeFinetti is extended so as to justify Koopman's axioms for SC.

(f) It is argued that the use of the Principle of Coherence to derive the axioms of confirmation is compatible with an objective theory of confirmation. This is the most important point of the present paper, since I believe that Ramsey's and DeFinetti's ingenious method of justifying the axioms of confirmation will be accepted more widely when it is separated from their subjectivistic epistemology.

2. THE CONCEPT OF COHERENCE

Coherence is a property of sets of beliefs. Beliefs held by a person to the total denial of any alternative constitute only a subset of the class of beliefs which are under consideration here. The more general class consists of beliefs which may be *partial* as well as total, and which may be held *relative* to the assumption that certain propositions are true. Partial relative beliefs will be discussed in both quantitative and comparative forms:

(i) X believes *h*, on assumption that *e* is true, to degree *r*.
(This statement will be symbolized as '$B_X(h/e) = r$'.)

(ii) X believes *h*, on assumption that *e* is true, to a lesser degree than he believes *h'* on assumption that *e'* is true.
(This will be symbolized as '$LB_X(h/e, h'/e')$'.)

(iii) X believes *h* on assumption that *e* is true to the same degree that he believes *h'* on assumption that *e'* is true.
(This will be symbolized as '$EB_X(h/e, h'/e')$'.)

In discussing the comparative form of partial relative beliefs I shall not assume that X is capable of specifying a linear order among his beliefs. Were such an assumption made, it would be possible to reduce the number of primitive concepts, for we could take the concept SB_X as primitive, where '$SB_X(h/e, h'/e')$' is to mean 'X believes *h* on assumption that *e* is true equally or less than he believes *h'* on assumption that *e'* is true', and then the concepts LB_X and EB_X could be defined respectively as

$SB_X(h/e, h'/e')$ & $\sim SB_X(h'/e', h/e)$ and
$SB_X(h/e, h'/e')$ & $SB_X(h'/e', h/e)$.

The concepts LB_X and EB_X as thus defined would then have the intended meaning. If linear ordering is not assumed, however, $\sim SB_X(h/e, h'/e')$ may indicate only that X is uncertain as to the relative strengths of his belief in *h* on assumption of *e* and his belief in *h'* on assumption of *e'*. Consequently, '$SB_X(h/e, h'/e')$ & $\sim SB_X(h'/e', h/e)$' does not adequately represent the intended meaning of '$LB_X(h/e, h'/e')$'. For this reason LB_X and EB_X will be taken as primitive concepts, and SB_X will be defined in terms of them as follows: '$SB_X(h/e, h'/e')$' for '$LB_X(h/e, h'/e')$ or $EB_X(h/e, h'/e')$'.

In order to define coherence, it is convenient to explicate the quantitative and comparative forms of partial belief in terms of willingness to bet. *Prima facie* this seems to be too narrow an interpretation of belief. Betting, however, need not be construed only as gambling in well-recognized games of chance. Any decision whose success or failure depends upon the actual truth or falsehood of the propositions which are the objects of belief is a bet. Thus the decision to cross a street is a bet whose success depends upon the truth or falsehood of the proposition that no accident will befall me while crossing, a proposition which in ordinary circumstances is believed to a much higher degree than its contrary. When betting is understood in such a wide sense, the following explications of partial relative belief are quite reasonable.

(a) $B_X(h/e) = r$ if and only if X would accept a bet on h on the following terms, or on terms more favorable to himself than these:

To pay rS and collect nothing for a net gain of $-rS$ in case e is true and
    h is false;
To pay rS and collect S, for a net gain of $(1-r)S$, in case e and h are both
    true;
To annul the bet, i.e., to have a net gain of 0, in case e is false.

In the above terms of the bet, S is known as the 'stake', and r, which is the ratio between the amount risked by X and the stake, is called 'the betting quotient'. It should be noted that the stake S must be measured in terms of units of utility.[22] Because of the diminishing marginal utility of money, and because there is utility both in the excitement of betting and in the security of risking little, the above explication would not be adequate to our intuitive notion of degrees of belief if S were measured in terms of money. It should also be noted that for ease in mathematical considerations, S is allowed to be either positive or negative. The interpretation of a negative stake is clear: 'to pay a negative quantity S' is equivalent to 'to collect a positive quantity $-S$', and conversely.

(b) $LB_X(h/e, h'/e')$ if and only if, for any r which X would accept as a betting quotient in a bet on h contingent on the truth of e for any positive or negative stake, there is an $r' > r$ which he would accept as a betting quotient in a bet on h' contingent on the truth of e' for any positive or negative stake. $EB_X(h/e, h'/e')$ if and only if any r which is acceptable to X as a betting quotient for a bet on h contingent on the truth of e for any stake, is also acceptable to him for a bet on h' contingent on the truth of e' for any stake, and conversely.

The comparative concepts LB_X and EB_X are clearly less idealized and more readily applicable than the quantitative concept B_X, for X may be able to assert that he would accept a higher (or equal) betting quotient for a bet on h' contingent on the truth of e' than for a bet on h contingent on the truth of e, even though he might be unable to specify quantitatively what he would consider an acceptable betting quotient in either case. The following terminology will be useful in subsequent discussion: The betting quotients p and p', for bets on h contingent on e and on h' contingent on e', respectively, are *compatible with* $LB_X(h/e, h'/e')$ if and only if $p < p'$, and are *compatible with* $EB_X(h/e, h'/e')$ if and only if $p = p'$. By a natural extension, we can speak of a set of betting quotients being compatible with a set of comparative beliefs.

With these preliminaries it is possible to give a precise definition for both the coherence of sets of quantitative beliefs and the coherence of

22. R. Carnap, *op. cit.*, section 51.

sets of comparative beliefs. In section 1 a tentative definition of coherence was formulated and attributed to Ramsey and DeFinetti. DeFinetti's own definition is this:

Lorsqu'un individu a évalué les probabilités des certains événements, deux cas peuvent se présenter: ou bien il est possible de parier avec lui en s'assurant de gagner à coup sûr, ou bien cette possibilité n'existe pas. Dans le premier cas on doit dire evidemment que l'évaluation de la probabilité donnée par cet individu contient une incohérence; dans l'autre cas, nous dirons que l'individu est cohérent.[23]

Ramsey uses the term 'consistent' with the same sense as DeFinetti's 'cohérent'. The clearest statement on this matter which I have been able to find in Ramsey's essay is the following:

Having degrees of belief obeying the laws of probability implies a further measure of consistency, namely such a consistency between the odds acceptable on different propositions as shall prevent a book being made against you.[24]

In usual gambling terminology 'a book being made against you' means that an opponent has made a series of bets with you such that, whatever the outcome may be concerning the propositions bet upon, he will make a net gain. Since this is what Ramsey understands by 'having a book made', the identification of his sense of 'consistency' with DeFinetti's 'coherence' is accurate.

Ramsey's and DeFinetti's notion of coherence seems too weak, however. There are sets of beliefs which are classified as coherent by their definition, but which intuitively we should classify as incoherent. Specifically, suppose X's beliefs are such that an opponent can propose a series of bets acceptable to X on the basis of his beliefs, and such that (i) X does not suffer a net loss in every eventuality, yet (ii) he makes a net gain in no eventuality, and in at least one possible eventuality he suffers a net loss. X's beliefs in this example are coherent, according to Ramsey's and DeFinetti's notion of coherence, although intuitively we are inclined to say that they are incoherent.

I propose, therefore, the following definition of 'a coherent set of quantitative beliefs', which is somewhat stronger than Ramsey's and DeFinetti's. Suppose that X entertains the beliefs $B_X(h_i/e_i) = r_i$, for $i = 1, \ldots, n$ (where n is finite but arbitrary). Now let a series of bets be proposed with the conditions

X will pay $r_i S_i$ and collect nothing, for a net gain of $-r_i S_i$, in case e_i is true and h_i is false;

23. "La prévision: ses lois logiques, ses sources subjectives," *Annales de l'Institut Henri Poincaré*, vol. 7 (1937), p. 7.
24. *Op. cit.*, pp. 182–83.

X will pay $r_i S_i$ and collect S_i, for a net gain of $(1 - r_i) S_i$, in case both h_i
and e_i are true;
X neither pays nor collects in case e_i is false.

In view of the explication above of '$B_X(h/e) = r$', X is willing to accept
this series of bets. Now there are two possibilities: (i) There exists a choice
of stakes S_i such that, if X accepts the series of bets at these stakes, then
no matter what the actual truth values of h_i and e_i may be, X can at best
lose nothing, and in at least one possible eventuality he will suffer a positive loss; (ii) such a choice of the S_i does not exist. In the latter case X's
system of beliefs is *coherent,* in the former it is *incoherent.*

The concept of coherence of a set of quantitative beliefs can now be
used in formulating a reasonable definition of 'a coherent set of comparative beliefs'. Suppose X holds the comparative beliefs $LB_X(h_i/e_i, h_i'/e_i')$
and $EB_X(h_j/e_j, h_j'/e_j')$ for $i = 1, \ldots, n$ and $j = n+1, \ldots, m$ (where m is finite but arbitrary). There are two possibilities: Either (i) there exists a
set of betting quotients p_i, p_i', p_j, p_j' for bets on h_i contingent on e_i, on
h_i' contingent on e_i', h_j contingent on e_j, and h_j' contingent on e_j', such
that the complete set of these betting quotients is compatible with the
above set of comparative beliefs, and such that the set of quantitative
beliefs consisting of $B_X(h_i/e_i) = p_i$, $B_X(h_i'/e_i') = p_i'$, $B_X(h_j/e_j) = p_j$, and
$B_X(h_j'/e_j') = p_j'$ constitute a coherent set of quantitative beliefs; or (ii) this
condition is not fulfilled. In the former case, X's set of comparative beliefs is *coherent,* in the latter case it is *incoherent.* It should be pointed
out that an equivalent definition of 'a coherent set of comparative beliefs'
can be formulated without employing the concept B_X. Such a formulation would have the advantage of not disguising the fact that the notion
of comparative belief is more fundamental than the notion of quantitative belief. However, such a formulation would be even more complex
and cumbersome than the definition given, and hence it was avoided.

3. THE PRINCIPLE OF COHERENCE

The proposition which asserts the relevance of considerations of the coherence of beliefs to confirmation is the Principle of Coherence. It may
be stated so as to apply neutrally both the comparative and to the quantitative concepts of confirmation: namely, *the concepts of confirmation
are such that sets of beliefs determined by correct confirmation judgments are coherent.* In order to derive the quantitative and comparative
axioms of confirmation, however, it is necessary to state precise specifications of the Principle of Coherence for both C and SC.

The Principle of Coherence for C: Suppose τ is a finite set of beliefs
$B_X(h_i/e_i) = r_i$ and suppose each of the (h_i, e_i) is in the range of C. A sufficient condition for τ to be coherent is that $B_X(h_i/e_i) = C(h_i/e_i)$.

The Principle of Coherence for SC: Suppose τ is a finite set of beliefs $LB_X(h_i/e_i, h_i'/e_i')$, $EB_X(h_j/e_j, h_j'/e_j')$. Then a sufficient condition for the coherence of τ is that $LC(h_i/e_i, h_i'/e_i')$ and $EC(h_j/e_j, h_j'/e_j')$. (Note: It evidently follows from this that a sufficient condition for the coherence of a set of beliefs $SB_X(h_i/e_i, h_i'/e_i')$ is that $SC(h_i/e_i, h_i'/e_i')$.)

One point must be clarified regarding the Principle of Coherence for C. The neutral formulation of the Principle of Coherence at the beginning of this section asserts only that beliefs "determined by" the concept of confirmation must be coherent, but it makes no mention of the precise mode of determination. Consequently, in the formulation of the Principle of Coherence in its quantitative specification the clause "$B_X(h_i/e_i) = C(h_i/e_i)$" could have been replaced by "$B_X(h_i/e_i) = f[C(h_i/e_i)]$", where f is any one–one function of the range of values of C onto the range of degrees of belief. This indicates that there is a certain amount of conventionalism in the choice of the confirmation function C (as there is in the specification of any metrical concept). No ambiguity results from this conventionalism, so long as it is explicitly indicated what convention is being adopted. In formulating the Principle of Coherence for C the simplest convention was adopted: namely, f was taken to be the identity function.

The Principle of Coherence for both C and SC seems to be analytic; i.e., it is true in virtue of relations which hold between the concepts of confirmation and the concepts of coherence, rather than in virtue of any contingent facts. It must be emphasized that these concepts are not linguistic entities, and that the Principle of Coherence itself is a proposition rather than a sentence. Consequently, we cannot explicate the analyticity of the Principle of Coherence as L-truth,[25] i.e., as truth in virtue of the semantical rules of a language. No complete explication exists of the non-semantical sense of 'analytic', but it is a concept with which we have some intuitive familiarity. For instance, if numbers and sets are considered to have some sort of Platonic existence, then the true propositions of arithmetic and set theory are analytic in the non-semantical sense.[26]

It may be objected that the analogy between the Principle of Coherence and the truths of mathematics is inaccurate, since our concepts of number and set are precise, whereas our concepts of confirmation are quite vague. Though there is some force in this objection, the difference in precision with which we grasp these various concepts seems to be a matter of degree. We do indeed have a remarkably precise grasp of the concept of number. The occurrence of the set-theoretical antinomies, and the failure to date of logicians to provide an axiomatization of the notion

25. R. Carnap, *op. cit.*, pp. 83–84.
26. K. Gödel, "Russell's mathematical logic," *The philosophy of Bertrand Russell* (edited by P. A. Schilpp), Menasha, Wisconsin, 1946, pp. 150–52.

of set which is both reasonably complete for the purposes of mathematics and clearly free from contradiction indicate, however, that our intuitive grasp of the concept of set is not precise.[27] Nevertheless, there is much that we do know about the concept of set, and we are probably justified in considering the basic axioms of set theory to be analytic.[28] Indeed, the example of the set concept shows that we are able to have a quite clear grasp of 'nuclear' aspects of certain crucial concepts, which suffices to establish the analyticity of some propositions involving these concepts, although our intuition of 'peripheral' aspects of the concepts may be vague.

In particular, it is reasonable to claim that the only aspects of the concept of confirmation upon which the truth of the Principle of Coherence depends are aspects which we do grasp clearly. First of all, whatever the detailed content of the concepts of confirmation and of rational belief may be, it is clear that C and SC are normative of rational quantitative and comparative beliefs respectively. In other words

(a) If a set of beliefs $B_X(h_i/e_i) = r_i$ or $LB_X(h_i/e_i, h_i'/e_i') \& EB_X(h_j/e_j, h_j'/e_j')$ is determined by correct confirmation judgments, then that set is rational.

Secondly, whatever the detailed content of the concept of rational belief may be, a necessary condition for a set of beliefs to be rational is that it be coherent. Since coherence has been defined only for finite sets of beliefs,[29] it is more accurate to say

(b) A necessary condition for a finite set of beliefs to be rational is that it be coherent.

It is difficult to see how any 'proof' could be given of the analyticity of either (a) or (b). Both assert such fundamental relations among the concepts involved in them that in defending them, one can only appeal to an intuitive grasp of these concepts.

It follows at once from (a) and (b) that any finite set of beliefs determined either by C or SC is coherent. This, however, is just the Principle of Coherence in its quantitative and comparative specifications. A proposition inferrible from two analytic propositions is analytic, and as we saw above, the analyticity of (a) and (b) seem intuitively evident. Hence the Principle of Coherence is analytic.

27. *Op. cit.*, p. 150.
28. K. Gödel, *op. cit.*, pp. 150–52; also his "What is Cantor's continuum hypothesis," *American mathematical monthly*, vol. 54 (1947), pp. 520–22.
29. The difficulty of extending the notion of coherence so as to apply to infinite sets will be pointed out in section 5.

4. THE CLOSURE RULE

An ordered pair of propositions (h, e) is *in the range of C* if and only if $C(h/e)$ has a well-determined value p. A pair of ordered pairs (h, e), (h', e') is *in the range of SC* if and only if either $SC(h/e, h'/e')$ or $SC(h'/e', h/e)$.[30] Little has been said so far as to the range of pairs of propositions over which C is defined, or the range of pairs of pairs over which SC is defined.[31] It is, indeed, a very difficult problem to specify the ranges over which these concepts are non-arbitrarily defined, but for the purpose of adequately discussing the axioms of confirmation, a detailed investigation of this problem is unnecessary. It is important, however, to show that certain minimum sets are included in the ranges of C and LC, respectively, and that these ranges are closed under certain operations.

An examination of axioms (2) or (2'), (3), and (4) shows that these axioms not only specify the relations between values of C for certain related arguments, but also implicitly impose certain requirements on the range of C. This is clear in the case of axioms (2) and (2'), for if $\Diamond e$ and $e \dashv h$ implies that $C(h/e) = 1$, it follows that the range of C includes every pair (h, e) such that $\Diamond e$ and $e \dashv h$. The imposition of conditions on the range of C is not quite so clear in the case of (3) and (4). Thus, (3) may be interpreted as saying that if $\Diamond e$ and $e \dashv \sim (h \& h')$ and C is well-defined for two of the three pairs (h, e), (h', e), $(h \lor h', e)$, then C is also well-defined for the third, and in addition $C(h \lor h'/e) = C(h/e) + C(h'/e)$; or it may be interpreted as saying that if $\Diamond e$ and $e \dashv \sim (h \& h', e)$ and C is well-defined for (h, e), (h', e), $(h \lor h', e)$, then $C(h \lor h'/e) = C(h/e) + C(h'/e)$. There is the same ambiguity in (4). The first of these two interpretations is evidently the stronger and does impose a condition on the range of C, whereas the second interpretation imposes no such conditions. The stronger interpretation of axioms (3) and (4) will be adopted in this paper.

To justify the weaker interpretation of axioms (3) and (4) the Principle of Coherence, together with the explications of partial belief and coherence, will suffice. To justify axioms (3) and (4) in their strong interpretations, and to justify axioms (2) and (2') and the comparative axioms, an additional principle, which I call 'the Closure Rule', will be used.

The Closure Rule for Quantitative Confirmation: Suppose that from $Q_1, \ldots, Q_n$ together with the premiss that (h, e) belongs to the range of C, it is possible to infer $C(h/e) = p$. Then it is permissible to infer categorically from $Q_1, \ldots, Q_n$ that (h, e) belongs to the range of C.

30. If this condition is satisfied we shall often use the expression '(h, e) and (h', e') are comparable'.
31. However, the assumption has been made that '$C(h/e)$' and '$SC(h/e, h'/e')$' are well-defined only if e and e' are logically possible propositions.

The Closure Rule for Comparative Confirmation: Suppose that from $Q_1, \ldots, Q_n$ together with the premiss that $((h, e), (h', e'))$ belongs to the range of SC, it is possible to infer $SC(h/e, h'/e')$. Then it is permissible to infer categorically from $Q_1, \ldots, Q_n$ that $((h, e), (h', e'))$ belongs to the range of SC.

The intuitive justification of the Closure Rule is similar to the following argument: Suppose X is determined to perform only actions which are objectively moral, yet he is uncertain whether there are any canons of objective morality. X does know, however, that *if* there are such canons, then they are such that in a particular situation where two alternative actions a and b are possible and exclusive and exhaustive, a is objectively moral and b is not. Clearly X should perform action a, for if there are no canons of objective morality, it is arbitrary which action is performed, whereas if there are such canons, then a is categorically the correct choice.

Now consider the Closure Rule in its comparative specification. Suppose that

(i) from $Q_1, \ldots, Q_n$ together with the premiss that $((h, e), (h', e'))$ belongs to the range of SC it is possible to infer $SC(h/e, h'/e')$.

Interpreted in terms of rational belief, this means

(ii) on the assumption that it is rational for X to hold one or both of the beliefs $SB_X(h/e, h'/e')$, $SB_X(h'/e', h/e)$, and on assumption that $Q_1, \ldots, Q_n$ are true, then the specific belief $SB_X(h/e, h'/e')$ is rational.

Now suppose that X, who is determined to act rationally, is faced with a practical decision which depends on whether $SB_X(h'/e', h/e)$ or $SB_X(h/e, h'/e')$, and suppose also that X knows $Q_1, \ldots, Q_n$ to be true. Finally suppose that X does not know whether there are grounds for classifying either $SB_X(h/e, h'/e')$ or $SB_X(h'/e', h/e)$ as rational, especially if 'rational belief' is explicated in a quite stringent sense. What ought X to do? Clearly he should prefer $SB_X(h/e, h'/e')$ and make the decision which this belief implies. For if there are no objective grounds for saying that either $SB_X(h/e, h'/e')$ or $SB_X(h'/e', h/e)$ is rational in the stringent sense, then the choice is arbitrary, and if there are such grounds, then by (ii) $SB_X(h/e, h'/e')$ is rational. Indeed, under these circumstances, we may say categorically that $SB_X(h/e, h'/e')$ is rational, where 'rational' is now used in a less stringent sense than before. There is, however, no injustice done to the intuitive notion of confirmation, if SC is explicated so that $SC(h/e, h'/e')$, if $SB_X(h/e, h'/e')$ is rational even in the less stringent sense. Consequently, it follows from (ii) that

(iii) on assumption that $Q_1, \ldots, Q_n$ are true, we can infer $SC(h/e, h'/e')$,

and *a fortiori*

(iv) on assumption of $Q_1, ..., Q_n$ we can infer that $((h, e), (h', e'))$ belongs to the range of *SC*.

Hence, by analysis of the concepts of rational belief and confirmation we have derived (iv) from (i), thus establishing the truth and analyticity of the Closure Rule in its comparative form. The justification of the Closure Rule in its quantitative form follows in the same way.

The name 'Closure Rule' was adopted because the rule does impose certain conditions of closure on the ranges of *C* and *SC*. For instance, suppose that axiom (3) is justified in its weak interpretation, and suppose also that $\Diamond e$, $e \dashv\vdash {\sim}(h \,\&\, h')$ and that both (h, e) and (h', e') belong to the range of *C*. Then *on assumption that* $(h \vee h', e)$ belongs to the range of *C*, axiom (3) in its weak interpretation implies that $C(h \vee h'/e)$ is uniquely determined. The Closure Rule implies then that $(h \vee h', e)$ does belong to the range of *C*, without any additional assumption to that effect. In other words, the range of *C* is closed under the kind of operation described in axiom (3). Likewise, the ranges of *C* and *SC* are closed under the operations involved in the other axioms, in virtue of the Closure Rule.

5. THE QUANTITATIVE AXIOMS

In this section the strong set of quantitative axioms (1), (2'), (3), and (4), which were listed in the first section and in footnote 12 of this paper, will be proved. The demonstration will be performed in two stages. First, several lemmas stating necessary conditions for the coherence of a set of quantitative beliefs will be proved. The only premises required for these lemmas are the explications of the quantitative concept of partial relative belief and of the concept of coherence. The second stage is the derivation of axioms (1), (2'), (3), (4) from the lemmas, by making use of the Principle of Coherence and the Closure Rule in their quantitative specifications.

The technique used in proving lemmas III and IV in this section and many of the lemmas of the following section is due to DeFinetti.

Lemma I. *If* $\Diamond e$, *then a necessary condition for the coherence of the set* τ *whose one member is* $B_X(h/e) = p$ *is that* $0 \leq p \leq 1$.

Proof. By the explication of quantitative belief in section 2, X will accept the following bet for any stake *S*:

To pay *pS* and collect nothing for a net gain of $-pS$, in case *e* is true and *h* is false;

To pay *pS* and collect *S*, for a net gain of $(1-p)S$, in case *e* and *h* are true;

To annul the bet, for a net gain of 0, in case *e* is false.

If $p > 1$ choose $S > 0$. Then clearly X suffers a net loss in the first two eventualities (at least one of which must be possible by the premiss $\Diamond e$), and neither gains nor loses in the third. There are no further distinct eventualities. Hence, by the explication of coherence for sets of quantitative beliefs, τ is incoherent. If $p < 0$ choose $S < 0$, and a similar argument can be carried out. Thus, a necessary condition for the coherence of τ is that $0 \leq p \leq 1$.

Lemma II. *If $\Diamond e$ and $e \dashv h$, then a necessary condition for the coherence of the set of beliefs τ whose one member is $B_X(h/e) = p$ is that $p = 1$.*

Proof. Suppose $p < 1$. According to the explication of quantitative belief, X would accept the following bet for any stake S:

To pay pS and collect nothing for a net gain of $-pS$ in case e is true and h is false;
To pay pS and collect S for a net gain of $(1-p)S$, in case e and h are true;
To annul the bet, for a net gain of 0, in case e is false.

Since $\Diamond e$ and $e \dashv h$, the second of these three eventualities can occur, but the first cannot. Let S be negative. Since $p < 1$, $(1-p)S$ is negative – i.e., X suffers a net loss in the second eventuality. In the third eventuality X neither wins nor loses. Hence, if $p < 1$, τ is incoherent; or conversely, a necessary condition for the coherence of τ is that $p \geq 1$. But by lemma I, a necessary condition for the coherence of τ is that $p \leq 1$. Consequently, a necessary condition for the coherence of τ is that $p = 1$.

Lemma III. *If $\Diamond e$ and $e \dashv \sim(h_1 \,\&\, h_2)$, then a necessary condition for the coherence of the set of beliefs τ whose members are $B_X(h_1/e) = p_1$, $B_X(h_2/e) = p_2$, $B_X(h_1 \lor h_2/e) = p_3$, is that $p_1 + p_2 = p_3$.*

Proof. Let 'h_3' be an abbreviation for '$h_1 \lor h_2$'. By the explication of quantitative belief X is willing to accept the following bets for any S_i, where $i = 1, 2, 3$:

To pay $p_i S_i$ and collect nothing, for a net gain of $-p_i S_i$, in case e is true and h_i is false;
To pay $p_i S_i$ and collect S_i, for a net gain of $(1-p_i)S_i$, in case both e and h_i are true;
To annul the bet, for a net gain of 0, in case e is false.

There are at most four distinct possible eventualities regarding e and the h_i:

(1) e true and h_1 true, h_2 false
(2) e true and h_1 false, h_2 true

(3) e true and both h_1 and h_2 false
(4) e false.

If G_j represents X's net gain in the jth of these eventualities, then evidently

$$G_1 = S_1 + S_3 - (p_1 S_1 + p_2 S_2 + p_3 S_3)$$
$$G_2 = S_2 + S_3 - (p_1 S_1 + p_2 S_2 + p_3 S_3)$$
$$G_3 = -(p_1 S_1 + p_2 S_2 + p_3 S_3)$$
$$G_4 = 0.$$

The first three of these equations constitute a set of three linear equations in the three variables S_1, S_2, S_3. If the G_i are given fixed values, it is a well-known algebraic theorem that simultaneous solutions can be found for these three equations unless the determinant $\Delta = 0$, where

$$\Delta = \begin{vmatrix} (1-p_1) & -p_2 & (1-p_3) \\ -p_1 & (1-p_2) & (1-p_3) \\ -p_1 & -p_2 & -p_3 \end{vmatrix}.$$

In particular, unless $\Delta = 0$, stakes S_i can be chosen so that all the G_i (for $i = 1, 2, 3$) have specified negative values. In other words, there exist stakes such that in each of the first three eventualities X will suffer a net loss. The propositions e, h_1, h_2, and h_3 may be such that one or two of the first three eventualities listed are logically impossible, but because $\Diamond e$ and $e \mathbin{-3} \sim (h_1 \& h_2)$, at least one of the three is possible. Hence, in all of the first three eventualities which are possible – and at least one is – X suffers a net loss. In the only other eventuality, the fourth, X neither wins nor loses, since $G_4 = 0$. Therefore, if $\Delta \neq 0$, τ is incoherent; or conversely, a necessary condition for coherence is that $\Delta = 0$. But on expanding the determinant we have $\Delta = -p_3 + p_1 + p_2$.

Hence a necessary condition for the coherence of τ is that $-p_3 + p_1 + p_2 = 0$, i.e., $p_3 = p_1 + p_2$.

Lemma IV. (a) *If $\Diamond e$ and $\Diamond(h \& e)$ and τ consists of the three beliefs $B_X(h \& h'/e) = p_1$, $B_X(h/e) = p_2$, $B_X(h'/h \& e) = p_3$, then a necessary condition for the coherence of τ is that $p_1 = p_2 p_3$. (b) If $\Diamond e$ and $\Diamond(h' \& e)$ and τ consists of the three beliefs $B_X(h \& h') = p_1$, $B_X(h'/e) = p_2$, $B_X(h/e \& h') = p_3$, then a necessary condition for the coherence of τ is $p_1 = p_2 p_3$.*

Proof. Similar to that of lemma III.

Lemma V. *If $\Diamond e$ and $B_X(h/e) = 1$, then a necessary condition for the coherence of the set of beliefs τ whose one member is $B_X(h/e) = 1$ is that $e \mathbin{-3} h$.*

Proof. The only distinct eventualities are (1) e false, (2) e and h true, (3) e true and h false. If it is false that $e \dashv h$, then at least the third of these eventualities is logically possible. In accordance with the explication of quantitative belief, a bet on h contingent on e at betting quotient 1 is acceptable to X, for any stake S. Choose $S > 0$. Then in the third eventuality X will lose S, whereas in the first and second he will neither gain nor lose. Hence, on assumption that $e \dashv h$ is false, τ is incoherent.

Lemmas I–V can now be used to demonstrate axioms (1), (2'), (3), and (4).

Axiom (1). $0 \le C(h/e) \le 1$.

Proof. Let τ consist of $B_X(h/e) = p$. By the Principle of Coherence, if (h, e) belongs to the range of C and $B_X(h/e) = C(h/e)$, then τ is coherent. Since an implicit assumption is $\Diamond e$, lemma I is applicable, and a necessary condition for the coherence of τ is that $0 \le p \le 1$. Hence $0 \le C(h/e) \le 1$. (Note: axiom (1) is the only quantitative axiom which imposes no condition on the range of C, but only on the values of C for those arguments for which C is well-defined.)

Axiom (2'). $C(h/e) = 1$ *if and only if* $e \dashv h$.

Proof. First assume that (h, e) belongs to the range of C and that $\Diamond e$ and $e \dashv h$. Let τ consist of $B_X(h/e) = p$. By the Principle of Coherence, if $B_X(h/e) = C(h, e)$, then τ is coherent. Lemma II implies, however, that if τ is coherent, then $p = 1$. Hence, if (h, e) belongs to the range of C and $\Diamond e$ and $e \dashv h$, then $C(h/e) = 1$. By using the Closure Rule it follows from this that $\Diamond e$ and $e \dashv h$ implies $C(h/e) = 1$. For the converse assume $\Diamond e$ and $C(h/e) = 1$. If τ is the unit class whose one member is $B_X(h/e) = 1$, then by the Principle of Coherence, τ is coherent. Then by lemma V, $e \dashv h$. (Note: The first part of this proof establishes axiom (2).)

Axiom (3) follows from lemma III and axiom (4) follows from lemma IV by arguments precisely parallel to the first part of the proof of axiom (2').

The only premises in the foregoing arguments are the Principle of Coherence and the Closure Rule, which we saw earlier to be analytic, and the definitions of quantitative belief and coherence, which are analytic propositions, since they only explicate the content of the respective concepts. Since a logical consequence of several analytic propositions is itself analytic, it follows that not only the truth but the analyticity of axioms (1), (2'), (3), and (4) has been established.

It should be noted that lemmas I–IV are demonstrable even if Ramsey's and DeFinetti's weak definition of coherence is adopted, provided only that the trivial eventuality of e being false is not counted as a distinct eventuality. (It is trivial because when e is false all bets are annulled.) Lemma V, however, can be demonstrated only if the strong sense of coherence is used. Consequently, axioms (1), (2), (3), and (4) are demonstrable using Ramsey's and DeFinetti's definition, but axiom (2'), which depends on lemma V, cannot be demonstrated on the basis of their definition.

One might conjecture that other principles of quantitative confirmation besides axioms (1), (2'), (3), and (4) could be established by the foregoing method. The falsity of this conjecture is shown by a remarkable theorem of Kemeny,[32] which can be formulated as follows: for the confirmation function C to be such that every finite set of quantitative beliefs determined by C is coherent (in the strong sense), it suffices that C satisfy axioms (1), (2'), (3), and (4).[33] In other words, the foregoing method cannot yield any principles of quantitative confirmation independent of axioms (1), (2'), (3), and (4).

It might be supposed, however, that by extending the concept of coherence so as to apply to infinite sets of beliefs, and by formulating the Principle of Coherence in terms of this extended concept, other principles of confirmation might be demonstrable by the above method. In particular, it would be desirable to justify the 'Principle of Complete Additivity' – i.e.,

If $\Diamond e$ and $e \supset \sim (h_i \,\&\, h_j)$ where $i \neq j$ and h_i and h_j are members of an infinite sequence $\{h_k\}$, then if h' is the proposition that one member of this sequence is true, $C(h'/e) = \sum_{i=1}^{\infty} C(h_i/e)$.[34]

Unfortunately, the following considerations throw doubt on the possibility of formulating a satisfactory concept of coherence applicable to both finite and infinite sets of quantitative beliefs. Let a finite set of quantitative beliefs which is coherent in the sense used previously in this section be called simply 'coherent', and a finite or infinite set of quantitative beliefs which satisfies the extended concept of coherence be called 'generally coherent'. It is natural to require (a) that if τ is generally coherent and σ is a finite subset of τ, then σ is coherent, and (b) that for every set (a_α, b_α) of ordered pairs of propositions such that $\Diamond a_\alpha$, $\Diamond b_\alpha$ there exist real numbers p_α such that the set of beliefs $\{B_X(a_\alpha/b_\alpha) = p_\alpha\}$ is generally coherent. Two examples will now be given, the first of which shows that if general

32. "Fair bets and degree of confirmation," *Journal of symbolic logic,* vol. 20 (1955), p. 263. Also R. S. Lehman, *ibid.,* p. 251.
33. Kemeny also proved the following: a sufficient condition for C to be such that every finite set of quantitative beliefs regulated by C is coherent in the usual sense of Ramsey and DeFinetti is that C should satisfy axioms (1), (2), (3), and (4).
34. A. Kolmogoroff, *The foundations of probability,* New York, 1950, p. 15.

coherence is applicable to non-denumerable sets of beliefs then a contradiction is deducible from (a) and (b), and the second of which shows that even if general coherence is restricted to finite and denumerably infinite sets an extremely counter-intuitive conclusion follows from (a) and (b).

(i) Suppose $\Diamond e$ and $\Diamond(h_\alpha \& e)$ where $\{h_\alpha\}$ is a non-denumerable set of propositions, and e logically implies that one of the $\{h_\alpha\}$ is true, and $e \dashv \sim (h_\alpha \& h_\beta)$ if $\alpha \neq \beta$. Let the set $\{h'_\beta\}$ consist of all finite disjunctions of members of $\{h_\alpha\}$; clearly $\{h_\alpha\}$ is a subset of $\{h'_\beta\}$. Suppose that – in accordance with (b) – real numbers p_β are chosen so that the set $\tau = \{B_X(h'_\beta/e) = p_\beta\}$ is generally coherent. By condition (a), each unit class σ_β consisting of a single $B_X(h'_\beta/e) = p_\beta$ is coherent. Since $e \dashv \sim h_\alpha$ is false, $e \dashv \sim h'_\beta$ is false. It follows then from lemmas I, II, III, and V that $p_\beta > 0$. Since $\{h_\alpha\}$ is non-denumerable, there exist some $\epsilon > 0$ and some infinite subset $\{h^\epsilon_\gamma\}$ of $\{h_\alpha\}$, such that $p^\epsilon_\gamma > \epsilon$ for all p^ϵ_γ with the property that $B_X(h^\epsilon_\gamma/e) = p^\epsilon_\gamma$ is in τ. Choose $n > 1/\epsilon$, and select an arbitrary set $\{h^\epsilon_{\gamma_1}, \ldots, h^\epsilon_{\gamma_n}\}$ of the $\{h^\epsilon_\gamma\}$. Then the set of beliefs

$$\phi = \{B_X(h^\epsilon_{\gamma_1}/e) = p^\epsilon_{\gamma_1}, \ldots, B_X(h^\epsilon_{\gamma_n}/e) = p^\epsilon_{\gamma_n}, B_X(h^\epsilon_{\gamma_1} \vee \cdots \vee h^\epsilon_{\gamma_n}/e) = p\}$$

is a finite subset of τ. Hence, by condition (a), ϕ is coherent, and therefore by lemma III and the above inequalities

$$p = \sum_{i=1}^{n} p^\epsilon_{\gamma_i} > n\epsilon > 1.$$

Since $p > 1$, by lemma I the unit class whose one member is $B_X(h^\epsilon_{\gamma_1} \vee \cdots \vee h^\epsilon_{\gamma_n}/e) = p$ is incoherent, and therefore by condition (a) τ is not generally coherent. But this is a contradiction.

(ii) Suppose $\Diamond e$ and $\Diamond h_i$, where $\{h_i\}$ is a denumerable sequence of propositions, and e logically implies one of the $\{h_i\}$ is true, and $e \dashv \sim (h_j \& h_k)$ for $j \neq k$. Then an argument similar to that of example (i) can be used, not to deduce a contradiction from (a) and (b), but to infer that unless $p_1 + p_2 + \cdots$ forms a series convergent to 1 the set $\{B_X(h_i/e) = p_i\}$ is incoherent. This implies, in particular, that whatever the content of the h_i and the e in this example may be, equal belief in all the h_i on evidence e leads to incoherence and hence is irrational. But such a stringent general condition seems counter-intuitive and artificial.

Thus a generalized concept of coherence satisfying (a) and (b) seems to be impossible, and without formulating such a concept precisely it is, of course, impossible to demonstrate any further principles of confirmation by the method used in this section. Furthermore, it is difficult to see how an intuitively satisfactory and fruitful concept of coherence could be formulated which did not satisfy both (a) and (b). Hence it seems that axioms (1), (2'), (3), and (4) are the totality of the principles of quantitative confirmation demonstrable by the method of Ramsey and DeFinetti.

An important problem remains, however. The same sort of reasoning as in example (i) above shows that the confirmation function C cannot satisfy the quantitative axioms (1), (2'), (3), and (4) if its domain of definition contains sets of pairs (h_α, e) which fulfill the conditions mentioned in either example (i) or example (ii). This seems to be a serious difficulty, since in considerations of probabilities in continua (so-called 'geometric probabilities') such sets of pairs are in fact included in the domain of C. There are several ways of answering this difficulty, none entirely satisfactory, perhaps, but sufficient to defend the general approach to confirmation which has been adopted in this paper.

(a) One way is to claim that in all scientifically and practically interesting inquiries we are concerned only with finite sets of alternatives – e.g., we never treat the continuum of the electro-magnetic spectrum as a mathematical continuum, but only as a finite set of bands of small but finite width. If this claim turns out to be consistent with physical theory and practice, then we need refer to infinite sets of alternatives only for the purpose of mathematical simplification, and the anomalies which we have observed are entirely ascribable to that simplification.

(b) A better method, I believe, is to admit infinite sets of alternatives, but to substitute for axiom (2') the weaker axiom (2). If this is done, then in a situation where h_α is one of an infinite set of mutually exclusive hypotheses each possible on assumption of e, the evaluation $C(h_\alpha, e) = 0$ is consistent. However, in view of the fact that (2') has been derived from the Principle of Coherence in the same manner as the other axioms were derived, it seems illegitimate to single out this axiom and to weaken it. To answer this objection, it is necessary to look more carefully at the concepts involved in the Principle of Coherence. The statement of the Principle of Coherence given earlier in this paper referred to the concept of partial relative belief, which in turn was explicated in terms of bets and betting quotients. These betting quotients were non-negative real numbers. Yet if h_α is one of an infinite set of mutually exclusive possibilities left open by e, neither a 0 betting quotient nor a finite betting quotient seems appropriate for a bet on h_α. An appropriate betting quotient would be an 'infinitesimal', which is neither 0 nor finite; but this is impossible because of the Archimedean property of the positive real numbers. For this reason it is impossible to formulate an entirely adequate explication of quantitative partial belief in a hypothesis which is one of an infinite set of mutually exclusive possible alternatives, if the explication is to be in terms of betting quotients. We have thus been working with an explication of partial relative belief which is only partially adequate, which in turn made the formulation of the Principle of Coherence somewhat inaccurate. The replacement of axiom (2) for axiom (2') in the infinite case is a compensation – and apparently the only one necessary in order to restore consistency – for this inaccuracy. Axiom (2′) may still be retained when h is one of a finite set of alternatives.

6. THE COMPARATIVE AXIOMS

There are two motives for demonstrating a set of axioms for the comparative concept of confirmation without bringing in considerations of the quantitative concept. In the first place, as indicated in section 1, the comparative concept is epistemologically prior to the quantitative, in the sense that we can often correctly judge that $SC(h/e, h'/e')$ without being able to evalute either $C(h/e)$ or $C(h'/e')$; and independent derivation of the comparative axioms avoids disguising this priority. In the second place, the demonstrations of the quantitative axioms depended upon the somewhat idealized notion of quantitative partial belief, whereas it suffices to use the less idealized concept of comparative belief in deriving the comparative axioms.

The set of axioms for SC to be discussed in this section is due to Koopman.[35] It is, to my knowledge, the most adequate and strongest axiomatization of the comparative concept of confirmation yet proposed. The notation used below differs from Koopman's, but his names for the axioms are retained.

Axiom I (Verified Contingency). *If $e' \multimap h'$ then $SC(h/e, h'/e')$.*

Axiom II (Implication). *If $SC(h/e, h'/e')$ and $e \multimap h$, then $e' \multimap h'$.*

Axiom III (Reflexivity). *If $e \equiv e'$ and $(h \,\&\, e) \equiv (h' \,\&\, e')$, then $SC(h/e, h'/e')$.*[36]

Axiom IV (Transitivity). *If $SC(h_1/e_1, h_2/e_2)$ and $SC(h_2/e_2, h_3/e_3)$, then $SC(h_1/e_1, h_3/e_3)$.*

Axiom V (Anti-Symmetry). *$SC(h/e, h'/e')$ implies $SC(\sim h'/e', \sim h/e)$.*

Axom VI (Composition). *If $\Diamond h_1$, $\Diamond h'_1$, $h_1 \multimap h_2 \multimap h_3$ and $h'_1 \multimap h'_2 \multimap h'_3$, then*
(i) *$SC(h_1/h_2, h'_1/h'_2) \,\&\, SC(h_2/h_3, h'_2/h'_3)$ implies $SC(h_1/h_3, h'_1/h'_3)$*
(ii) *$SC(h_1/h_2, h'_2/h'_3) \,\&\, SC(h_2/h_3, h'_1/h'_3)$ implies $SC(h_1/h_3, h'_1/h'_3)$.*

Axiom VII (Decomposition). *If $\Diamond h_1$, $\Diamond h'_1$, $h_1 \multimap h_2 \multimap h_3$, $h'_1 \multimap h'_2 \multimap h'_3$, and $SC(h'_1/h'_3, h_1/h_3)$, then*

35. Cf. footnote 19.
36. '$p \equiv q$' means 'p is logically or strictly equivalent to q'. See F. B. Fitch, *op. cit.,* pp. 77–78.

(i) $SC(h_1/h_2, h_1'/h_2')$ implies $SC(h_2'/h_3', h_2/h_3)$
(ii) $SC(h_2/h_3, h_1'/h_2')$ implies $SC(h_2'/h_3', h_1/h_2)$
(iii) $SC(h_1/h_2, h_2'/h_3')$ implies $SC(h_1'/h_2'/, h_2/h_3)$
(iv) $SC(h_2/h_3, h_2'/h_3')$ implies $SC(h_1'/h_2', h_1/h_2)$.

Axiom VIII (Alternative Presumption). *If $SC(h/e_1 \& e_2, h'/e')$ and $SC(h/e_1 \& \sim e_2, h'/e')$, then $SC(h/e_1, h'/e')$.*

Axiom IX (Subdivision). *For an integer n let the propositions $h_1, \ldots, h_n$, $h_1', \ldots, h_n'$ be such that $\sim\Diamond(h_i \& h_j)$ and $\sim\Diamond(h_i' \& h_j')$ for $i \neq j$ and $i, j = 1, \ldots, n$. Also suppose $\Diamond h$ and $\Diamond h'$ where h is $h_1 \vee \cdots \vee h_n$ and h' is $h_1' \vee \cdots \vee h_n'$, and that $SC(h_1/h, h_2/h), \ldots, SC(h_{n-1}/h, h_n/h), SC(h_1'/h', h_2'/h'), \ldots, SC(h_{n-1}'/h', h_n'/h')$. Then $SC(h_1/h, h_n'/h')$.*

These axioms, like the quantitative axioms, will be demonstrated in two stages. First, a number of lemmas stating necessary conditions for the coherence of sets of comparative beliefs will be proved, the proofs depending only upon the explications given in section 2 for comparative belief and for coherence of sets of comparative beliefs. Second, the axioms for *SC* will be derived from these lemmas, by making use of the Principle of Coherence and the Closure Rule in their comparative specifications.

The proofs of most of the lemmas are lengthy and therefore are omitted. The omitted proofs follow the same pattern as the proofs of lemma I of this section and lemma III of the previous section.

Lemma I. *If $\Diamond e_1$, $\Diamond e_2$ and $e_1 \,\lightning\, h_1$, then the set τ whose one member is $LB_X(h_1/e_1, h_2/e_2)$ is incoherent.*

Proof. Let p_1 and p_2 be any betting quotients for bets on h_1 contingent on e_1 and on h_2 contingent on e_2, respectively, such that p_1 and p_2 are compatible with $LB_X(h_1/e_1, h_2/e_2)$. Then $p_1 < p_2$. Consider bets made with these betting quotients at stakes S_i ($i = 1, 2$):

To pay $p_i S_i$ and collect nothing for a net gain of $-p_i S_i$, in case e_i is true and h_i false

To pay $p_i S_i$ and collect S_i for a net gain of $(1 - p_i)S_i$, in case e_i and h_i are both true

To annul the bet, for a net gain 0, in case e_i is false.

Since $e_1 \,\lightning\, h_1$, the only distinct eventualities are

(1) e_1, e_2, h_1, h_2 true
(2) e_1, e_2, h_1 true and h_2 false
(3) e_1, h_1 true, e_2 false
(4) e_1 false, e_2 and h_2 true

(5) e_1 false, e_2 true and h_2 false

(6) e_1 and e_2 false.

At least one of the first two eventualities is logically possible. There are two cases.

Case (i): $p_1 = 1$. In this case $p_2 > 1$. Let $S_1 = 0$ and $S_2 > 0$. Then there is net loss in (1), (2), (4), and (5) and neither gain nor loss in (3) and (6).

Case (ii): $p_1 \neq 1$. Then let $S_2 = 0$ and S_1 be positive or negative according as $p > 1$ or $p < 1$. Then there is net loss in (1), (2), and (3) and neither gain nor loss in (4), (5), and (6). Since p_1 and p_2 are arbitrary betting quotients compatible with τ, it follows from the explication of coherence for sets of comparative beliefs that τ is incoherent.

Lemma II. *If $\Diamond e$ and $e \preccurlyeq h$, then a necessary condition for the coherence of the set of beliefs τ whose one member is $SB_X(h/e, h'/e')$ is that $e' \preccurlyeq h'$.*

Lemma III. *If $e \equiv e'$ and $(h \,\&\, e) \equiv (h' \,\&\, e')$, then the set of beliefs τ whose one member is $LB_X(h'/e', h/e)$ is incoherent.*

Lemma IV. *The set of beliefs τ whose members are $SB_X(h_1/e_1, h_2/e_2)$, $SB_X(h_2/e_2, h_3/e_3)$, $LB_X(h_3/e_3, h_1/e_1)$ is incoherent.*

Proof. A set of betting quotients p_i for bets on h_i contingent on e_i ($i = 1, 2, 3$) must satisfy the conditions $p_1 \leq p_2$, $p_2 \leq p_3$, $p_3 < p_1$, in order to be compatible with τ. But this is impossible. Hence there is no set of betting quotients compatible with τ, and *a fortiori* no set of betting quotients which satisfies all the conditions necessary for the coherence of τ.

Lemma V. *The set of beliefs τ whose members are $SB_X(h/e, h'/e')$ and $LB_X(\sim h/e, \sim h'/e')$ is incoherent.*

Lemma VI. *If $\Diamond h_1$, $\Diamond h_1'$, $h_1 \preccurlyeq h_2 \preccurlyeq h_3$, and $h_1' \preccurlyeq h_2' \preccurlyeq h_3'$, then (i) the set of beliefs τ whose members are $SB_X(h_1/h_2, h_1'/h_2')$, $SB_X(h_2/h_3, h_2'/h_3')$, and $LB_X(h_1'/h_3', h_1/h_3)$ is incoherent, (ii) the set of beliefs whose members are $SB_X(h_1/h_2, h_2'/h_3')$, $SB_X(h_2/h_3, h_1'/h_3')$, and $LB_X(h_1'/h_3', h_1/h_3)$ is incoherent.*

Lemma VII. *If $\Diamond h_1$, $\Diamond h_1'$, $h_1 \preccurlyeq h_2 \preccurlyeq h_3$, $h_1' \preccurlyeq h_2' \preccurlyeq h_3'$, then (i) the set of beliefs τ whose members are $SB_X(h_1'/h_3', h_1/h_3)$, $SB_X(h_1/h_2, h_1'/h_2')$, and $LB_X(h_2/h_3, h_2'/h_3')$ is incoherent, (ii) the set of beliefs whose members are $SB_X(h_1'/h_3', h_1/h_3)$, $SB_X(h_2/h_3, h_1'/h_2')$, and $LB_X(h_1/h_2, h_2'/h_3')$ is incoherent, (iii) the set of beliefs whose members are $SB_X(h_1'/h_3', h_1/h_3)$, $SB_X(h_1/h_2, h_2'/h_3')$, and $LB_X(h_2/h_3, h_1'/h_2')$ is incoherent, and (iv) the* set of beliefs whose members are $SB_X(h_1'/h_3', h_1/h_3)$, $SB_X(h_2/h_3, h_2'/h_3')$, and $LB_X(h_1/h_2, h_1'/h_2')$ is incoherent.

Lemma VIII. *If $\Diamond e$ and τ consists of $SB_X(h/e_1 \& e_2, h'/e')$, $SB_X(h/e_1 \& {\sim}e_2, h'/e')$, and $LB_X(h'/e', h/e)$, then τ is incoherent.*

Lemma IX. *For arbitrary n, suppose the propositions $h_1, \ldots, h_n, h_1', \ldots, h_n'$ are such that $\sim\Diamond(h_i \& h_j)$, and $\sim\Diamond(h_i' \& h_j')$ for $i \neq j$ and $i, j = 1, \ldots, n$. Also suppose $\Diamond h$ and $\Diamond h'$, where h is $h_1 \vee \cdots \vee h_n$ and h' is $h_1' \vee \cdots \vee h_n'$. Then the set of beliefs τ whose members are $SB_X(h_1/h, h_2/h), \ldots, SB_X(h_{n-1}/h, h_n/h), SB_X(h_1'/h', h_2'/h'), \ldots, SB_X(h_{n-1}'/h', h_n'/h'), LB_X(h_n'/h', h_1/h)$ is incoherent.*

Proof. Let $p_1, \ldots, p_n, p_1', \ldots, p_n'$ be betting quotients for bets on h_1 contingent on $h, \ldots, h_n$ contingent on h, h_1' contingent on $h', \ldots, h_n'$ contingent on h', respectively. To be compatible with τ these betting quotients must satisfy the conditions $p_n' < p_1$, $p_1 \leq p_2 \leq \cdots \leq p_{n-1} \leq p_n$ and $p_1' \leq p_2' \leq \cdots \leq p_{n-1}' \leq p_n'$. Suppose bets are made in accordance with these betting quotients with stakes $S_1, \ldots, S_n, S_1', \ldots, S_n'$. The only distinct eventualities are

> (1) h_1 and h_1' true, h_i and h_j' false for $i \neq 1$ and $j \neq 1$
> $\vdots$
> (n) h_1 and h_n' true, h_i and h_j' false for $i \neq 1$ and $j \neq n$
> ($n+1$) h_1 true and h' false and h_i false for $i \neq 1$
> $\vdots$
> (n^2+2n) h_n' true and h false and h_i' false for $i \neq n$
> (n^2+2n+1) h and h' false.

Now suppose $p_1 > 1/n$. Then $p_i > 1/n$ for $i = 1, \ldots, n$. Let $S_1 = \cdots = S_n > 0$ and $S_1' = \cdots = S_n' = 0$. In each of the first n^2+n eventualities there is a net gain of $S - (p_1 + p_2 + \cdots p_n)S$, which is negative, and at least one of the eventualities is logically possible. In the remaining eventualities there is neither gain nor loss. Hence it must be possible for p_1 to be chosen $\leq 1/n$ if τ is coherent. Similarly, if τ is coherent it must be possible to choose $p_n' \geq 1/n$. But if $p_1 \leq 1/n$ and $p_n' \geq 1/n$, then the condition $p_n' < p_1$ cannot be satisfied. Hence τ is incoherent.

We can now use lemmas I–IX, together with the Principle of Coherence and the Closure Rule, to justify the comparative axioms.

Axiom I. *If $e \mathbin{-\!3} h$, then $SC(h'/e', h/e)$.*

Proof. Suppose $e \mathbin{-\!3} h$ and that (h, e) and (h', e') are comparable. If $\sim SC(h'/e', h/e)$, then by the assumption of comparability $SC(h/e, h'/e')$, and hence by the Principle of Coherence the set of beliefs whose one member is $LB_X(h/e, h'/e')$ is coherent. This contradicts lemma I, however, since we implicitly assume $\Diamond e$ and $\Diamond e'$ as antecedents in all the confirmation statements. Hence, by indirect proof, $SC(h'/e', h/e)$ follows. We have used the suppositions $e \triangleleft h$ and the comparability of (h, e) and (h', e') in the foregoing derivation, but by the Closure Rule we can drop the explicit assumption of comparability and still derive $SC(h'/e', h/e)$.

Axiom II. *If $SC(h/e, h'/e')$ and $e \triangleleft h$, then $e' \triangleleft h'$.*

Proof. Suppose that $SC(h/e, h'/e')$ and $e \triangleleft h$. By the Principle of Coherence the set whose one member is $SB_X(h/e, h'/e')$ is coherent. By lemma II, however, $e' \triangleleft h'$. (Note: axiom II is the only comparative axiom for which the Closure Rule is not needed.)

Axioms III–IX follow from lemmas III–IX precisely as axiom I follows from lemma I.

The deductive strength of axioms I–IX is examined by Koopman[37] and will not be discussed in detail here. A sense of the content of these axioms may be gained, however, by considering the consequences of adding two further principles to axioms I–IX.

The Axiom of n-Scales. *For every natural number n there is a set of propositions $e, h_1, \ldots, h_n$, such that $\Diamond e$, $e \triangleleft (h_1 \vee \cdots \vee h_n)$, $e \triangleleft \sim (h_i \& h_j)$ for $i \neq j$ and $1 \leq i, j \leq n$, and $EC(h_i/e, h_j/e)$ for $1 \leq i, j \leq n$.*

The Axiom of Comparability. *If $\Diamond e$ and $\Diamond e'$, then either $SC(h/e, h'/e')$ or $SC(h'/e', h/e)$.*

Adding the Axiom of n-Scales makes it possible to define the quantitative concept C in terms of SC, in a manner analogous to the usual definition of the metrical concept of length in terms of the comparative concept *longer than* and an infinitely divisible unit of length.[38] Furthermore, from axioms I–IX together with the Axiom of n-Scales it can be proved that the concept C as thus defined satisfies the quantitative axioms (1), (2), (3), and (4).[39] $C(h/e)$ cannot be shown to be well-defined for every ordered pair (h, e) such that $\Diamond e$, but it can be shown that C as thus defined satisfies the Closure Rule,[40] and hence that the range of C is closed under

37. *Op. cit.,* footnote 4.
38. Cf. the discussion of weight measurements in M. Cohen and E. Nagel, *An introduction to logic and scientific method,* New York, 1934, pp. 297–98.
39. B. O. Koopman, *op. cit.,* pp. 290–91, especially theorems 20, 24, 25.
40. *Ibid.*

the operations involved in axioms (1), (2), (3), and (4). If, in addition, the Axiom of Comparability is assumed, then it can be proved that $C(h/e)$ is well-defined for every pair of propositions (h, e) such that $\Diamond e$; and, incidentally, axiom VIII (Alternative Presumption) and axiom IX (Subdivision) become superfluous.

The effect of axioms I–IX, in brief, is to impose the classical conditions of probability on pairs of propositions (h, e), up to the point of comparability and metrization. (The pair (h, e) and (h', e') are comparable if either h is more confirmed on e than h' is on e', or the converse is true, or the two degrees of confirmation are equal.) Metrization is partially achieved by the Axiom of n-Scales, though not completely, because it cannot be demonstrated on the basis of this axiom together with axioms I–IX that every (h, e) such that $\Diamond e$ is in the range of C. Comparability is imposed by the Axiom of Comparability, thereby also completing the metrization.

The questions arise as to whether and how the Axioms of n-Scales and Comparability are justified. I hope to discuss these questions in subsequent papers, and shall only outline the main considerations here.

Koopman accepts the Axiom of n-Scales, yet does not group it with the axioms I–IX, since it is not a "law of consistency", but rather the postulation of "conceptual experiments" of the sort familiar in theoretical physics.[41] A more precise justification, I believe, can be given on the basis of the Principle of Indifference.[42] Let e be the proposition that one entity is selected at random from a collection of n entities which accurately resemble each other in all relevant respects, and let h_i be the hypothesis that the ith entity is selected, where the numbering is assigned in any fashion compatible with the randomness of selection. Clearly e is logically possible, even though for very large n it may be physically impossible. Also $e \prec (h_1 \lor \cdots \lor h_n)$ and $e \prec \sim(h_i \,\&\, h_j)$ for $i \neq j$ and $1 \leq i, j \leq n$. Finally, by even a quite stringent formulation of the Principle of Indifference, h_i and h_j are equally confirmed on e for $1 \leq i, j \leq n$. Hence, on the assumption of a stringent form of the Principle of Indifference, the Axiom of n-Scales is demonstrated.

Koopman rejects the Axiom of Comparability, as did von Kries[43] and Keynes[44] previously. They present impressive arguments against this axiom, mainly stressing that for (h, e) and (h', e') in different domains there

41. "The bases of probability," *Bulletin of the American Mathematical Society,* vol. 46 (1940), p. 769.
42. For a discussion of this principle, see J. M. Keynes, *op. cit.,* pp. 41–64. In my dissertation, "A theory of confirmation" (submitted to the Graduate School of Yale University in 1953), I attempted to establish the analyticity of a strict formulation of the principle.
43. *Op. cit.,* pp. 29ff.
44. *Op. cit.,* pp. 34–40.

may be no objective grounds for comparing the degree of confirmation of h on e with that of h' on e'. Although I agree in general with these criticisms, I think it may be possible to justify the Axiom of Comparability if its function is properly interpreted: namely, that it is a rule for making conventions. When objective grounds are absent for judging either that h is more confirmed on e than h' is on e', or conversely, or that they are equally confirmed, than any one of these alternatives can be chosen arbitrarily as a basis for action; no error is committed thereby, because there are no objective grounds for a unique correct choice. Likewise no positive error is committed in assuming that an unspecified one of the alternatives $SC(h/e, h'/e')$, $SC(h'/e', h/e)$ is considered by convention to be true. The Axiom of Comparability asserts that a systematic set of such conventional choices can be made, so that either on objective grounds, or if these grounds are absent then conventionally, one of the two propositions $SC(h/e, h'/e')$, $SC(h'/e', h/e)$ holds for all logically possible e and e'. The advantage of adopting the Axiom of Comparability lies in the simplification of some of the mathematical discussion of confirmation. For instance, $C(h/e)$ can be treated as well-defined for every pair (h, e) such that $\Diamond e$ – though, of course, in view of the conventional character of the Axiom of Comparability it is impossible in general to find objective grounds for determining the value of $C(h/e)$ beyond a certain degree of precision.

The question arises, however, as to the consistency of taking the Axiom of Comparability as a convention. Suppose M is the set of all objectively grounded propositions $SC(h/e, h'/e')$. The Axiom of Comparability asserts the existence of a set M' such that M' is an extension of M (i.e., $M \subseteq M'$), and such that if $\Diamond b$ and $\Diamond b'$ then M' contains either $SC(a/b, a'/b')$ or $SC(a'/b, a/b)$. The set M of objectively grounded SC-propositions satisfies axioms I–IX, but what guarantees that any choice of M' will also satisfy axioms I–IX? Is it not possible that the structure of M is tangled in such a way that any systematic extension of M which satisfies the Axiom of Comparability will at some point violate one of the axioms I–IX? A partial answer to this question is provided by a theorem of E. Szpilran: "For every relation π establishing a partial ordering in a domain E there exists a relation ρ containing π and establishing a complete ordering in E."[45] It can be shown on the basis of this theorem that M is extendable to a set M' which satisfies the Axiom of Comparability and also axioms II (Reflexivity) and IV (Transitivity). I have neither proved nor disproved, however, the hypothesis that *any* system which

45. "Sur l'extension de l'ordre partiel," *Fundamenta mathematicae*, vol. 16 (1930), pp. 386–89. This theorem was independently proved by Professor E. Begle of the Department of Mathematics of Yale University.

satisfies axioms I–IX can be extended to a system which satisfies both I–X and the Axiom of Comparability. It seems likely that the hypothesis is false. But a negative result would not in itself preclude the consistency of adopting the Axiom of Comparability as a convention, for the set M of objectively grounded confirmation propositions may be extendable, even if all systems satisfying I–IX are not. Whether M is extendable to a system M' which satisfies both axioms I–IX and the Axiom of Comparability depends upon the details of the structure of M, concerning which we have only partial knowledge. Consequently, the consistency of adopting the Axiom of Comparability as a convention must for the present remain a conjecture.

7. COHERENCE AND THE OBJECTIVITY OF CONFIRMATION

The method of justifying the axioms of confirmation by considerations of coherence has previously been used only by Ramsey and DeFinetti, who deny that there exist objective grounds for confirmation judgments. The following argument might be given to show that this is not a historical accident, but that Ramsey's and DeFinetti's method of deriving the axioms of confirmation presupposes a subjectivistic theory of confirmation: If confirmation judgments are only subjective, one can consciously adjust them so as to avoid incoherence; whereas, if there are objective grounds for correct C and SC propositions, they cannot be adjusted at will, and thus there may be correct sets of C and SC propositions which determine incoherent sets of beliefs. (This argument is analogous, though obviously not in all respects, to Kant's argument that *a priori* principles of mathematics and natural science can be known to hold without exception only if the mind prescribes them to nature.[46])

I propose to answer this argument in rather Platonic terms. The correctness of a confirmation proposition $C(h/e) = r$ or $SC(h/e, h'/e')$ is determined by the logical content of the propositions h, e, h', e' and the content of C and SC respectively. S and SC – like such concepts as set and number – are Platonically considered to have objective status and are independent of being thought or designated by human beings. Consequently, the correctness of $C(h/e) = r$ and $SC(h/e, h'/e')$ is an objective matter. As we saw in section 3 and section 4, the concepts C and SC are partially explicated in formulating the Principle of Coherence and the Closure Rule. These principles suffice, as we also saw, for deriving the axioms of confirmation. In order to evaluate $C(h/e)$ and to judge $SC(h/e, h'/e')$ in specific instances, however, further principles are required (e.g., an adequate formulation of the Principle of Indifference).

46. *Prolegomena to any future metaphysics,* translated by Paul Carus, Chicago, 1912, section 36.

The essential point is that these further principles are like the Principle of Coherence and the Closure Rule in being partial explications of C and SC, each of which is a complex but single concept. For this reason, the use of considerations of coherence for deriving the axioms of confirmation is compatible with an objective determination of the correctness of specific propositions $C(h/e)$ and $SC(h/e, h'/e')$.

Two particular considerations strengthen this conclusion. The first is that intuitive confirmation judgments in practical situations (e.g., in gambling and in evaluating scientific hypotheses on the basis of experimental evidence) are uniformly found to conform to the Principle of Coherence and hence to the axioms of confirmation. This indicates that when adequate general rules are formulated for making specific confirmation judgments, they too will be compatible with the Principle of Coherence. The second consideration is that no general principle for making C and SC judgments has been proposed which satisfies our intuitions into the nature of confirmation and also conflicts with the Principle of Coherence and the Closure Rule. A thorough defense of this claim is not possible here, because it requires a complete analysis of the rules which have been proposed by inductive logicians for making specific confirmation judgments. My defense, however, would be along the following lines: The only principle for making specific confirmation judgments which is clearly analytic in virtue of the concepts C and SC is a carefully formulated version of the Principle of Indifference, and this certainly seems to be compatible with the Principle of Coherence and the Closure Rule. Other principles are indeed required in the theory of confirmation – refined versions, perhaps, of Mill's Uniformity of Nature[47] or Keynes' Principle of Limited Variety[48] or Jeffreys' Simplicity Postulate.[49] Such principles, however, are not partial explications of C or SC, but synthetic propositions which are incorporated into the evidence e of confirmation judgments. Since their truth does not depend on explicating an aspect of C or SC, *a fortiori* it does not depend on an aspect incompatible with the truth of the Principle of Coherence and the Closure Rule.

It is not inconceivable that some day a highly intuitive principle for making confirmation judgments will be proposed which does conflict with the Principle of Coherence and the Closure Rule. We should then be confronted with 'an antinomy of confirmation theory'. Just as the antinomies of set theory forced logicians to revise their intuitive grasp of the concept of set, such an antinomy would force us to reconsider our intuitive conception of confirmation. I suspect, however, that the discovery of a serious

47. *A system of logic,* BK. III, ch. III.
48. *Op. cit.,* chapters XX–XXII.
49. *Op. cit.,* p. 113.

antinomy of confirmation is unlikely, and that the main effort of inductive logicians must be to augment rather than to revise our knowledge of the very elusive concept of confirmation.

COMMENT

The first edition of Carnap's *Logical Foundations of Probability,* referred to in footnote 14, did not offer a justification of the axioms of confirmation. Kemeny's review of this book (cited in footnote 15) questioned the universal validity of Axiom (4) and claimed to have a counterexample. Although Carnap was not convinced by the claim of a counterexample, he did see the need for some kind of derivation of the axioms from more fundamental considerations. When I began doctoral research on inductive logic in the summer of 1952, I relied heavily on the bibliography of Carnap's book, and soon read two articles by Bruno DeFinetti. The formal principles of subjective probability were established by DeFinetti as necessary conditions for what he called "coherence." Although Carnap was aware of DeFinetti's work and the similar work of F. P. Ramsey, he sees not to have considered their results to be relevant to his program. The most important thesis of my paper – point (f) of Section 1 – is that DeFinetti's and Ramsey's justification of the axioms applies as well to the logical concept of probability as to the subjective concept. Carnap was convinced, and he incorporated arguments from coherence into his later treatments of logical probability: *Logical Foundations of Probability,* 2nd ed. (Chicago: University of Chicago Press, 1962); "My Basic Conceptions of Probability and Induction" in *The Philosophy of Rudolf Carnap,* ed. P. A. Schilpp (La Salle, IL: Open Court, 1963), pp. 966–73; and *Studies in Inductive Logic and Probability,* vols. 1 and 2, in collaboration with R. Jeffrey (Berkeley and Los Angeles: University of California Press, 1961 and 1980).

As pointed out in Section 2, my definition of "a coherent set of quantitative beliefs" is somewhat stronger than that of DeFinetti and Ramsey. I met DeFinetti in 1971, and he told me that he had considered the definition that I proposed as an alternative to the one he chose; he rejected the former because it leads to the conclusion that $C(h/e) = 1$ only if e logically implies h (which is part of Axiom (2′) of Section 5), and this conclusion is troublesome if one is dealing with an infinite set of beliefs. I am skeptical, however, of the appropriateness of betting considerations except in situations where there is a finite set of outcomes, and in these situations there is no good reason for abstaining from the strong concept of coherence.

In Axiom (4) of Section 1 the evidential proposition of the second factor on the right-hand side of the first equation is written as $h \& e$, but in the second equation as $e \& h'$. With this asymmetrical way of writing the axiom it is straightforward to deduce from the axioms, without an additional assumption, that if e and e' are logically equivalent then $C(h/e) = C(h/e')$; but the derivation does not go through if the evidential proposition in the second equation is written as $h' \& e$, as in the original printing of this article. I wish to thank Hugues Leblanc for bringing this point to my attention. The coherence argument of Section 5 justifies either version of Axiom (4) and can also trivially be used to justify $C(h/e) = C(h/e')$ when e and e' are logically equivalent, as well as $C(h/e) = C(h'/e)$ when h and h' are logically equivalent.

Throughout, the word "range" was used in a nonstandard way, to mean "domain".

7

An Adamite derivation of the principles of the calculus of probability*

I. A RECONSTRUCTION OF ADAM'S REASONING

If Adam was a rational man even before he had garnered much experience of the world, and if the ability to reason probabilistically is an essential part of rationality (as Bishop Butler maintained when he wrote that "But, to us, probability is the very guide of life"[1]), then Adam must at least tacitly have known the Principles of the Calculus of Probability. Specifically, Adam must have known that the epistemic concept of probability – probability in the sense of "reasonable degree of belief" – satisfies these Principles, for it is the epistemic concept, rather than the frequency concept or the propensity concept, which enters into rational assessments about uncertain outcomes.[2] But what warrant did Adam have for either an explicit or a tacit assertion that epistemic probability satisfies the Principles of the Calculus of Probability?

Today, the best known and most widely accepted justification of this assertion is the "Dutch Book" Theorem, originally proved independently

This work originally appeared in J. Fetzer (ed.), *Probability and Causality,* Dordrecht. © 1988 by D. Reidel Publishing Company. All rights reserved. Reprinted by permission of Kluwer Academic Publishers.

*This paper is based upon a lecture given at the University of Pittsburgh in January 1982. A similar thesis is developed independently by Bas van Fraassen in "Calibration: A Frequency Justification for Personal Probability," in *Physics, Philosophy, and Psychoanalysis: Essays in Honor of Adolf Grünbaum,* ed. by R. S. Cohen and L. Laudan (D. Reidel: Dordrecht and Boston, 1983). I dedicate the paper to Wesley C. Salmon, because he explored with penetration and devotion the epistemic applicability of the frequency concept of probability.

1. Butler, Bishop Joseph, *The Analogy of Religion, Natural and Revealed, to the Constitution and Course of Nature,* ed. by G. R. Crooks (Harper: New York, 1868; originally published in 1736), third paragraph of the Introduction.
2. Some philosophers have maintained, however, that the frequency concept of probability can be applied epistemically, for example, Hans Reichenbach, *The Theory of Probability* (U. of California: Berkeley, 1949); Wesley C. Salmon, *The Foundations of Scientific Inference* (U. of Pittsburgh: Pittsburgh, 1966), pp. 83–96; and Wesley C. Salmon, "Statistical Explanation," in *The Nature and Function of Scientific Theories,* ed. by R. G. Colodny (U. of Pittsburgh: Pittsburgh, 1970), pp. 173–231.

by F. P. Ramsey[3] and F. DeFinetti[4] for a subjectivist or personalist version of epistemic probability, but applicable also to non-subjectivist or not-entirely-subjectivist versions.[5] It is most implausible, however, to attribute awareness of this theorem to Adam, partly because it requires some mathematics that is not completely trivial, partly because of the rarity of gambling in the Garden of Eden, and partly because in the one case in which Adam did indulge in a gamble he exhibited little skill at decision theory. A less well known but quite ingenious proof that epistemic probability satisfies the standard Principles was given independently by R. T. Cox[6] and I. J. Good[7] and refined by J. Aczél.[8] But mathematically, this type of proof is even more complicated than the Dutch Book Theorem, and furthermore the proof is actually incomplete without some recourse to considerations of betting.[9] It is my contention that Adam must have had a much simpler warrant for the Principles of the Calculus of Probability than either of these, and the purpose of this paper is to reconstruct his simple and sturdy reasoning.

The crux of the reconstructed reasoning is the rough idea that epistemic probability is somehow an estimate of relative frequency. In R. Carnap's list[10] of the informal concepts which can be identified with epistemic probability, or probability$_1$, as he calls it, we find first "a measure of evidential support," second "a fair betting quotient," and third "an estimate of relative frequency," but I am supposing that the third is the most primitive. This idea is particularly attractive and appropriate if the hypothesis *h* asserts that an individual drawn from a particular collection has the property *M*, while the evidence *e* gives some information about the collection and about the randomness of the mode of drawing. Then the epistemic

3. Ramsey, Frank P., "Truth and Probability," in *The Foundations of Mathematics and Other Logical Essays* (Routledge and Kegan Paul: London, 1931). Reprinted in *Studies in Subjective Probability,* ed. by H. Kyburg and H. Smokler (Wiley: New York, 1964).
4. DeFinetti, Bruno, "La prévision: ses lois logiques, ses sources subjectives," *Annales de l'Institut Henri Poincaré* 7, 1–68 (1937). Reprinted in English translation in *Studies in Subjective Probability,* ed. by H. Kyburg and H. Smokler (Wiley: New York, 1964).
5. Shimony, Abner, "Coherence and the Axioms of Confirmation," *Journal of Symbolic Logic* 20, 1–28 (1955).
6. Richard T. Cox, "Probability, Frequency, and Reasonable Expectation," *American Journal of Physics* 14, 1–13 (1946).
7. Good, I. J., *Probability and the Weighing of Evidence* (C. Griffin: London, 1950).
8. Aczél, J., *Lectures on Functional Equations and their Applications* (Academic Press: New York, 1966).
9. Shimony, Abner, "Scientific Inference," in *The Nature and Function of Scientific Theories,* ed. by R. G. Colodny (U. of Pittsburgh: Pittsburgh, 1970), pp. 79–172, especially pp. 108–110.
10. Carnap, Rudolf, *Logical Foundations of Probability* (U. of Chicago: Chicago, 1950), pp. 168–175.

probability statement $P(h/e) = r$ can reasonably be construed as stating an estimate of the relative frequency of individuals with the property M in the collection. To say this is not tantamount to identifying the epistemic concept of probability with the frequency concept, because the latter refers to the *actual frequency* of M in the collection, whereas the former refers to an *estimate of the actual frequency.*

Estimation thus plays a central role in the reconstruction of Adam's probabilistic reasoning. It will not do to explicate estimation in the standard manner of sophisticated probability theory, viz., the estimate upon evidence e of a quantity A which may have one of the values $a_1, ..., a_n$ is

$$E(A/e) = \sum_{i=1}^{n} a_i P(h_i/e),$$

where h_i is the hypothesis that A has the value a_i. This standard explication is out of place in our reconstruction, because it would generate a circularity. I suggest, therefore, that estimation be taken as a primitive operation, essentially that which is conveyed by "most reasonable guess of a quantity." (In Section 2 we shall be more cautious and represent the estimate of relevant quantities by an interval of the real line rather than by a single real number.)

The explication of epistemic probability as an estimate of relative frequency now permits an extremely simple demonstration of the four basic Principles of the Calculus of Probability, which we state as follows:[11]

  (i) If $P(h/e)$ is a well defined real number, it lies in the interval $[0, 1]$,
 (ii) if e is not the impossible proposition 0 and e logically implies h, then $P(h/e) = 1$.
(iii) if $P(h/e)$, $P(h'/e)$ and $P(h \vee h'/e)$ are well defined real numbers, then $P(h \vee h'/e) = P(h/e) + P(h'/e)$.
 (iv) if $P(h/e)$, $P(h'/h \& e)$, $P(h/e \& h')$, and $P(h \& h'/e)$ are well defined real numbers, then $P(h \& h'/e) = P(h/e) P(h'/h \& e) = P(h'/e) P(h/e \& h')$.

If the epistemic probability expression $P(h/e)$ is replaced by an appropriate expression for probability in the frequency sense, then Principles (i)–(iv) would hold trivially, since all would be simple statements about ratios of the cardinalities of specified sets. But with little modification, a similar justification of Principles (i)–(iv) holds when $P(h/e)$ is interpreted, as we have done, as epistemic probability. A useful notation in presenting the argument is to let C designate the collection referred to in

11. Only in Principle (ii) has it been explicitly stated that e is not the impossible proposition, but this restriction is a necessary condition for clauses like "$P(h/e)$ is a well defined real number" which occur as antecedents in Principles (i), (iii), and (iv).

the evidential proposition e and let C_M designate the subset of C with the property M, while $n(C)$ and $n(C_M)$ respectively are the cardinalities of C and C_M. The Principles are now justified as follows:

(i) Since $n(C_M)/n(C)$ is necessarily a real number in the inverval $[0, 1]$ if it is defined at all (i.e., if $n(C) \neq 0$), the estimate of this ratio also lies in this interval.

(ii) If e logically implies h then $n(C_M) = n(C)$, and hence if $n(C) \neq 0$ the ratio $n(C_M)/n(C)$ equals 1.

(iii) If e implies the falsity of $h \& h'$, and h and h' attribute the properties M and M' respectively to the randomly chosen individual, then C_M and $C_{M'}$ are disjoint sets, and hence $n(C_M \cup C_{M'}) = n(C_M) + N(C_{M'})$. But $P(h \vee h'/e)$ is the estimate of the relative frequency $n(C_M \cup C_{M'})/n(C)$, and if the denominator $n(C) \neq 0$ then this ratio is well defined and equals the sum of $n(C_M)/n(C)$ and $n(C_{M'})/n(C)$, which are estimated by $P(h/e)$ and $P(h'/e)$ respectively.

(iv) If h and h' attribute the properties M and M' respectively to the randomly selected individual, then $P(h \& h'/e)$ is the estimate of $n(C_M \cap C_{M'})/n(C)$, $P(h'/h \& e)$ is the estimate of $n(C_M \cap C_{M'})/n(C_M)$, and $P(h/e \& h')$ is the estimate of $n(C_M \cap C_{M'})/n(C_{M'})$. But if $n(C_M) \neq 0$ (and *a fortiori* $n(C) \neq 0$), then $n(C_M \cap C_{M'})/n(C) = [n(C_M \cap C_{M'})/n(C_M)][n(C_M)/n(C)]$. It is reasonable then that the estimate of the lhs equals the product of the estimates of the two factors of the rhs, establishing the first half of the Principle. The second half follows in the same way.

One more step is necessary for the reconstruction of Adam's reasoning. In many cases the pair (h, e) in $P(h/e)$ neither specifies a definite collection nor attributes a definite property to a randomly selected member of the collection; for h need not refer to an individual at all, or it may refer to an individual by a proper name without indicating a reference class, and there is a great variety of forms of evidential propositions. The foregoing justification of Principles (i)–(iv) for epistemic probability, which clearly was parasitic upon the standard and trivial justification of the corresponding Principles for the frequency concept of probability, can be extended by the appropriate construction of a reference class associated with the pair (h, e). Adam can consider a very large class of pairs (h_i, e_i), each referring to a different and independent situation, but such that $P(h_i/e_i) = P(h/e)$ for each i. The ground for this equation may be a subjective judgment of indifference, if Adam was a personalist, or it may be something additional, if he was a logical probabilist or a tempered personalist;[12]

12. See Ref. 9, especially Section III.

no commitment need be made on this point for the purposes of the argument to come. Now the statement $P(h/e) = r$ can be explicated straightforwardly as an estimate of an appropriate relative frequency as follows. Let C be the set of the pairs (h_i, e_i) in which e_i is true, and C_M the subset of C in which h_i is also true. Then $P(h/e) = r$ means that the estimate of the quotient $n(C_M)/n(C)$ is r. (Even if Adam was cautious and used an interval estimate when $n(C)$ is small, the independence of the situations eliminates the grounds for suspicion of correlations among the outcomes of the h_i, and hence the estimate of $n(C_M)/n(C)$ should diminish practically to a point as $n(C)$ becomes very large.[13]) The arguments above for Principles (i)–(iv) can now be paralleled with little modification.

I submit that this line of reasoning, which is informal and unrigorous, but simple and sturdy, can be considered to be a reconstruction of Adam's explicit or (more likely) tacit justification that epistemic probability satisfies the four Principles of the Calculus of Probability.

2. A REFINEMENT OF THE ADAMITE DERIVATION

It is interesting to give a rigorous version of the Adamite derivation, in order to allay the suspicion that illegitimate steps may lurk in the informal reasoning, and to make explicit the mild but nontrivial assumptions upon which the reasoning depends.

We shall take as primitive the concept of *equiprobability* or *indifference,* but in order to avoid the appearance of an antecedent commitment to the existence of a quantitative probability $P(h/e)$ for an arbitrary pair of propositions (h, e), we shall use the notation $I(h, e; h', e')$ to mean "h is as probable on evidence e as h' is on evidence e', and conversely."

In our argument we use the term "situation," and for our purposes it will suffice to take situations to be finite Boolean algebras of propositions. A situation is thus not specified by a factual assertion but rather by a finite set of possible facts, the set being closed under negation and disjunction and consequently under conjunction. A situation is thus a mode of discriminating possibilities.

The concept of "independent situations" will be taken as primitive. We have good intuitive judgments that certain situations are independent of each other or irrelevant to each other, e.g., B_1 consisting of all Boolean combinations of a finite set or propositions concerning the outcome of the state lottery, and B_2 consisting of all Boolean combinations of a finite

---

13. For a discussion of the effect of correlations on the character of the estimate see Arthur Hobson, "The Interpretation of Inductive Probabilities," *Journal of Statistical Physics* **6**, 189–193 (1972), and Abner Shimony, "Comment on the interpretation of inductive probabilities," *Journal of Statistical Physics* **9**, 187–191 (1973).

set of propositions concerning the decay of a nucleus in a distant nebula. It would be a major enterprise to explicate clearly and judiciously the concept of "independent situations," but for our purposes there is no need to attempt even a crude explication. The reason is that we shall never actually have to decide whether two situations are independent, and hence we do not have to be equipped with a criterion for such a decision. Instead, we shall be engaged in thought experiments in which an arbitrarily large class of independent situations is available for the purpose of making estimates of truth frequencies, each of which is equivalent – in a sense to be specified – to a given situation of interest, but the detailed content of each is irrelevant.

By the *equivalence of situations* B_1 and B_2 we mean the following: B_1 and B_2 are isomorphic as Boolean algebras, and if h_1 and e_1 are the counterparts of h_2 and e_2 respectively under the isomorphism, then $I(h_1, e_1; h_2, e_2)$.

If B is a situation and $\{B_i\}$ is a finite set of independent situations equivalent to B, then we shall take as primitive the concept

$$E(h/e; \{B_i\}),$$

which is an interval estimate of the relative truth frequency of counterparts of $h \& e$ to counterparts of e in the set $\{B_i\}$. In other words, if h_i and e_i are the counterparts in B_i of h and e, and if $n(e; \{B_i\})$ and $n(h \& e; \{B_i\})$ are respectively the numbers of true e_i and true $h_i \& e_i$ in the $\{B_i\}$, then

$$E(h/e; \{B_i\}) = [r_1, r_2]$$

means "r_1 and r_2 are the glb and lub respectively for reasonable values of the quotient $n(h \& e; \{B_i\})/n(e; \{B_i\})$."

We now make two mild assumptions which connect the concept of estimation to mathematical properties of relative truth frequencies.

Assumption E_1. If e, h belong to situation B, e is distinct from the impossible proposition 0, $\{B_i\}$ is a finite set of independent situations equivalent to B, and h_i and e_i are respectively the counterparts in B_i of h and e, then if it is necessary that the quotient $n(h \& e; \{B_i\})/n(e; \{B_i\})$ belongs to the interval $I \subseteq \mathbb{R}$ whenever the denominator $n(e; \{B_i\})$ is non-zero, it follows that $E(h/e; \{B_i\}) \subseteq I$.

Assumption E_2. Let $\{B_i\}$ be a finite set of independent situations equivalent to B, h^α and e^α be propositions in B ($\alpha = 0, 1, \ldots, k$), and h_i^α and e_i^α be the counterparts of h^α and e^α in B_i. If q^α is defined as $n(h^\alpha \& e^\alpha; \{B_i\})/n(e^\alpha; \{B_i\})$ and if the functional relation $q^0 = Q(q^i, \ldots, q^k)$ holds whenever all the q^α are well defined (i.e., when each $n(e^\alpha; \{B_i\})$ is non-zero), then

$$E(h^0/e^0, \{B_i\}) \subseteq [R_1, R_2],$$

where

$$R_1 = \text{glb}\{Q(x_1, \ldots, x_k) \mid x_1 \in I_1, \ldots, x_k \in I_k\},$$

$$R_2 = \text{lub}\{Q(x_1, \ldots, x_k) \mid x_1 \in I_1, \ldots, x_k \in I_k\},$$

and

$$I_\alpha = E(h^\alpha/e^\alpha, \{B_i\}).$$

We also make the following assumption about the existence of independent situations equivalent to a given situation.

Assumption S_1. If B is a situation and e^α ($\alpha = 1, \ldots, k$) is a set of propositions in B, each different from the impossible proposition 0, then there exists a countable sequence $\{B_i\}$ of independent situations each equivalent to B in which there are infinitely many true propositions among the counterparts e_i^α of the e^α.

[*Note:* Assumption S_1 guarantees arbitrarily large reference classes for each of the e^α, but this does not mean that the relative frequency of true e_i among all the e_i of the $\{B_i\}^N$ must approach a non-zero value as N approaches ∞ ($\{B_i\}^N$ being the initial segment of length N in the infinite sequence $\{B_i\}$).]

We now have the materials at hand to define epistemic probability and to demonstrate Principles (i)–(iv).

Definition. If h is an arbitrary proposition and e is a proposition distinct from the impossible proposition 0, then $P(h/e) = r$ (r a real number) holds if and only if for every situation B to which h and e belong and every countable sequence $\{B_i\}$ of independent situations equivalent to B and containing infinitely many true counterparts of e the following two limiting relations hold:

$$\lim_{N \to \infty} \text{lub}[E(h/e; \{B_i\}^N)] = r,$$

$$\lim_{N \to \infty} \text{glb}[E(h/e; \{B_i\}^N)] = r.$$

Comment. Because of assumption S_1 there exists a sequence of independent situations equivalent to every situation B containing h and e, with infinitely many true counterparts of e, and therefore the class of countable sequences mentioned in the *Definition* is not empty. As a result, the warrant of $P(h/e) = r$ cannot be the trivial one of the emptiness of the class of sequences; and consequently it is impossible that $P(h/e) = r$ and $P(h/e) = r'$ if $r \neq r'$. Clearly, however, the *Definition* and assumptions do not guarantee that there exists a real number r such $P(h/e) = r$ for each e distinct from the impossible proposition 0.

Proof of Principle (i). Suppose $P(h/e) = r$, and let $\{B_i\}$ be one of the sequences mentioned in the *Definition*. Then for every N large enough that $\{B_i\}^N$ contains at least one true counterpart of e the quotient $n(h \& e; \{B_i\}^N)/n(e; \{B_i\}^N)$ is well defined and lies in the interval $[0, 1]$. Hence, by assumption E_1 $E(h/e; \{B_i\}^N) \subseteq [0, 1]$, and therefore the limit r, which is asserted to exist by the *Definition*, must also lie in $[0, 1]$.

Proof of Principle (ii). If e logically implies h, then in any situation (i.e., finite Boolean algebra) containing h and e the conjunction $h \& e$ is identical with e. Hence, in any situation B_i equivalent to B the counterpart of $h \& e$ is identical with the counterpart of e. Hence, for any N large enough that the denominator of the quotient $n(e \& h; \{B_i\}^N)/n(e; \{B_i\}^N)$ is non-zero this quotient is 1, and hence by assumption E_1 both the lub and the glb of the interval $E(h/e; \{B_i\}^N)$ are 1. By the *Definition* it follows that $P(h/e) = 1$.

Proof of Principle (iii). Let e be distinct from the impossible proposition 0 and let it logically imply $\sim (h \& h')$. Generate a Boolean algebra B from e, h, and h'. Suppose $P(h/e) = r$, $P(h'/e) = r'$, and $P(h \lor h'/e) = r''$. Then given any countable sequence $\{B_i\}$ of independent situations equivalent to B with infinitely many true counterparts of e, and given $\epsilon > 0$, there exists an integer N_ϵ such that all of the following relations hold:

$$|\mathrm{lub}[E(h/e; \{B_i\}^{N_\epsilon})] - r| < \epsilon,$$

$$|\mathrm{glb}[E(h/e; \{B_i\}^{N_\epsilon})] - r| < \epsilon,$$

$$|\mathrm{lub}\, E(h'/e; \{B_i\}^{N_\epsilon}) - r'| < \epsilon,$$

$$|\mathrm{glb}\, E(h'/e; \{B_i\}^{N_\epsilon}) - r'| < \epsilon,$$

$$|\mathrm{lub}\, E(h \lor h'/e; \{B_i\}^{N_\epsilon}) - r''| < \epsilon,$$

$$|\mathrm{glb}\, E(h \lor h'/e; \{B_i\}^{N_\epsilon}) - r''| < \epsilon.$$

By assumption S_1 there exists such a sequence $\{B_i\}$, and for $\epsilon > 0$ let N_ϵ be the appropriate integer for this sequence with the properties displayed above. Then if we write

$$E(h/e; \{B_i\}^{N_\epsilon}) = [s_1, s_2],$$

$$E(h'/e; \{B_i\}^{N_\epsilon}) = [t_1, t_2],$$

and

$$E(h \vee h'/e; \{b_i\}^{N_\epsilon}) = [u_1, u_2],$$

we have

$$r - \epsilon < s_1 \leq s_2 < r + \epsilon,$$

$$r' - \epsilon < t_1 \leq t_2 < r' + \epsilon,$$

$$r'' - \epsilon < u_1 \leq u_2 < r'' + \epsilon.$$

Since e implies $\sim (h \& h')$, it follows that $e \& h$ is the same proposition as $e \& h \& \sim h'$, and $e \& h'$ is the same proposition as $e \& h' \& \sim h$. Hence, $e_i \& h_i$ and $e_i \& h_i \& \sim h_i'$, which are the counterparts in B_i of $e \& h$ and $e \& h \& \sim h'$ respectively, are identical propositions; and likewise the counterparts in B_i of $e \& h'$ and $e \& h' \& \sim h$ are identical propositions. It follows that the sets of true counterparts of $e \& h$ and of $e \& h'$ in $\{B_i\}^{N_\epsilon}$ are mutually exclusive, and since the set of counterparts of $e \& (h \vee h')$ is the union of these two sets, we obtain

$$n(h/e; \{B_i\}^{N_\epsilon})/n(e; \{B_i\}^{N_\epsilon}) + n(h'/e; \{B_i\}^{N_\epsilon})/n(e; \{B_i\}^{N_\epsilon})$$

$$= n(h \vee h'/e; \{B_i\}^{N_\epsilon})/n(e; \{B_i\}^{N_\epsilon}),$$

provided that N_ϵ is taken large enough that the denominators in all these quotients is non-zero. By assumption 2 we infer

$$[u_1, u_2] \subseteq [r + r' - 2\epsilon, r + r' + 2\epsilon].$$

Hence,

$$r + r' - 3\epsilon \leq r'' \leq r + r' + 3\epsilon,$$

and letting $\epsilon \to 0$ we obtain

$$r'' = r + r'.$$

Proof of Principle (iv). Suppose that $P(h/e) = r$, $P(h'/e \& h) = r'$, $P(h \& h'/e) = r''$. Let B be the Boolean algebra generated from e, h, and h' and $\{B_i\}$ a countable sequence of independent situations equivalent to B with infinitely many true counterparts of e and $e \& h$. Then by the *Definition* if $\epsilon > 0$ there exists an integer N_ϵ such that the following relations hold:

$$|\text{lub}[E(h/e; \{B_i\}^{N_\epsilon})] - r| < \epsilon,$$

$$|\text{glb}[E(h/e; \{B_i\}^{N_\epsilon})] - r| < \epsilon,$$

$$|\text{lub}[E(h'/h \& e; \{B_i\}^{N_\epsilon})] - r'| < \epsilon,$$

$$|\text{glb}[E(h'/h \& e; \{B_i\}^{N_\epsilon})] - r'| < \epsilon,$$

$$|\text{lub}[E(h \& h'/e; \{b_i\}^{N_\epsilon})] - r''| < \epsilon,$$

$$|\text{glb}[E(h \& h'/e; \{B_i\}^{N_\epsilon})] - r''| < \epsilon.$$

If

$$E(h/e; \{B_i\}^{N_\epsilon}) = [s_1, s_2],$$

$$E(h'/h \& e; \{B_i\}^{N_\epsilon}) = [t_1, t_2],$$

$$E(h \& h'/e; \{B_i\}^{N_\epsilon}) = [u_1, u_2],$$

then

$$r - \epsilon < s_1 \leq s_2 < r + \epsilon,$$

$$r' - \epsilon < t_1 \leq t_2 < r' + \epsilon,$$

$$r'' - \epsilon < u_1 \leq u_2 < r'' + \epsilon.$$

By assumption E_2

$$[u_1, u_2] \subseteq [rr' - \epsilon r' - \epsilon r - \epsilon^2, rr' - \epsilon r' - \epsilon r + \epsilon^2]$$

(where the lower bound on the rhs is specified conservatively). Hence

$$rr' - \epsilon r' - \epsilon r - \epsilon^2 - \epsilon \leq r'' \leq rr' - \epsilon r' - \epsilon r + \epsilon^2 + \epsilon.$$

Letting $\epsilon \to 0$ we obtain $r'' = rr'$. Parallel reasoning yields the second part of Principle (iv).

Our rigorous derivation of Principles (i)–(iv) is now complete. This derivation follows the general strategy of the Adamite proof given in Section 1, but in view of the unavoidable epsilontics, perhaps it should be called a "Solomonic" derivation.

The most disagreeable part of the derivation is the existence assertion in assumption S_1. This assumption may appear palatable if one considers the possibility of taking each B_i to be a finite Boolean algebra of propositions concerning subsets of a region R_i of space. Neither the infinity of different situations (if space is infinite) nor the independence of the various B_i is troublesome; and the equivalence of each B_i to B may be achievable by an appropriate partitioning of R_i. An alternative to this strategy would be to rephrase the *Definition* of $P(h/e)$ counterfactually, asserting the limiting relations conditionally upon the existence of the infinite sequence $\{B_i\}$ and, of course, construing the counterfactual not to be automatically true if the condition is not factually fulfilled. We should not wish to attribute such counterfactual reasoning to Adam, but it would not be beyond the powers of Solomon.

COMMENT

A major motivation of this paper was to justify the axioms of epistemic probability without any consideration of utilities, on the ground that these are prima facie irrelevant to purely theoretical investigations. A contrary thesis is that theoretical knowledge itself is a human value. The consequences of this thesis for inductive logic are explored in detail by Isaac Levi, *Gambling with Truth* (New York: Knopf, 1967; Cambridge, MA: MIT Press, 1973); see also Jaakko Hintikka and Juhani Pietarinen, "Semantic Information and Inductive Logic" in *Aspects of Inductive Logic,* ed. J. Hintikka and P. Suppes (Amsterdam: North-Holland, 1961), pp. 96–112.

8

The status of the Principle of
Maximum Entropy

The Principle of Maximum Entropy (hereafter referred to as "PME") was proposed by E. T. Jaynes [1957, 1963, 1967, 1968, 1976, 1979] as a fundamental instrument of inductive inference. Jaynes's ideas have been enthusiastically accepted and applied by numerous scientists, engineers, and philosophers, but there have also been critics and skeptics. Because of several papers over the past decade [Friedman and Shimony 1971; Friedman 1973; Shimony 1973; Dias and Shimony 1981] my collaborators and I are considered to belong to the second group. Our effort, however, has not been entirely negative. We have attempted to discover tacit presuppositions of PME and thereby to determine the circumstances under which it may legitimately be applied. The present paper is a summary of our results, with only a sketch of some of the mathematical arguments.

We share with Jaynes the conviction that an epistemic concept of probability plays a central role in inductive inference. If h is a hypothesis and d is an evidential proposition (not self-contradictory), then we suppose that it is meaningful to speak of "the probability of h upon assumption that d is the total body of evidence" and to consider this probability to be a definite real number denoted by "$P(h\,|\,d)$". The statement "$P(h\,|\,d) = r$" is to be interpreted as "the reasonable degree of belief in h, if d is the total evidence, is r". At the beginning of the study of inductive logic, the words used in this interpretation are vague. In particular, there is a fundamental question about the objectivity or subjectivity of the probability function p. The name "epistemic concept of probability" is deliberately chosen to be neutral on this question, and it could be employed both by objectivists like Keynes, Jeffreys, and Carnap, and by subjectivists like Ramsey, DeFinetti, and Savage. On this question Jaynes's terminology is nonstandard and can mislead the unattentive reader. He says, "To the subjectivist, the purpose of probability theory is to help us in forming plausible conclusions in cases where there is not enough information available to lead to certain conclusions" [Jaynes 1957, p. 622]; but he also says that "two persons with the same relevant prior information should assign the same prior probabilities" [Jaynes 1968, p. 228], and that once the evidence is given the probability assignment is "independent of the personality of the user" [*ibid.*]. It is clear that Jaynes calls himself a subjectivist only to emphasize his rejection of a frequency concept of probability, and not to deny that there is an impersonal (and in this sense objective) "reasonable degree of belief" in a given hypothesis upon given evidence.

A novel element in Jaynes's view of inductive inference is his employment of information theory precisely for the purpose of obtaining a unique, rational, impersonal probability assignment. From C. Shannon [1948] he accepts the following reasoning. Suppose that at the beginning of an investigation a person possesses a body of evidence e, according to which one of a set of mutually exclusive hypotheses $h_1, \ldots, h_n$ must be true, and such that the probabilities of the respective hypotheses are $p(h_i|e) = p_i$. At the conclusion of the investigation it is found that h_j is true. How much information has the person gained? The question is obviously badly posed, because no definition has been given of "quantity of information". Shannon, however, makes two natural requirements which suffice to supply the requisite definition almost uniquely.

(a) The quantity of information gained is a function $I(p_j)$ – i.e., it is independent of the content of h_j and depends only upon the antecedent probability.

(b) If the investigator makes two independent investigations (among $h_1, \ldots, h_n$ on the one hand, and among $h'_1, \ldots, h'_m$ on the other), the quantity of information which he gains is the sum of the quantities which he would gain from the two separate investigations.

Suppose now that the results in the two investigations are h_j and h'_k. Because of independence

$$p(h_j \& h'_k|e) = p(h_j|e) \cdot p(h'_k|e) = p_j p'_k.$$

Hence, conditions (a) and (b) yield the functional equation

$$I(p_j p'_k) = I(p_j) + I(p'_k),$$

the only regular solutions of which are

$$I(p) = C \log p, \quad \text{where } C \text{ is a constant.}$$

(C is usually chosen to be negative, so that the information gained in the investigation is a positive quantity.) Finally, the *information theoretical entropy* S is defined as

(1) $$S = \sum p_i I(p_i) = C \sum p_i \log p_i.$$

S can be interpreted as the *average value (given e) of the information gained in the investigation,* or equivalently (and more usefully for Jaynes's purposes) as *the average value of the information which is missing at the beginning of the investigation.*

The *peripeteia* of Jaynes's theory is to use Shannon's expression (1) for information theoretical entropy for the purpose of determining the probabilities $p(h_i|e) = p_i$. Shannon himself says little about these probabilities, although his illustrations indicate that he is content with a frequency interpretation of them. Since Jaynes is committed to an epistemic interpretation of probability, he needs a way of determining the p_i. To do so, he enunciates PME:

The p_i have those values, consistent with the evidence e, which maximize the entropy S.

Jaynes's justification of PME is epistemological and moral: when one is ignorant of something, one ought to recognize one's ignorance, and can do so by admitting that the quantity of missing information is the maximum compatible with the evidence. Jaynes claims that PME "provides the most honest description of what we know" [1967, p. 97] and that the maximizing probability distribution "is uniquely determined as the one which is maximally noncommittal with regard to missing information" [1957, p. 623]. The acknowledgment of ignorance thus provides the means of overcoming subjectivity in the evaluation of epistemic probabilities.

Here are two typical applications of PME. If e is nothing but the background information b that $h_1, \ldots, h_n$ are mutually exclusive and exhaustive, then the only constraints on the p_i are the ones provided by probability theory – that they are nonnegative and sum to unity. Then PME yields the "prior probabilities"

$$(2) \qquad\qquad p(h_i|b) = 1/n.$$

by a standard analysis using a single Lagrange multiplier. If, however, e consists of the evidence that the mean value of some quantity A is α, where h_i implies $A = a_i$, there is an additional constraint

$$(3) \qquad\qquad \langle A \rangle = \sum a_i p_i = \alpha.$$

Then PME yields the "posterior probabilities"

$$(4) \qquad\qquad p(h_i|e) = \frac{e^{-\beta a_i}}{\sum_{k=1}^{n} e^{-\beta a_k}},$$

where β is a constant determined by α. The probability distribution of Eq. (4) is the Boltzmann distribution, although the interpretation of probability intended by Jaynes is different from that of Boltzmann. The facility of Jaynes's derivation, its independence from difficult and dubious physical assumptions like ergodicity, and its avoidance of fictions like "virtual ensemble" have impressed many students of statistical mechanics and inductive logic.

Nevertheless, doubts remain about the justification of PME. No one can reasonably dispute the maxim that one should be honest about the extent of one's ignorance, which is the core of Jaynes's epistemological and moral argument for PME. Yet the exact determination of probabilities on the basis of ignorance is somewhat suspicious: it is a kind of *docta ignorantia*. This objection is not decisive, however, just because epistemic probability is not a matter of fact but is only reasonable degree of belief; PME can be regarded as a refinement of the Principle of Indifference, which is the classical rule for evaluating epistemic probabilities on the basis of ignorance.

A more serious objection is an anomaly discovered by Friedman and Shimony [1971] concerning the connection between Eqs. (2) and (4), which are respectively the "prior" and the "posterior" applications of PME. The evidence e adds to the background information b the information that the mean value of the quantity A is α, as expressed in Eq. (3). The parameter β in the probability distribution (4) is an invertible function $\beta = \beta(\alpha)$, so that $\alpha = \alpha(\beta)$, and the content of the evidence e can equally well be expressed by fixing α or β. The evidence with this context will be designated either by $\hat{e}_\alpha$ or e_β. In a discussion of PME at a high level of generality it is not necessary to give a detailed discussion of the nature of the evidence e, although an illustration will be given later. For the present it suffices to say that the quantities α and β are unknown at the beginning of investigation (i.e., not specified by the background information b), that they can be specified by appropriate empirical data, and that a person who has no other information than b may have reasonable degrees of belief concerning the possible values of α and β. Specifically we assume that there are additive, nonnegative, normalized functions μ and $\hat{\mu}$ on the set $\mathcal{B}$ of Borel subsets of the real line R, such that for any $U \in \mathcal{B}$

$\hat{\mu}(U)$ equals the probability (in the sense of reasonable degree of belief),
        given background information b, that $\alpha \in U$,
$\mu(U)$ equals the probability given b that $\beta \in U$,
$\hat{\mu}(R) = \mu(R) = 1$,

and

$$\mu(U) = \hat{\mu}(\alpha(U)) = \hat{\mu}(\beta^{-1}(U)),$$
$$\hat{\mu}(U) = \mu(\beta(U)) = \mu(\alpha^{-1}(U)).$$

If the probability space is taken large enough to include the $\{h_i\}$ and also the $\{e_\beta\}$ (or, equivalently, the $\{\hat{e}_\alpha\}$), then the general principles of probability theory give

(5) $$p(h_i \mid b) = \int p(h_i \mid b \,\&\, e_\beta)\, d\mu.$$

The prior and posterior applications of PME in Eqs. (2) and (4) then yield

(6) $$\frac{1}{n} = \int \frac{e^{-\beta a_i}}{\sum_{k=1}^{n} e^{-\beta a_k}}\, d\mu.$$

Suppose now that not all of the a_i are equal and that one of them, a_m, is the average of all:

(7) $$a_m = \frac{1}{n} \sum_{i=1}^{n} a_i.$$

Friedman and Shimony derived from Eqs. (6) and (7) that

(8)
$$\mu(\{0\}) = \mu(R),$$
$$\mu(U) = 0 \quad \text{if } 0 \notin U.$$

(A proof is given in Appendix A.) In other words, on the basis of the background information b there is a reasonable degree of belief 1 (virtual certainty) that the evidence e will be found to be e_0, the evidence which specifies that the parameter β has value 0. This is an astonishingly strong result. The range of possible values of β is the whole real line, and it is therefore very surprising to find that a procedure which emphasizes honesty about one's ignorance should require one a priori to have virtual certainty concerning the outcome of an observation.

In order to make the foregoing result more vivid, let us consider the situation studied by Boltzmann, in which A is the energy of a physical system and a_i is the energy of the system when it is in the ith state. The average energy α of the system is constrained in some way, e.g., by placing it in contact with a reservoir of definite temperature. Some empirical procedure is envisaged for determining α, although the rule for applying PME once α is known is independent of the details of this procedure. The result expressed in Eq. (8) is that one has virtual certainty on the basis of the background information b that a parameter β will be found to be 0. It is easy to show that the value of α which corresponds to $\beta = 0$ is the a_m of Eq. (7), and therefore one has virtual certainty given b that this is the average energy which will be found empirically. Finally, the customary analysis of statistical thermodynamics shows that the parameter satisfies

$$\beta = 1/kT,$$

where k is Boltzmann's constant and T is the Kelvin temperature. It follows then, from the result expressed in Eq. (8), that on background information b it is reasonable to believe with degree 1 (virtual certainty) that *the reservoir temperature is infinite.* This conclusion can be interpreted as showing that *we are in hell,* which seems to be the first a priori argument to this effect that has ever been offered!

Someone might suggest that the disagreeable conclusions we have reached are avoidable by dropping the assumption that the probability measure $\mu(R)$ is normalized. For the purpose of calculating expectation values $\mu(R)$ need not be finite, since one could use a limiting procedure like the following:

$$(9) \qquad \langle F \rangle = \lim_{K \to \infty} \left[\frac{\int_{[-K,K]} F(\beta)\, d\mu}{\mu([-K,K])} \right].$$

The use of non-normalized probability measures does not, however, provide a loophole for PME from the foregoing objection. Dias and Shimony [1981] assumed only that $\mu(S) < \infty$ for any bounded Borel subset S of R, and they replaced Eq. (6) by

$$(10) \qquad \frac{1}{n} = \lim_{k \to \infty} \frac{\int_{[-K,K]} \dfrac{e^{-\beta a_i}}{\sum e^{-\beta a_k}}\, d\mu}{\mu([-K,K])}.$$

They still were able to demonstrate Eq. (8), and incidentally to show that $\mu(R) < \infty$; in other words, normalization is a result rather than a premiss.

In the foregoing analysis Eq. (5) was used without special comment. It is clear upon inspection that Eq. (5) is closely related to Bayes's theorem, and in fact is a natural generalization of Bayes's theorem when the space of evidential propositions $\{e_\beta\}$ is a continuum. Hence, the anomaly that has been presented is *almost* a demonstration that PME is inconsistent with Bayesian probability theory. The qualification "almost" is obligatory, since PME can indeed be incorporated into a Bayesian framework if one is willing to pay the high price of accepting Eq. (8). An excellent general discussion of the difficulties of enriching Bayesianism by deploying PME and other principles is given by Seidenfeld [1979]. He concludes that these principles yield

mathematically convenient idealizations wherein specified distributions are elevated to the roles of "ignorance" and "partial information" distributions. But the cost that goes with the idealization is a violation of conditionalization [*op. cit.,* pp. 434–35].

Jaynes has answered the foregoing objection by claiming that the information used to provide a constraint in PME cannot be used as evidence in an application of Bayes's theorem:

If a statement *d* referring to a probability distribution in space *S* is testable (for example, if it specifies a mean value $\langle f \rangle$ for some function $f(i)$ defined on *S*), then it can be used as a constraint in PME; but it cannot be used as a conditioning statement in Bayes' theorem because it is not a statement about any event in *S* or any other space.

Conversely, a statement *D* about an event in the space S^n (for example, an observed frequency) can be used as a conditioning statement in applying Bayes' theorem, whereupon it yields a posterior distribution on S^n which may be contracted to a marginal distribution on *S*; but *D* cannot be used as a constraint in applying PME in space *S*, because it is not a statement about any event in *S*, or about any probability distribution over *S*, i.e., it is not testable information in *S*. [1979, p. 54]

A similar reply was made by Cyranski [1979, p. 298]. Skyrms [1985] has systematically replied to Jaynes by discussing how a probability space can be enriched in order to conditionalize on the information contained in a PME constraint, and he argues that PME can be regarded as a special case of conditionalization. My answer to Jaynes will be less systematic and less ambitious. I shall simply present an example constructed by Dias and Shimony [1981, section IV] of a statement which can unequivocally be used both as a constraint in PME and as a conditioning statement in Bayes's theorem.

The example will be constructed in a language $\mathcal{L}$ studied by Carnap [1965], a language with *N* individual names $x_1, \ldots, x_N$ and a family of *n* mutually exclusive and exhaustive predicates $P_1, \ldots, P_n$, and logical connectives "~" (negation), "∧" (conjunction), and "∨" (disjunction). The sentences of $\mathcal{L}$ are the atomic sentences of the form "$P_j x_i$" and all finite molecular sentences constructed from these by means of logical connectives. Carnap constructs and studies a family c_λ of probability functions, where the index λ can be any non-negative real number or ∞. The family $\{c_\lambda\}$, which Carnap calls "the lambda continuum", is the complete set of functions satisfying both the standard probability axioms and certain additional reasonable assumptions (a fact which was independently discovered by Johnson [1932], Carnap [1952], and Kemeny [1953]). The sentence "$c_\lambda(h \mid e) = r$" asserts that "the probability of the hypothesis *h*, given the evidence *e*, is *r*", but it makes a tacit commitment to the choice c_λ from among the continuum of possible probability functions. The probability sentence is expressed in a metalanguage $\mathfrak{M}$, not in the language $\mathcal{L}$, which lacks the resources for making statements involving the real numbers (or even the integers); in other words, *h* and *e* are names in $\mathfrak{M}$ for sentences in $\mathcal{L}$, *r* is a name in $\mathfrak{M}$ of a real number, and c_λ is a function in $\mathfrak{M}$ which takes pairs of names of sentences as arguments. (Note that in the notation "$p(h \mid e) = r$" used earlier in this paper the letters *h* and *e* designated propositions, which are nonlinguistic entities, rather than sentences.)

The strongest possible consistent sentences in $\mathcal{L}$ are the *state descriptions,* each of which has the form

$$P_{j_1} x_1 \wedge P_{j_2} x_2 \wedge \cdots \wedge P_{j_N} x_N,$$

where j_i is one of the integers $1, \dots, n$ for each of the indices $i = 1, \dots, N$. In other words, a state description D assigns a definite one of the family of predicates $P_1, \dots, P_n$ to each of the individuals. The n-tuple $\{N_1, \dots, N_n\}$ of the state description is an ordered set of n non-negative integers such that N_i is the number of individuals which are assigned the predicate P_i by D. Obviously the N_i sum to the number N of individuals. A *structure description* $D_{\{N_1, \dots, N_n\}}$ is the disjunction of all state descriptions having the same n-tuple $\{N_1, \dots, N_n\}$. Suppose that a structure description is taken as an evidential statement, and an atomic sentence $P_j a_i$ is taken to be the hypothesis. Then it is easy to show that for each value of λ,

(11) $$c_\lambda(P_j x_i \mid D_{\{N_1, \dots, N_n\}}) = N_j / N.$$

Although I believe that Jaynes's method of evaluating probabilities would give results in agreement with Eq. (11), this concurrence is not very interesting. We proceed, therefore, to construct a statement which clearly plays both the role of being a conditioning statement and a constraint in PME.

Consider a statement $E_{[\alpha_1, \alpha_2]}$ which is the disjunction of all those structure descriptions $D_{\{N_1, \dots, N_n\}}$ such that

(12) $$\alpha_1 \le \frac{1}{N} \sum_{j=1}^n j N_j \le \alpha_2,$$

where α_1 and α_2 are real numbers such that $1 \le \alpha_1 \le \alpha_2 \le n$. There may be no structure descriptions that satisfy Inequality (12), in which case $E_{[\alpha_1, \alpha_2]}$ is not a consistent sentence, but it is still well defined as the null disjunction.

The foregoing translation of $E_{[\alpha_1, \alpha_2]}$ has, of course, been written in a metalanguage rich enough to talk about real numbers; but $E_{[\alpha_1, \alpha_2]}$ itself is a sentence in the extremely simple language $\mathcal{L}$, since it is the disjunction of a finite number of structure descriptions, each of which is a sentence in $\mathcal{L}$. Furthermore, if there is at least one structure description in this disjunction (which depends upon the real numbers α_1 and α_2), then it is a consistent sentence which can be inserted in the second argument place in any of the functions c_λ – in other words, $E_{[\alpha_1, \alpha_2]}$ can be used as a statement in Bayesian conditionalization.

But $E_{[\alpha_1, \alpha_2]}$ can also be used as a constraint in PME. Consider the probability space S of n points designated by the n sentences $P_1 x_i, P_2 x_i, \dots,$ $P_n x_i$, and let the quantity A mentioned in Eq. (3) have the value $a_j = j$

if $P_j x_i$ is true. If one's evidence consisted of a structure description $D_{\{N_1, ..., N_n\}}$, one could calculate a mean value for A, namely

(13) $$\langle A \rangle = \frac{1}{n} \sum_{j=1}^{n} j N_j.$$

It would not be reasonable, however, to use this value of $\langle A \rangle$ as a constraint in PME in order to compute the probabilities of the $P_j x_i$. The reason, of course, is that the structure description provides much more data than just the mean value of A. In fact, because of this additional information, the reasonable evaluations of the probabilities are given in Eq. (11). If, by contrast, the evidence is described not by a structure description but by $E_{[\alpha_1, \alpha_2]}$, and if the interval $[\alpha_1, \alpha_2]$ is large enough to permit a large number of structure descriptions to satisfy Eq. (12), then the evidence is exhausted by information about the mean value of A. Furthermore, if the interval $[\alpha_1, \alpha_2]$ is narrow compared with the possible range of averages of A over structure descriptions, which is from 1 to n, then the circumstances are surely appropriate for applying PME. The posterior probabilities can be written in the form of Eq. (4), but the parameter β is specified only as lying within a small interval $[\beta(\alpha_2), \beta(\alpha_1)]$ instead of being given by a single real number. The narrowness of the interval $[\alpha_1, \alpha_2]$ is consistent with the condition that a large number of structure descriptions satisfy Eq. (12) only if the number of individuals N is very large, which is typically the case in statistical mechanics, where the Boltzmann distribution of Eq. (4) has enjoyed its greatest triumphs. For example, if N is of the order of 10^{24}, then there are immensely many structure descriptions satisfying Eq. (12) with $|\alpha_2 - \alpha_1| = 10^{-10}$.

The conclusion is that the sentence $E_{[\alpha_1, \alpha_2]}$ provides a counterexample to the claim made by Jaynes in the quotation above. The sentence $E_{[\alpha_1, \alpha_2]}$ not only can be used as a constraint in PME, but is a statement about an event in the space L, the points of which are the designata of the state descriptions of $\mathcal{L}$. It is true that $E_{[\alpha_1, \alpha_2]}$ is not about an event in the space S, the points of which are the designata of the n sentences $P_1 x_i, ..., P_n x_i$. But contrary to Jaynes, $E_{[\alpha_1, \alpha_2]}$ is about an event in *some* space, and furthermore, that space L is an enrichment of S, in the sense that the points of S are subspaces of L.

This discussion of the dual role played by $E_{[\alpha_1, \alpha_2]}$ also indicates that a very interesting and attractive suggestion of Skyrms is not always applicable. In the passage about to be quoted "Pr_f" is the "final probability," i.e., the probability computed by PME when the mean value of some quantity is given as evidence. He writes:

The foregoing suggests that the natural interpretation of maximum entropy inference may be found when Pr_f is interpreted as chance. We do not think of Pr_f there as simply some degree of belief resulting from assimilating some new evidence, but rather as the limiting upshot of assimilating an infinite sample [Skyrms, 1985, p. 69].

Contrary to Skyrms's statement, the $E_{[\alpha_1, \alpha_2]}$ constructed above is a sentence in a language about a finite (though very large) collection of individuals. Of course, the price which was paid for using a finite population was that the mean value $\langle A \rangle$ was constrained to lie in an interval, rather than to be a point.

We may now pose a very instructive question. For what value of Carnap's parameter λ, if any, is there agreement between $c_\lambda(P_j x_i | E_{[\alpha_1, \alpha_2]})$ and $p(P_j x_i | E_{[\alpha_1, \alpha_2]})$ computed by PME with the constraint (12)? The answer was given by Dias and Shimony [1981, section 4]: for any finite value of λ there are $\delta_\lambda > 0$ and integer N_λ such that for N larger than N_λ the discrepancy between the two sets of probability evaluations is greater than δ_λ; whereas, the discrepancy between the PME probabilities and $c_\infty(P_j x_i | E_{[\alpha_1, \alpha_2]})$ goes to 0 as the population size N goes to infinity and the interval $[\alpha_1, \alpha_2]$ shrinks to 0.

We do not claim that this result shows PME to be a covert version of Carnap's c_∞, because, in fact, it is only under rather special circumstances that Carnap's inductive methods and PME can both be applied. Nevertheless, we do assert that there is an affinity between c_∞ and PME.

There are three ways of setting forth the peculiar character of the probability function c_∞. (i) Suppose the evidential sentence e asserts that in a sample of M individuals, M_j of them have the property designated by P_j, and suppose that x_i names an individual which is not in the sample. Then for any finite λ,

$$(14) \qquad c_\lambda(P_j x_i | e) = \frac{M_j + \lambda/n}{M + \lambda}.$$

Clearly, the smaller is λ, the more is the probability of Eq. (14) determined by the empirical ratio M_j/M, and the less is the probability affected by the number n. In fact, for $\lambda = 0$ the probability equals M_j/M. Hence, λ is an index of the weight put upon the logical structure of the language, specifically upon the number n of predicates in the family $P_1, ..., P_n$. In the limit of $\lambda \to \infty$, Eq. (14) yields

$$(15) \qquad c_\infty(P_j x_i | e) = \frac{1}{n}.$$

(ii) Eq. (15) shows that c_∞ is inadequate for the purpose of inductively learning from experience, since any empirical evidence about individuals other than the one designated by x_i is irrelevant to an atomic sentence $P_j x_i$. (iii) Eq. (15) also implies that all state descriptions have the same prior probability. If t is a tautology (null evidence, stating no more than can be inferred from the rules of $\mathcal{L}$), then for any state description D

(16) $c_\infty(D\,|\,t) = (\text{number of state descriptions})^{-1} = n^{-N}.$

Eq. (16) throws some interesting light upon the anomaly derived earlier, and expressed in Eq. (8). If the state description D were given, then its n-tuple would be determined simply by counting, and $\langle A \rangle$ could be computed by Eq. (13). But upon background information one does not know what the true state description is. If Eq. (16) is assumed, then all state descriptions have equal a priori probability. However, combinatorial considerations ensure that for large N nearly all of the state descriptions have n-tuples that are approximately $\{N/n, \ldots, N/n\}$. But this means that there is high probability a priori that

(17) $$\langle A \rangle \simeq \frac{1}{n}(1 + 2 + \cdots + n)$$

or, equivalently, high probability a priori that the parameter β is near 0. The exact meaning of "high probability" depends upon the size of N and the latitude permitted in satisfying Eq. (17). But when these technicalities are settled, "high probability" means $1 - \omega(N)$, where $\omega(N)$ goes to 0 as N goes to infinity. In other words, the anomalous Eq. (8), which asserts that one has virtual certainty a priori about the value of the parameter β, is essentially what one expects if one works with the probability function c_∞.

This affinity betwen c_∞ and PME was independently discovered by Hadjisavvas [1981], who concluded that PME is thereby justified but given a derivative status, since it is "implied from Laplace's principle of equiprobabilities." This justification of PME is questionable, however, precisely because of the connection between assertions (ii) and (iii) made above concerning c_∞. The price of the a priori equiprobability of the state descriptions (statement iii) is that c_∞ is inadequate for learning from experience (statement ii).

Even though c_∞ is not a reasonable choice of a probability function, it may be reasonable to employ a relativized version of c_∞ in certain circumstances. The evidence E may restrict the possible state descriptions to a class $\mathcal{S}$, the cardinality of which is $\mathfrak{N}(\mathcal{S})$. It may then be reasonable to use the relativized probability function c_∞^{rel}, defined only for evidential sentences of the form $E \wedge f$, and such that

(18) $c_\infty^{\text{rel}}(D\,|\,E) = 1/\mathfrak{N}(\mathcal{S})$ if $D \in \mathcal{S}$,

                                    $= 0$ if $D \notin \mathcal{S}$.

If the class $\mathcal{S}$ is defined by Eq. (12), then c_∞^{rel} yields probability evaluations in good agreement with PME, as we have seen previously. A commitment to c_∞^{rel} is much more restricted than a commitment to c_∞, and in particular does not require that one make inductive inferences in accordance with the disagreeable Eq. (15) and (8). Furthermore, if the simple language $\mathcal{L}$ is replaced by a language sufficiently sophisticated for scientific and practical discourse, it may still be possible to find evidential statements E for which some analogue of Eq. (18) is reasonable.

In appropriate situations, gambling devices, random noise generators, collections of radioactive particles, gases in thermal equilibrium, and many other systems seem to permit the employment of a relativized probability function with a uniformity property similar to Eq. (18). In many of these situations, as has been indicated, a relativized probability function which is uniform over a restricted class of state descriptions is in good agreement with PME. Hence, there is some validity in the claim of Jaynes and his followers that PME has an abundance of applications. But assent to this extent is far from concurrence with Jaynes's claim that PME is a general instrument of inductive inference which is grounded entirely upon epistemological considerations. Usually a nontrivial investigation is required in order to establish a uniformity like that of Eq. (18), and in the absence of such an investigation there is a nontrivial assumption. For example, an analogue of Eq. (18) in statistical mechanics is the uniform probability distribution on an energy hypersurface (the microcanical distribution), and the effort to justify this distribution – by showing ergodic behavior of a many-particle system or by other means – has occupied statistical mechanicians from the time of Boltzmann to the present.

It thus becomes a factual question, and often a quite intricate one, whether PME is an appropriate method of inference to use in a given situation. This conclusion lends support to Skyrms's suggestion, quoted above, that "the natural interpretation of maximum entropy inference may be found when Pr_f is interpreted as chance," where "chance" signifies a nonepistemic (or propensity) concept of probability. Although I did not fully accept Skyrms's suggestion I suspect that it is true in most cases of legitimate use of PME. However, I am not thereby retracting the statement made at the beginning of the paper that an epistemic concept of probability plays a central role in inductive inference. Rather, I am saying that the epistemic and the propensity concepts of probability are often closely connected, as the role of gambling devices in the history of probability clearly indicates. If e asserts that a roulette wheel has a *chance or propensity* 0.51 of showing "red" when it is normally spun, then

$$\text{Prob(red will show on the next turn } e) = 0.51$$

is an excellent example of *epistemic probability*. Upon what empirical evidence we are entitled to assign high epistemic probability to this e is another matter. A systematic treatment of the relation between the epistemic and propensity concepts of probability is beyond the scope of this paper, but it is significant that reflection on PME has brought us to the threshold of this problem.

APPENDIX A

To prove the theorem of Friedman and Shimony, insert a_m of Eq. (7) into Eq. (6):

(A1)
$$\frac{1}{n} = \int \frac{e^{-\beta a_m}}{\sum_{k=1}^{n} e^{-\beta a_k}} \, d\mu = \int \left[\sum_{k=1}^{n} e^{\beta(a_m - a_k)} \right]^{-1} d\mu$$

$$\equiv \int I(\beta) \, d\mu.$$

We assume that $\mu(R) = 1$. The derivative of the integrand $I(\beta)$ in Eq. (A1) is

(A2)
$$\frac{dI(\beta)}{d\beta} = - \left[\sum_{k=1}^{n} e^{\beta(a_m - a_k)} \right]^{-2} \sum_{k=1}^{n} (a_m - a_k) e^{\beta(a_m - a_k)},$$

which vanishes at $\beta = 0$, since

(A3)
$$\sum_{a_m < a_k} (a_m - a_k) = - \sum_{a_m > a_k} (a_m - a_k),$$

by Eq. (7). But $dI(\beta)/d\beta \neq 0$ for $\beta > 0$, since in

$$\sum_{k=1}^{n} (a_m - a_n) e^{\beta(a_m - a_k)}$$

all the terms occurring in the lhs of Eq. (A3) enter with a weighting factor $e^{\beta(a_m - a_k)} < 1$, while all those on the rhs of Eq. (A3) enter with a weighting factor > 1. Likewise, $dI/d\beta \neq 0$ if $\beta < 0$. Hence, the only extremum of $I(\beta)$ is at $\beta = 0$. (We can easily check that this extremum is a maximum, but we have no need for this fact here.) It follows that Eq. (A1) can be satisfied only if $\mu(\{0\}) = 1$ and $\mu(R - \{0\}) = 0$. Informally, we may say that the a priori probability density of the parameter β is the Dirac delta function $\delta(\beta)$.

APPENDIX B

In private communications Prof. Isaac Levi made several important comments on the papers of Friedman and Shimony and of Dias and Shimony.

1. Both papers use "the general principles of probability theory" to assert

$$P(h_i \,|\, b) = \int P(h_i \,|\, b \,\&\, e_\beta) \, d\mu,$$

which was Eq. (5) of the present paper. Levi pointed out, however, that Eq. (5) is derivable only if "the general principles" include countable additivity, which may not be justifiable for epistemic probability [DeFinetti 1975, pp. 91–93]. An answer to Levi is provided by the following theorem, which assumes that a certain kind of interpolation of PME is permitted but dispenses with countable additivity.

Suppose that R is decomposed into a finite set of intervals $(-\infty, \beta_1]$, $(\beta_1, \beta_2], \ldots, (\beta_{z-1}, \infty)$, and let D_j be the proposition that the parameter β (which is the parameter introduced in Eq. (4)) falls in the jth of these intervals. The interpolation of PME which will be assumed is

(B1) $$P(h_i \,|\, b \,\&\, D_j) \in \left(\frac{e^{-\beta_{j-1} a_i}}{\sum_k e^{-\beta_{j-1} a_k}}, \frac{e^{-\beta_j a_i}}{\sum_k e^{-\beta_j a_k}} \right).$$

Condition (B1) is related to the "principle of conglomerability": if for each element e_j of the denumerable partition of E

$$c' \le P(A \,|\, e_j) \le c'' \quad (i = 1, \ldots),$$

then

$$c' \le P(A \,|\, E) \le c''$$

[DeFinetti 1975, section 5]. Condition (B1), however, is based on a continuous partition of the jth interval, but it is applied only to the special case of probabilities evaluated by PME when β is known within an interval rather than exactly. It is hard to see how Jaynes could avoid using condition (B1) if PME is to be applied in practical circumstances, when one does not have exact knowledge of the value of β.

Since we are restricting our attention to a finite decomposition $D_1, \ldots, D_z$, we can assume normalization of probability to unity:

(B2) $$\sum_{j=1}^{z} P(D_j \,|\, b) = 1.$$

Finite (not countable) additivity yields the following analogue to Eq. (5):

(B3) $$P(h_i \,|\, b) = \sum_{j=1}^{z} P(h_i \,|\, b \,\&\, D_j) \cdot P(D_j \,|\, b).$$

Eq. (B1) implies that for each i there exists a set of real numbers $\hat{\beta}_1^i, \ldots, \hat{\beta}_z^i$, with $\hat{\beta}_j^i$ lying in the jth interval in the decomposition, such that

(B4) $$P(h_i \,|\, b) = \sum_{j=1}^{z} \frac{e^{-\hat{\beta}_j^i a_i}}{\sum_{k=1}^{n} e^{-\hat{\beta}_j^i a_k}} \cdot P(D_j \,|\, b).$$

The index i on $\hat{\beta}_j^i$ indicates that the interpolation rule (B1) is applied separately for each h_i. We focus now particularly on the hypothesis h_m, where m is defined by Eq. (7), and we shall use the simplified notation $\hat{\beta}_j = \hat{\beta}_j^m$. Then for $i = m$ Eq. (B4) can be rewritten as

(B5) $\quad P(h_m | b) = \sum_{j=1}^{z} \frac{e^{-\hat{\beta}_j a_m}}{\sum_{k=1}^{n} e^{-\hat{\beta}_j a_k}} \cdot P(D_j | b) = \int_{-\infty}^{\infty} \frac{e^{-\beta a_m}}{\sum e^{-\beta a_k}} f(\beta)\, d\beta,$

where

(B6) $\qquad\qquad\qquad f(\beta) = \sum_{j=1}^{z} P(D_j | b)\,\delta(\beta - \hat{\beta}_j).$

Now if $P(h_m | b) = 1/n$, as required by the application of PME in Eq. (2), then the same argument which led from Eq. (A1) to Eq. (A3) yields

(B7) $\qquad\qquad\qquad\qquad f(\beta) = \delta(\beta).$

Hence,

(B8a) $\qquad P(D_j | b) = 0$, if 0 is not contained in the jth interval,

(B8b) $\qquad P(D_j | b) = 1$, if 0 is contained in the jth interval.

Since the interval containing $\beta = 0$ can be chosen as narrow as one wishes without abandoning a finite decomposition of the real line, a conclusion almost as strong as that of Friedman and Shimony has been obtained, but without Eq. (5) and hence without countable additivity.

2. Levi's second comment is that the theorem of Dias and Shimony was proved only when a special procedure was used to take the limits to $\pm\infty$, namely that indicated in Eqs. (9) and (10). Will it be possible to prove that $\mu(\{0\}) = \mu(R) < \infty$ if, instead of Eq. (10), one writes

(B9) $\qquad\qquad \dfrac{1}{n} = \lim_{t \to \infty} \dfrac{\displaystyle\int_{-K_1(t)}^{K_2(t)} \frac{e^{-\beta a_i}}{\sum e^{-\beta a_k}}\, d\beta}{\mu([-K_1(t), K_2(t)])},$

where $K_1(t)$ and $K_2(t)$ go monotonically and continuously from 0 to ∞ in an arbitrary fashion, possibly reaching ∞ at finite values of t? One might suspect that the answer is negative, since, of course, for an arbitrary function F the limit as $t \to \infty$ of $F(K_1(t), K_2(t))$ is not independent of the paths $K_1(t)$ and $K_2(t)$. The function in the brackets on the rhs of Eq. (9), however, is quite special, and therefore the answer to the question raised by Levi turns out to be positive. This claim can be proved by paralleling the argument in section 2 of the paper of Dias and Shimony. There are essentially two cases: (i) both $K_1(t)$ and $K_2(t)$ remain finite for the same range of values of t, and (ii) one of $K_1(t)$, $K_2(t)$ attains the value ∞ while the other is still finite. These two cases must be treated slightly differently from each other, but in both cases the argument is straightforward.

3. Levi's third comment is that Dias and Shimony use nonstandard terminology when they call a measure function μ "sigma-finite" if $\mu(S) < \infty$ for any bounded Borel subset S of R. Levi's comment is correct, and we request that readers take note that Dias and Shimony use "sigma-finite" in a stronger than standard sense. The theorem in section 2 of their paper, however, is correct if μ is taken to satisfy their strong and nonstandard sense of "sigma-finite."

Acknowledgments. The author is indebted to Profs. Isaac Levi and Teddy Seidenfeld for very stimulating discussions and correspondence. The research for this paper was supported by the National Science Foundation, Grant No. SES-8309118.

REFERENCES

Carnap, R.: 1952, *The Continuum of Inductive Methods,* Univ. of Chicago Press, Chicago.

Carnap, R.: 1965, 'A Basic System of Inductive Logic, Part I', in R. Carnap and R. Jeffrey, (eds.), *Studies in Inductive Logic and Probability,* Vol. I, Univ. of California Press: Berkeley, Los Angeles.

Cyranski, J. F.: 1979, 'Measurement, Theory, and Information', *Information and Control* 33, 275–304.

DeFinetti, B.: 1975, *Theory of Probability: A Critical Introductory Treatment,* Vol. II, John Wiley, London.

Dias, P. and Shimony, A.: 1981, 'A Critique of Jaynes' Maximum Entropy Principle', *Advances in Applied Mathematics* 2, 172–211.

Friedman, K.: 1973, 'Replies to Tribus and Motroni and to Gage and Hestenes', *Journal of Statistical Physics* 9, 265–69.

Friedman, K. and Shimony, A.: 1971, 'Jaynes's Maximum Entropy Prescription and Probability Theory', *Journal of Statistical Physics* 3, 381–84.

Hadjisavvas, N.: 1981, 'The Maximum Entropy Principle as a Consequence of the Principle of Laplace', *Journal of Statistical Physics* 26, 807–15.

Jaynes, E. T.: 1957, 'Information Theory and Statistical Mechanics', I, *Phys. Rev.* 106, 620–30; II, *Phys. Rev.* 108, 171–90.

Jaynes, E. T.: 1963, 'Information Theory and Statistical Mechanics', in K. W. Ford (ed.), *Statistical Physics,* Brandeis University Summer Institute Lectures in Theoretical Physics 3, Benjamin, New York.

Jaynes, E. T.: 1967, 'Foundations of Probability Theory and Statistical Mechanics', in M. Bunge (ed.), *Delaware Seminar in the Foundations of Physics,* Springer, New York.

Jaynes, E. T.: 1968, 'Prior Probabilities', *IEEE Trans. Syst. Sci. Cybernet,* SSC-4, 227–41.

Jaynes, E. T.: 1976, 'Confidence Intervals vs. Bayesian Intervals', in W. Harper and C. Hooker (eds.), *Foundations of Probability Theory, Statistical Inference, and Statistical Theories of Science,* Vol. II, Reidel, Dordrecht.

Jaynes, E. T.: 1979, "Where Do We Stand on Maximum Entropy?' in R. Levine and M. Tribus (eds.), *The Maximum Entropy Formalism,* MIT Press, Cambridge, Mass.

Johnson, W. E.: 1932, 'Probability', *Mind* **41**, 1–16, 281–96, 408–23.

Kemeny, J., 1953: 'A Contribution to Inductive Logic', *Philosophy and Phenomenological Research* **13**, 371–74.

Seidenfeld, T.: 1979, 'Why I Am Not an Objective Bayesian: Some Reflections Prompted by Rosenkrantz', *Theory and Decision* **11**, 413–40.

Shannon, C.: 1948, 'A Mathematical Theory of Communication', *Bell System Technical Journal* **27**, 379–523.

Shimony, A.: 1973, 'Comment on the Interpretation of Inductive Probabilities', *Journal of Statistical Physics* **9**, 187–91.

Skyrms, B.: 1985, 'Maximum Entropy Inference as a Special Case of Conditionalization', *Synthese* **63**, 55–74.

COMMENT

In E. T. Jaynes's "Some Random Observations" [*Synthese* 63 (1985), pp. 115–38], there is a brief notice of my critical argument (pp. 134–37). He asserts that "Errors in this argument have now been pointed out five times, by Tribus and Motroni (1972), Hobson (1972), Gage and Hestenes (1973), Jaynes (1978), and Cyranski (1979), and yet he persists in publishing that same argument over and over again" (p. 135). All of these rebuttals have been carefully answered, however, in articles mentioned in the references for this essay: Tribus and Motroni and Gage and Hestenes by my collaborator Friedman (1973), Hobson by Shimony (1973), Jaynes by Dias and Shimony (1981), and Cyranski by Skyrms (1985). I have also had extensive correspondence with Hobson and Cyranski and a conversation with Hestenes.

Jaynes supplements his reference to earlier literature with the following brief argument:

In the maximum entropy problem, the quantity β has no previous existence; it is a Lagrange multiplier that is treated only in the process of entropy maximization. But it appears only for mathematical convenience; the problem could be solved also by direct algebraic reduction without ever introducing it. β is not "estimated", but *defined,* by the PME formalism. . . .

It does not make sense, therefore, to speak of having prior knowledge of β, much less of honestly representing that knowledge. A Lagrange multiplier does not have a probability distribution; it is no different, in principle, from a normalization constant that also appears in a probability distribution. That too is not estimated, but defined; and indeed, to infinite accuracy (p. 136).

This argument focuses upon the mathematical operation that is performed when the information-theoretical entropy is maximized subject to a constraint, like that of Eq. (3):

$$\langle A \rangle = \sum a_i p_i = \alpha.$$

When α is given, the value of the Lagrange multiplier β can be computed; or alternatively, as Jaynes says, the p_i can be expressed directly in terms of α. But the quotation from Jaynes plays down the obvious fact that the value of α must be obtained from some kind of observational data, since it surely is not known a priori. Consequently, even though β is nothing but a parameter defined by the PME formalism, its value depends indirectly upon observational data. From a general Bayesian standpoint, the probabilities p_i calculated by the distribution of Eq. (4), when β is known, are posterior probabilities – they are the probabilities of the various h_i, given the data that fix α and hence β. I grant that my terminology "estimate" is not appropriate if the data are sufficient to fix the value of α unequivocally, but if they are not strong enough for that purpose the terminology is appropriate. Finally, Bayesians generically are willing to speak of the prior probabilities of the possible data from which α, and hence β, are inferred. For this reason, it does make sense for a Bayesian to speak of the prior probability distribution of the Lagrange multiplier β. If this conclusion is distasteful to Jaynes, he has the option of retrenching his commitment to Bayesianism. The criticism of Jaynes by Friedman, Dias, and me indicated the need for a retrenchment *somewhere,* either of PME or of Bayesian probability theory, but it does not say where the retrenchment should be made.

PART D

Inductive inference: The dialectic of experience and reason

9
Scientific inference

There's a divinity that shapes our ends,
Rough-hew them how we will.
Shakespeare, *Hamlet*

I. INTRODUCTION

A

My epistemological position can be described as naturalistic, or, with obvious latitude, as Copernican.[1] It recognizes that a human being is a minute part of a universe which existed long before his birth and will survive long after his death; it considers human experience to be the result of complex

This work originally appeared as "Scientific Inference" in *The Nature and Function of Scientific Theories,* Robert G. Colodny, editor. Published in 1970 by the University of Pittsburgh Press. Used by permission of the publisher.

This essay is an expanded version of a paper presented to a philosophy of science workshop at the University of Pittsburgh in May, 1965. The comments of the other participants in the workshop were very helpful in revising the original paper. Conversations with Marx Wartofsky, Joseph Agassi, and Imre Lakatos were also very stimulating. For many years Richard Jeffrey has patiently listened to my conjectures concerning induction and has given expert criticism. Howard Stein read several drafts of the essay with great care and made a large number of valuable suggestions on matters of principle, on the structure of the argument, and on style; furthermore, conversations with him over many years have strongly influenced my philosophical outlook. I am grateful to the National Science Foundation for a Senior Postdoctoral Fellowship, during the tenure of which most of this essay was written. I also wish to thank the Harvard Physics Department and the M.I.T. Humanities Department for providing secretarial assistance in preparing the typescript.

My former teacher, Rudolf Carnap, was always very generous and tolerant in giving encouragement and technical advice, even though my approach to induction has diverged from his. I dedicate this essay to his memory.

1. To my knowledge the first use of this term in an epistemological sense was Kant's "Copernican Revolution," but in view of his doctrine that "the order and regularity in the appearances, which we entitle *nature,* we ourselves introduce" (Critique A 125), his epistemology is anti-Copernican in my sense of the term. Two contemporary philosophers who use "Copernican" as I do are Feigl (1950: pp. 40–41) and Smart (1963: p. 151), but I surmise that nineteenth-century precedents can be found. It is reasonable to understand "naturalistic" as having a wider extension than "Copernican" and applying to any epistemological theory which is based upon a study of man's place in nature. Thus Aristotle's epistemology is naturalistic, though I would not classify it as Copernican. In characterizing my own point of view, I shall use both descriptions: "naturalistic" to emphasize

interactions of human sensory apparatus with entities having careers independent of their being perceived; and it acknowledges the probability that the fundamental principles governing the natural order will seem extremely strange from the standpoint of ordinary human conceptions. In order to proclaim oneself a Copernican at the present stage in history, one admittedly does not have to be radical and nonconformist. Copernicanism is part of the generally accepted scientific world view, and it has been the doctrine of a number of philosophical schools, such as the American naturalists, the eighteenth-century materialists, the Lockean empiricists, and (anachronistically) the Greek atomists. Nevertheless, this familiar point of view is worthy of restatement, partly because it provides a perspective which can prevent narrowness in the technical investigations of philosophy of science and partly because its implications have by no means been exhaustively explored.

From the Copernican point of view it is natural to see two grand philosophical problems, which may be described by borrowing the vivid phrases of Heraclitus. The first problem is "the way up": to show how it is that human beings, whose experience is extremely restricted in space and time and is conditioned by physiological and psychological peculiarities, can obtain an understanding of the universe beyond themselves. The second is "the way down": to understand in terms of fundamental principles how there can be entities such as ourselves in nature, capable of the kind of experience and endowed with the kind of faculties which make natural knowledge possible. "The way up" is essentially the subject matter of methodology and epistemology, whereas "the way down" is roughly the concern of philosophical psychology, though it overlaps the domains of many sciences. A complete solution to either of these two grand problems will surely depend upon a complete solution of the other, and even partial solutions can be expected to be mutually dependent. As a result, the Copernican point of view involves some circularity, although, I believe, the circularity need not be vicious.[2] In this essay I shall not examine the question of circularity in full generality, nor give detailed criticisms of epistemological theories, like those of Descartes and Russell, which purport to escape circularity by developing the structure of knowledge along an architectural plan of successive stories resting upon a firm foundation. I shall, however, discuss circularity in the theory of inductive inference, where it is particularly acute, as Hume pointed out, and I hope that the

the continuity of philosophical and scientific investigations and "Copernican" to emphasize the insignificance of human beings in the universe.

2. The possibility of nonvicious circularity in the structure of knowledge was first suggested to me by Weiss (1938: chap. 1).

proposals made in this context will be capable of generalization to a wider range of philosophical problems.

B

I shall not be concerned with all varieties of inductive inference but only with scientific inference. If "induction" is construed broadly as meaning any valid nondeductive reasoning, then all but the logical and mathematical components of scientific inference are inductive. However, there are problems of induction which are not subsumed under scientific inference, especially problems of making practical decisions in a rational manner under conditions of uncertainty or partial knowledge. My treatment of scientific inference is instrumentalistic, in the sense that the justification for certain prescriptions is their conduciveness to achieving certain ends, but the ends admitted are only the theoretical ones of learning the truth about various aspects of the world; accordingly, these prescriptions are characterized by open-mindedness, caution, and a long-range point of view.[3] In making practical decisions there are other ends than pure knowledge, and a short-range point of view is often unavoidable. Although I believe that my treatment of scientific inference can be embedded in a more general theory of induction, in which the interrelations of theory and practice are properly taken into account, I shall not undertake to formulate such a theory.

The treatment of scientific inference in the following sections will be divided into two stages. In the first stage (sections II, III, and IV) the examination will proceed as far as possible by methodological considerations. Scientific inference will be formulated in terms of probability, and a conception of probability (essentially due to Harold Jeffreys) which I call "tempered personalism" will be presented. It will be argued that scientific inference, thus formulated, is a method of reasoning which is sensitive to the truth, whatever the actual constitution of the universe may be. In this stage very little will be assumed about the place of human beings in nature, but the result of such austerity is that scientific method is only minimally justified, as a kind of counsel of desperation.[4] In the second stage (section V) the justification is enlarged and made more optimistic by noting the biological and psychological characteristics of human beings which permit them to use induction as a discriminating and efficient instrument. Thus the second stage considers scientific methodology from

3. But not *infinitely* long-range. See sec. IV.
4. To paraphrase Goethe:
   Grau ist alle Theorie,
   Doch grauer Methodologie.

the standpoint of "the way down," and it evidently presupposes a considerable body of scientific knowledge about men and their natural environment, thereby raising the issue of circularity.

The outline of this treatment of scientific inference is due to Peirce. The first stage derives from Peirce's conception of the scientific method as a deliberate and systematic "surrendering" to the facts, whatever they may be (for example, Peirce 5.581, 7.78). The second stage derives from Peirce's proposition that all men, and the great discoverers preeminently, possess "an inward power, not sufficient to reach the truth by itself, but yet supplying an essential factor to the influences carrying their minds to the truth" (1.80). He was the first philosopher to understand fully that a study of instinct and intelligence in the light of evolutionary biology could provide a surrogate for classical rationalism, thereby reconciling a Copernican skepticism about the reliability of intuition with a recognition of the role of innate ideas in human knowledge.

There are two important respects, both in the first stage, in which my treatment of scientific inference departs from the main line of Peirce's argument. First, my treatment makes essential use of a concept of probability which is entirely distinct from the only one considered by Peirce to be legitimate, namely, the frequency concept (for example, 2.673–85). Second, Peirce usually claims that in the infinitely long run it is certain that scientific inference will asymptotically approach the truth, whereas the claims I put forth are qualified and conditional, and concern the results of investigations in finite intervals of time. However, there are some interesting passages in Peirce's later writings, which will be cited in section IV, in which he appears to depart from his usual position in the direction that I am proposing.

In a broad sense the structure[5] of the theory of scientific inference which I am proposing, and indeed of my entire program of naturalistic epistemology, is Platonic. I try above all to retain the characteristics of Plato's dialectic of beginning *in medias res* with hints and clues rather than with certainties, and of regarding knowledge of fundamental principles as the terminus of inquiry. The interplay of considerations of "the way up" and "the way down," by invoking some of the propositions of natural science at the same time that the validity of scientific procedures is being examined, fits into the dialectic pattern, even if the wording is not typically Platonic. Even Plato's doctrine of recollection is found in modern form in the conjecture of Peirce, which is incorporated into my theory, that the phylogenetic heritage of ideas and dispositions which

5. Structure is emphasized here partly because in sec. II I reject the Platonic conception of probability proposed by Keynes, according to which probabilities depend only upon certain internal relations among propositions and concepts.

result from evolutionary adaptation are of great heuristic value for scientific investigations of the nature of things.

C

I hope that in time the themes mentioned above will be developed in a comprehensive naturalistic epistemology and applied in detailed studies of many special problems. There are, however, great difficulties in trying to carry out this program. One is the obvious need for a critical examination of the relationship between philosophy and the natural sciences. A naturalistic epistemology is supposed to be based upon scientific knowledge, but the term "based upon" is elusive. After all, there are few if any modern philosophers, even among those whom I would classify as non-Copernican (such as Berkeley, Hume, Kant, the phenomenologists, the logical positivists, and the ordinary language analysts) who do not admit the validity, on an appropriate level, of the scientific propositions to which a Copernican appeals. Since philosophical systems have "many degrees of freedom," differing in their treatments of experience, of evidence, of confirmation, of existence, etc., they can accommodate the results of natural science in various ways (cf. McKeon 1951). Nevertheless, it may be possible to go beyond a relativism of systems by carefully applying criteria of philosophical adequacy which are recognized (to be sure, with varying interpretations) from all points of view – for example, coherence, fineness of reasoning, comprehensiveness, openness to evidence, and richness of content. Furthermore, the application of the last three criteria requires serious and detailed attention to the results of natural science. In subtle ways, then, the natural sciences (and, to the extent that these sciences are veridical, the facts of the world) would exercise a guiding and perhaps even a controlling role in philosophical investigations. However, this methodological supposition, which is Platonic in character, can be made convincing only by a detailed examination of the relevance of scientific material to philosophical controversies.[6] This indicates a second difficulty in carrying out the program of a naturalistic epistemology: it requires extensive knowledge of the sciences, particularly of psychology, far exceeding the amount that philosophers have usually deemed necessary for their purposes. (A case in point is the problem of the relation between observational and theoretical terms, which has often been treated with considerable logical ingenuity, but seldom with more than

---

6. My recommendation of the use of the hypothetico-deductive method in philosophical investigations (1965: p. 261) now seems to me to be an oversimplification. An excellent example of the kind of detailed study which is needed is Stein's examination of classical physical geometry (1967).

armchair information about empirical psychology.) A third difficulty is the most formidable but also the most interesting: the currently accepted body of scientific propositions does not provide all the information which a naturalistic epistemologist requires; nor are accepted propositions immune to critical probing by philosophers. For example, a naturalistic epistemology can offer no more than speculative proposals concerning some problems of perception, because the mind–body relationship is a vast desert of ignorance in the sciences of psychology and physiology. As a result, a naturalistic epistemology should not merely be an appendage to natural science, but rather the relation should be one of mutual criticism, stimulation, and illumination. If the demarcation between philosophy and natural science is thereby blurred, this will be regarded as healthy by anyone who appreciates the seventeenth-century sense of "natural philosophy."

In view of these difficulties a program of naturalistic epistemology must be carried out cumulatively and with contributions from many workers. I find it encouraging that in recent years a number of epistemologists have been working, although diversely and with little coordination, along the lines that I envisage.[7] The present essay is intended to be a contribution to a comprehensive program of naturalistic epistemology in two ways: first, as a treatment of one important problem from a naturalistic standpoint, and second, as an illustration of a set of themes, particularly concerning the structure of knowledge, which may be capable of incorporation into a more general theory. I feel strongly that my proposals are steps in the right direction, but I am fully aware that at many points they need to be made more precise, augmented, and corrected.

II. BAYESIAN FORMULATIONS OF SCIENTIFIC INFERENCE

A

The general theses of Copernicanism in epistemology are too broad to determine the character of scientific inference in any detail, but they do impose some demands. Above all, they require that scientific inference not be apodictic, since Copernicanism doubts the existence of human powers (such as those claimed in Aristotle's theory of intuitive induction) for attaining certainty with regard to propositions which are not entailed by the evidence. Also, they require that the kind of "inverse" reasoning which is exhibited in the hypothetico-deductive method be given a central place in scientific methodology, since Copernicanism recognizes that

7. A few with whose work I am acquainted are Fodor (1966), Harris (1965), Hirst (1959), Mandelbaum (1965), Piaget (1950), Popper (1962), Quine (1957), Sellars (1963), and Smart (1963). A near contemporary whose work is too little noticed by later naturalistic epistemologists is Dewey (1929 and elsewhere).

important scientific truths may be remote from direct experience and, therefore, unattainable by methods (like those of Bacon) which test only straightforward empirical generalizations.

A natural way to satisfy these demands is to formulate scientific inference in "Bayesian" terms. This means (a) using as the central concept in all nondeductive reasoning a probability function $P(h|e)$ (read "the probability of the hypothesis h upon evidence e"); (b) ensuring that P satisfies the standard axioms of probability,[8] namely,

(i) $0 \leq p(h|e) \leq 1$,
(ii) $P(h|e) = 1$ if e entails h,
(iii) $P(h \vee h'|e) = P(h|e) + P(h'|e)$, if e entails $\sim(h \& h')$,
(iv) $P(h \& h'|e) = P(h|e) \cdot P(h'|h \& e) = P(h'|e) \cdot P(h|e \& h')$;

and (c) allowing as admissible values for h not only statistical hypotheses, referring to the composition of very large reference classes, but arbitrary hypotheses, including such radically nonstatistical ones as singular propositions and scientific theories. The name "Bayesian" is applied to a method of inference which satisfies (a), (b), and (c) because it permits the use of Bayes's theorem

$$P(h|e \& a) = \frac{P(h|a) \cdot P(e|h \& a)}{P(e|a)}$$

(an immediate consequence of axiom [iv]), in order to evaluate the *posterior probability* $P(h|e \& a)$ in terms of the *prior probability* $P(h|a)$ and the *likelihood* $P(e|h \& a)$. Other treatments of inductive inference, such as the "orthodox" statistical theories of Fisher and of Neyman and Pearson, disallow the application of Bayes's theorem to arbitrary hypotheses, on the grounds that prior probability is in general not well defined.

A Bayesian is able to construe the hypothetico-deductive method as an application of Bayes's theorem in special circumstances. Suppose that there is a class of admissible hypotheses $h_1, h_2, \ldots$, a body of auxiliary information and assumptions a, and a body of evidence e such that for every integer i either e or $\sim e$ is entailed by $h_i \& a$. (This is a great idealization, since in realistic situations an unequivocal relationship of entailment or inconsistency between the hypothesis, together with the auxiliary assumptions, and the evidence is very rare. Also, there is a deep methodological problem, related to the central problem of section III, in demarcating the class of admissible hypotheses; but this will be set aside in the present formal analysis of hypothetico-deductive inference.) Suppose further that $h_1 \& a$ entails e, and that if h_j is any other hypothesis which

---

8. For a powerful mathematical development of probability theory it is necessary to replace axiom (iii) by the strong axiom of complete additivity: $\sum P(h_i|e) = P(H|e)$, where e entails $\sim(h_i \& h_j)$, for $i \neq j$, and H is the infinite disjunction of the h_i.

shares this property with h_1, then $P(h_1|a) \gg P(h_j|a)$. It follows from Bayes's theorem that the posterior probability $P(h_1|a \& e)$ is much greater than the posterior probability of any other member of the admissible class, and this conclusion can be presented by a Bayesian as the essential content of the hypothetico-deductive confirmation of h_1. (A Bayesian may wish to construe the hypothetico-deductive confirmation of h_1 more strongly, for example, as asserting that $P(h_1|a \& e) > 1 - \epsilon$, where ϵ is some preassigned small positive number; but to do this the suppositions regarding the prior probabilities must be strengthened.) That the usual formulations of the hypothetico-deductive method seldom compare the prior probabilities of those hypotheses which share with h_1 the property of "saving the appearances" can be regarded by a Bayesian as merely a deficiency in explicitness of analysis. Furthermore, a Bayesian can correctly claim that in many cases the hypothetico-deductive method cannot be used without drastically idealizing and distorting the conditions of investigation, whereas other inferences using Bayes's theorem are valid (cf. Salmon 1966: pp. 250–51).

There exist non-Bayesian treatments of scientific inference which attempt to satisfy the demands imposed by a Copernican point of view, the most important perhaps being that of Popper, who construes the hypothetico-deductive method as a procedure of elimination or refutation. In this paper, however, I shall not try to evaluate such treatments,[9] since my purpose is the constructive one of exhibiting a specific Bayesian formulation of scientific inference which is both thoroughly Copernican in spirit and defensible against the usual objections to the application of Bayes's theorem. It is generally granted, even by opponents, that the Bayesian approach is systematic, treating a great variety of special modes of inference, from the hypothetico-deductive method to statistical decision procedures, as cases of a few general principles (in contrast to orthodox statistical theory, which proceeds in a much more piecemeal fashion).[10] But the opponents maintain that this systematization is specious, since it is based upon the concept of prior probability, which they claim is either empty or arbitrary or otherwise useless for scientific inference, depending upon its interpretation. Their criticism is partially valid, I believe, when it is directed against the two conceptions of probability which most Bayesians have espoused, namely, the "logical concept" and the "personalist concept." Again, since my purpose is constructive rather than critical, I shall only briefly discuss these two concepts, in order to

9. A judicious evaluation of Popper's position from a Bayesian point of view is made by Salmon (1966: pp. 252–55).
10. See, for example, Hacking's remarks on personalism, which is a version of Bayesian probability theory (1965: p. 208).

provide the motivation for a version of Bayesianism which is based upon the "tempered personalist" concept of probability. It will be seen that this is essentially the concept of probability which Harold Jeffreys develops in his treatise, *Theory of Probability,* his verbal allegiance to the logical concept notwithstanding.

B

The leading exponents of the logical concept of probability are Keynes and Carnap, but the concepts which they present are different because of the divergence of their views of logic.

For Keynes, logic is concerned with relations between propositions, which are characterized as "the objects of knowledge and belief" (1921: p. 12) and evidently are conceived by him to be nonlinguistic entities. Probability theory is part of logic because "probability-relations" are a kind of relation between propositions: "if a knowledge of h justifies a rational belief in a of degree α, we say that there is a *probability-relation* of degree α between a and h" (*ibid.:* p. 4). (The "degrees" are not assumed to be ordered as the real numbers or even to constitute a simply ordered set [p. 34].) These relations are asserted to be "real" and "objective" (p. 5), independent of what human beings happen to believe (p. 4), and incapable of analysis in terms of simpler ideas (p. 8). Human beings are attributed the power to have direct knowledge of "secondary propositions" that probability relations hold between various pairs of propositions (p. 16), but the extent and accuracy of this power varies from person to person.[11] "The perception of some relations of probability may be outside the powers of some or all of us" (p. 18). Keynes thus conceives of probability relations in much the same way as a Platonist in mathematics, such as Gödel, conceives of the propositions of set theory – as true in virtue of internal relations among sets, which are real entities independent of human thought, language, or axiom systems.

It is just this comparison between mathematics and the logical theory of probability which exhibits the weakness of Keynes's position. For a Platonist in mathematics can point to a rich set of principles which have very strong intuitive appeal and which are sufficient for the "construction" (in the set theoretical sense rather than in the sense of recursive function theory) of all the structures ordinarily investigated by mathematicians. Although mathematical intuition becomes "astigmatic" (Gödel 1946: pp. 150–52) concerning esoteric matters, such as the continuum hypothesis, the edifice of set theory is sufficiently impressive to provide good

---

11. Passmore (1957: p. 348) perceptively notes Keynes's debt to Moore, who interprets ethical concepts in this manner.

cases both for the ontological independence of mathematical entities and for the excellence of human intuition concerning them. By contrast, the intuitive knowledge which Keynes cites about probability is slight, the only systematic components being: (a) a set of axioms regarding comparisons of degrees of probability (later strengthened by Koopman [1940]) and (b) a weak form of the principle of indifference, which in spite of the caution of its statement is nevertheless too strong to be defensible as it stands (cf. Russell 1948: p. 375). With trivial exceptions (for example, when e entails h or $\sim h$), the axioms only permit the derivation of probability propositions from other probability propositions, and these may be compared with those elementary principles of set theory which assert that the results of Boolean operations upon sets are sets – a very small part of set theory! Furthermore, the axioms of probability can be justified without any assumption about objective probability relations (cf. sections II.C and III.D). An extremely weak version of the principle of indifference, asserting only that $P(h_i \mid e) = P(h_j \mid e)$ if h_i and h_j are symmetrical in all respects relative to e, is indeed intuitively irresistible, but this version is of no use in any actual scientific inference without supplementary judgments to the effect that the symmetry-breaking features of the alternatives (which are always present in actual cases) are probabilistically irrelevant. Even though I am sympathetic with Platonism in mathematics, I find it hard to withstand Ramsey's argument

that there really do not seem to be any such things as the probability relations he describes. He supposes that, at any rate in certain cases, they can be perceived; but speaking for myself I feel confident that this is not true. I do not perceive them, and if I am to be persuaded that they exist it must be by argument; moreover I shrewdly suspect that others do not perceive them either, because they are able to come to so very little agreement as to which of them relates any two given propositions. All we appear to know about them are certain general propositions, the laws of addition and multiplication; . . . and I find it hard to imagine how so large a body of general knowledge can be combined with so slender a stock of particular facts. It is true that about some particular cases there is agreement, but these somehow paradoxically are always immensely complicated; we all agree that the probability of a coin coming down heads is 1/2, but we can none of us say exactly what is the evidence which forms the other term for the probability relation about which we are then judging. If, on the other hand, we take the simplest possible pairs of propositions such as 'This is red' and 'That is blue' or 'This is red' and 'That is red', whose logical relations should surely be the easiest to see, no one, I think, pretends to be sure what is the probability relation which connects them. (1931: pp. 161–62)[12]

For Carnap a sentence is logically true if its truth is a consequence of the syntactical and semantical rules of the language in which it occurs.

12. Even the very limited Platonism in probability theory which I once accepted (1955: pp. 11, 27–28) no longer seems tenable to me.

Statements of "probability" in the sense of "degree of confirmation" (as contrasted with the frequency sense) are metalinguistic statements of the form "$c(h, e) = r$," where h and e are sentences in some object language and r is a real number; and the true sentences of this form are consequences of metalinguistic rules governing the function c. (In an unpublished work [1968] Carnap continues to refer to some language or class of languages in discussing inductive logic, but his probability statements are now of the form "$C(H | E) = r$," where "H" and "E" denote propositions which are nonlinguistic entities.) In order to guide the selection of a c-function, Carnap lays down a number of conditions of adequacy (1963: pp. 974–76) (which are converted into axioms for the function C [1968]). These have been criticized both on the ground that some of them have too much content to be imposed a priori (for example, Nagel 1963: pp. 797–99 and Salmon 1961: p. 249) and on the ground that they are so weak that they admit too large a range of functions (for example, Lenz 1956). Carnap has also been criticized for saying that a c-function is an instrument (specifically, "an instrument for the task of constructing a picture of the world on the basis of observational data and especially of forming expectations of future events as a guidance for practical conduct" [1963: p. 55]) and, therefore, that the choice of a c-function may properly be evaluated by its performance, "that is, the values it supplies and their relation to later estimates" (*ibid.*); according to critics, it is only future performance which is relevant, and any assertion about that must be based upon induction, the rules of which are in question when deliberating upon the choice of c (for example, Lenz 1956). On these disputed points, however, I believe that the general direction of Carnap's thought is more correct and more suggestive than his critics have allowed. There may well be a place in the theory of induction for principles which are not justifiable a priori, but which are entrenched deeply though without an unconditional commitment (cf. section V.D). It may very well be that any reasonable set of general rules will fall short of determining numerically the probabilities of most hypotheses of interest upon most bodies of evidence, and yet the need to rely upon subjective decisions to supplement such rules may preclude neither eventual consensus among reasonable men nor sensitivity to the truth (cf. section III). And the need to appeal to induction in order to evaluate the performance of an inductive method may be both an inescapable consequence of the human situation and, with proper qualifications, an opportunity to use a legitimate and nonvicious kind of circular reasoning (cf. section V.E). (Carnap [1968: sec. 4] makes an analogy to geometry and distinguishes "pure" from "applied" inductive logic, placing principles which are justifiably a priori in the first class and those which are not in the second. The accuracy of this analogy is questionable, since inductive logic has a kind of normative character which is alien to geometry. Nevertheless, the general suggestion of dichotomizing inductive principles may be fruitful.)

In my opinion, the crucial error in Carnap's published works on induction is his taking the choice of a c-function, which is tantamount to one sweeping decision regarding the evaluation of all probabilities, as a desideratum. There are two major objections to this procedure: (1) the rules governing a c-function are purely syntactical and semantical and take no account of whether a given hypothesis h has been considered worthy of serious consideration by any one. But treating on the same footing those hypotheses which have been seriously proposed and those which have not has the effect of drastically decreasing the value set upon intelligent guessing and insight into the subject matter of investigation. In other words, undiscriminating impartiality toward hypotheses which no one has seriously suggested is a veiled kind of skepticism, since it would leave very little prior probability to be assigned to any specific proposal. Putnam presents in detail an objection along these lines in his contribution to the volume on Carnap in *The Library of Living Philosophers* (1963: pp. 770–74). Carnap's reply is remarkably receptive: "it is worthwhile to consider the possibility of preserving an interesting suggestion which Putnam offers, namely to make inductive results dependent not only on the evidence but on the class of actually proposed laws" (1963: p. 916). This reply can be made consistent with a purely logical definition of a c-function by increasing the number of arguments of c from two to three, that is, by defining a function $c(h, e, l)$, which is the degree of confirmation of the hypothesis h on the basis of evidence e, if the set of actually proposed laws is l. If this modification were worked out in detail, it would provide an answer to objection (1). I can also see an attractive alternative modification, which has the advantage of not increasing the number of arguments of c: the serious proposal of a hypothesis h by someone could be recognized as a relevant piece of evidence, expressed in a sufficiently rich language by the sentence s; and if e expresses all other evidence, there is no reason to suppose that $c(h, e \cdot s)$ is even approximately equal to $c(h, e)$. It would be difficult to work out this answer in detail, however, for according to the multiplication axiom (axiom iv),

$$c(h, e \cdot s) = \frac{c(h \cdot s, e)}{c(s, e)},$$

and the technical work so far presented by Carnap gives no help in evaluating the ratio on the right-hand side of this equation.

(2) Even if there is a "best" c-function, which is supposed somehow to be chosen, it would by no means be clear how to use it in practice. According to Carnap's "requirement of total evidence," a person applying

inductive logic in a given situation "must use as evidence *e* his total observation knowledge" (1963: p. 972). But how can this total knowledge be marshaled and expressed in a sentence suitable for substitution in the function $c(h, e)$, when by far the larger part of it is only dimly remembered or has been transformed into beliefs, habits, and adjustments? In order to conform to the requirement of total evidence, an inductive method is needed which effectively deploys the sharp part of our observational knowledge, such as might be recorded in an experimenter's notebook, but also somehow takes account of the massive background of experience.

The difficulties associated with a single sweeping decision regarding a *c*-function are largely avoided in Carnap's unpublished work. There he suggests that each separate kind of scientific investigation be taken as a separate domain of "applied inductive logic" (1968: secs. 4 and 5). He does not discuss the extent or the autonomy of the various domains. Nevertheless, his suggestion is at least partially in agreement with the principle of locality, recommended in section III below.

C

The personalist concept of probability was proposed independently by Ramsey and DeFinetti and has been elaborated recently by a number of Bayesian statisticians, notably Savage, Good, Raiffa, and Schlaifer. They find it convenient to measure a person's credence (that is, the inten- of his belief) in *h*, conditional upon the truth of *e*, by finding the largest fraction of a unit of utility which he is willing to stake in order to receive one unit if *h* is true, the bet being cancelled if *e* turns out to be false. Personal probabilities are the intensities, thus measured, of a person's subjective beliefs, provided that these are examined so as to eliminate lapses of *coherence* (a reasonable extension of consistency, which will be explained in the next paragraph). Sometimes "subjective probability" is used as synonymous with "personal probability," but it seems preferable to distinguish these terms, applying the former to actual beliefs, which may violate the condition of coherence (Edwards, Lindman, and Savage 1963: p. 197). The personal probability of *h*, conditional on *e*, for *X* is denoted by "$P_X(h|e)$", where "*X*" refers to a person at a definite time (the temporal qualification being required because of fluctuations in a man's beliefs); the subscript *X* is often omitted if it is clear from context whose beliefs are being considered. It should be noted that *e* is not assumed to be the total body of evidence as in the similar notation for the logical concept of probability.

The only normative constraint upon *X*'s personal probabilities, according to the personalists, is the condition of coherence: that no set of bets can be proposed which are acceptable to X according to his probability evaluations, but which cause him to suffer a net loss regardless of the truth values of the hypotheses bet upon (that is, no "book" can be made against X).[13] This is a nontrivial constraint, however, since, as DeFinetti has demonstrated, it implies that X's personal probabilities satisfy the standard axioms of probability (axioms [i] through [iv] in section II.A). And from these axioms follows Bayes's theorem, which asserts that in a certain sense X is constrained to learn from experience. For if the ratio $P_X(e\,|\,a\,\&\,h)/P_X(e\,|\,a)$ is either much greater or much less than 1, then the posterior probability differs sharply from the prior probability; and even though this ratio is a matter of subjective belief, once e is specified, the evidence e actually obtained is not under X's control, and to this extent the posterior probability is shaped by experience. The personalists claim that in this way the system of personal probabilities is adequate for guiding our practical expectations and for checking the theoretical hypotheses of natural science. They say, furthermore, that all attempts to evade the subjective basis of induction are deceptive, and they cite in particular the subjective character of choices of significance levels and of confidence intervals which are made by statisticians committed to an "objectivist" concept of probability (that is, probability in the sense of relative frequency).[14] The critics of personalism are parodied as saying: "We see that it is not secure to build on sand. Take away the sand, we shall build on the void" (paraphrase of DeFinetti in Edwards, Lindman, and Savage 1963: p. 208).

It is difficult to see, however, that the personalists have adequately explicated the concept of probability which is involved in scientific inference. A man's beliefs are intimately bound up with his emotional attachments, his subconscious associations, and the ideals of his culture, and it is likely that the belief systems of almost everyone in most cultures, and of the majority of men even in "advanced" cultures, are not conducive to performing good inductions on theoretical matters beyond the concerns of ordinary life. I do not find an adequate discussion of these complicating features of belief systems by members of the personalist school of probability. The only kind of irrationality which they explicitly discuss is incoherence, but there are surely other properties of belief systems which are characterizable as irrational and which would inhibit the progress of natural science if they were universal. In particular, the assignment of extremely small prior probabilities to unfamiliar hypotheses is compatible

13. This is DeFinetti's definition of "coherence." I have proposed a slightly stronger definition (1955: p. 9).
14. This argument is elaborated by Polanyi (1958: chap. 2), whose personalist epistemology is in the spirit of the personalist probability theorists but is of much wider scope.

with coherence, but may be irrational in the sense of being the result of prejudice or of obeisance to authority or of narrowness in envisaging possibilities. But unless a hypothesis is assigned a nonnegligible prior probability, Bayes's theorem does not permit it to have a large posterior probability even upon the grounds of a very large body of favorable observational data, and thus the theorem does not function as a good instrument for learning from experience. Perhaps the personalists make an implicit assumption that those who employ their methods are reasonable in some broad sense; after all, anyone who is willing to use the machinery of probability theory in order to adjust his opinions is unlikely to be a hardened fanatic. It is interesting that one occasionally finds personalists stating obiter dicta which seem to qualify their official attitude toward subjective belief. Thus

a prior distribution which has a region of zero probability is therefore undesirable unless you really consider it impossible that the true parameter might fall in that region. Moral: Keep the mind open, or at least ajar. (Edwards, Lindman, and Savage 1963: p. 211)

And

more generally, two people with widely divergent prior opinions but reasonably open minds will be forced into arbitrarily close agreement about future observations by a sufficient amount of data. (*Ibid.:* p. 201)

Implicit in these passages is a normative principle supplementing coherence, which ought to be made explicit if inductive reasoning is to be understood.

D

Despite the differences between Carnap's logical concept of probability and the personalist concept, both seem to be attempts to explicate the same informal concept (in Carnap's terminology, the same explicandum), namely, "rational credibility function" (1963: p. 971). Furthermore, Carnap, like the personalists, imagines betting situations in order to treat belief quantitatively: "$P(h, e)$ is a fair betting quotient for X with respect to a bet on h" (*ibid.:* p. 967).[15] The essential difference between their conceptions is that the personalists admit no criterion for rationality or fairness other

15. In Carnap's notation "P" represents the explicandum of which a c-function is an explicatum. It should be noted that Carnap regards the explanation of $P(h, e)$ in terms of a betting quotient as accurate only when the betting stakes are small compared with X's total fortune, and he gives another explanation, in terms similar to but not exactly the same as betting (namely, estimating the subjective utility of a benefit bestowed in case h is true), which he regards as more accurate (1963: p. 967).

than coherence, whereas Carnap formulates a series of additional criteria and hopes to find yet more.

There are reasons for doubting that the concept of probability as it occurs in scientific inference can be properly explained in terms of betting quotients, or even that it can be equated with the concept of a rational credibility function. The source of the difficulties is the indispensability of general theories in natural science as it is now studied and, indeed, as it conceivably could be studied if the ideal of general insight into the nature of things is not to be abandoned. Although probability is a valuable instrument in reasoning about scientific theories, it does not seem to make sense to explain the probability of a theory as a fair betting quotient. As Putnam argues, if the bet concerns the truth or falsity of a scientific theory, then how should the outcome of the bet be decided? Even if we can agree to call the theory false in certain definite cases, under what circumstances would we decide that the theory is certainly true and that the stakes should be paid to the affirmative bettor (1963a: p. 3)? There is no doubt that bets with definite payoff conditions can be made *concerning* a theory, for instance, on whether strongly unfavorable evidence will be found by the year 2000 or on whether the theory will be accepted by working scientists and enshrined in textbooks by that date, but such bets are evidently not about the *truth* or *falsity* of the theory. Moreover, even if one does not explain "credence" in terms of "acceptable betting quotient" and, therefore, "rational credibility" in terms of "fair betting quotient," the identification of the probability of a theory with its rational credibility is problematic, at least in the case of theories which are both general and precise and, therefore, very strong in their assertions (as in fundamental physics).[16] For it can be argued that upon any body of supporting evidence the rational degree of belief that such a theory is literally true, without qualification in conditions far different from those in which experiments have been conducted (for example, at much higher energies, or in much smaller space–time regions, or at much higher or lower temperatures than in the experiments) and without exception throughout all space and time, is zero or extremely small. Most scientists who are aware of the history of scientific revolutions can be expected to have very small credence in the literal and unqualified truth of theories of great strength. But if probability evaluations are to be guides to the acceptance (even tentatively) of general theories, the smallness of the probabilities of theories would be troublesome. Although this argument does not appear to me to be as decisive as the one against reasoning about theories in terms of bets, since conditions of acceptance might be

16. I am indebted to Prof. Joseph Agassi for a discussion on this point.

formulated in terms of ratios of posterior probabilities (provided these are not all zero) rather than in terms of absolute values, it nevertheless provides a motivation for reexamining the probability of theories and for seeking an explicandum different from rational credibility.

Some suggestions for an alternative explicandum can be gathered from the history of science. It has often happened that a scientific theory which was once successful, in the sense of being the only one among the commonly known rival theories to survive severe tests by independent experimenters, later had to be displaced. However, one of the reasons for believing that the scientific enterprise is progressive, despite the limitations of human powers which Copernicans emphasize, is that the displacement of once successful theories is seldom ignominious. (Evidently, this claim is cogent only if "successful" is used in the sense specified above, which drastically limits the reference class of theories, and not in the sense of "generally accepted.") Usually something, at least in the form of suggestions, is salvaged from an old theory. A select few (such as classical mechanics and Euclidean physical geometry) were displaced under such honorable circumstances that they may be regarded as promoted to the rank of "theory emeritus": they not only yielded the same observational predictions as the new theories over a wide domain, but the new theories were in a certain sense conceptual generalizations of them. This suggests that a person whose belief in the literal truth of a general proposition h, given evidence e, is extremely small may nevertheless have a nonnegligible credence that h is related to the truth in the following way: (i) within the domain of current experimentation h yields almost the same observational predictions as the true theory; (ii) the concepts of the true theory are generalizations or more complete realizations of those of h; (iii) among the currently formulated theories competing with h, there is none which better satisfies conditions (i) and (ii).[17] If a word is needed for this modality of belief, I suggest "commitment." If "rational degree of commitment" is identified as the explicandum of "probability" in the context of scientific inference, there will be no prima facie obstacle to finding that even very strong theories upon appropriate evidence have probabilities close to 1. Furthermore, the use of "probability" in this way would permit a formulation of scientific inference which fits the contours of the actual thinking of investigators, by focussing attention upon the progress of knowledge rather than upon the ultimate truth of theories. In the case of singular propositions and other special hypotheses, the compunctions stated above about a high credence are not applicable, and, therefore,

17. When conditions (i) and (ii) hold, the true theory is related to h by a kind of "correspondence principle."

credence and degree of commitment should be the same or very close. In these cases, then, "rational degree of commitment" virtually coincides with Carnap's explicandum of "probability," namely, "rational credibility function."

There is a difficulty concerning the concept of rational degree of commitment which arises from the occurrence of several vague expressions ("almost the same observational predictions" and "more complete realizations") in the first two of conditions (i) to (iii). This vagueness is troublesome in treating the "catchall" hypothesis h_n, which is the negation of the disjunction of all the specific hypotheses $h_1, \ldots, h_{n-1}$ considered in a scientific investigation. The ratio of the probabilities, upon given evidence, of h_i and h_j, where neither i nor j equals n, does not seem to depend upon the exact meanings of these expressions. On the other hand, the ratio of the probability of h_i ($i \neq n$) to that of h_n is sensitive to their exact meanings: the more stringently "almost the same observational predictions" and "more complete realizations" are construed, the stronger is the proposition that h_i will be displaced only under conditions (i) to (iii), and hence the smaller is the ratio of its probability to that of h_n. I can see three possible ways of dealing with this difficulty.

(a) Conditions (i) and (ii) might be reformulated so as to eliminate vagueness. In the case of condition (i) this reformulation would be relatively straightforward, consisting in a quantitative specification of "almost the same." In the case of condition (ii) I do not know a straightforward procedure of reformulation, since "more complete realization" does not lend itself in an obvious way to quantitative treatment. Furthermore, there is not merely a question of quantitative specification, since the character of the relationship between a moderately successful old theory and a more successful revolutionary theory cannot be known in advance of the scientific revolution.

(b) It might be possible to formulate scientific inference in a way that avoids treating the catchall hypothesis on the same footing with the specific hypotheses $h_1, \ldots, h_{n-1}$. Numerical weights might be attached to the specific hypotheses but not to h_n, and the assignment of these weights might be governed by principles similar to the standard probability axioms. However, there would necessarily be some departure from Bayesianism, for h_n is equivalent to $\sim(h_1 \vee \cdots \vee h_{n-1})$, and hence the disallowance of a numerical weight to h_n implies that the domain of the weight function is not closed under Boolean operations. Because of the arguments already given in section II.A and also because of the justification of the axioms of probability to be presented in section III.C, I am very reluctant to abandon a Bayesian formulation of scientific inference. Nevertheless, solution (b) would be attractive if two conditions could be fulfilled: the development of a reasonable set of principles, comparable to the standard axioms of probability, governing the assignment of numerical weights; and the provision of a reasonably open-minded treatment of the possibility that none of the specific hypotheses $h_1, \ldots, h_{n-1}$ is a good approximation to the truth. I conjecture that all the methodological proposals of the remainder of this paper (the "localization" of problems, the role assigned to subjective judgment, the preference accorded to seriously proposed hypotheses, the rejection of a priori ordering of hypotheses according to simplicity, and the derivation of methodological guidelines from the tentative body of scientific knowledge) would be compatible with a partial modification of Bayesianism.

(c) The procedure for evaluating the rational degree of commitment to a hypothesis h relative to evidence e may be such that the vagueness of the concept of commitment is methodologically innocuous. One thesis of section III is that a major component of the *rationality* of a rational degree of commitment is conduciveness to the progress of knowledge, and this criterion in turn leads to a prescription of open-mindedness toward seriously proposed hypotheses (that is, giving each of them a chance to be accepted into one's tentative body of knowledge if supported by sufficiently favorable evidence). But the application of this prescription in numerically weighing hypotheses is compatible with the vagueness of the concept of commitment. It will be argued in section III.C that persons who disagree sharply regarding prior probability evaluations and yet conform to the prescription of open-mindedness (which is there called "the tempering condition") are able to achieve rough consensus regarding posterior probabilities relative to a moderate amount of experimental data; and this argument can be adapted to show that rough consensus regarding posterior probabilities can be achieved even if quite different senses of "commitment" are used. Consequently, despite arbitrariness in the ratios of the prior probability of h_i ($i \neq n$) to that of h_n, it is possible to make numerical probability evaluations fruitfully without a prior clarification of the concept of commitment. In fact, the converse can be expected to occur: as knowledge progresses, and in particular as we learn how crude but partially successful theories are replaced by more refined theories, we may be able to clarify conditions (i) and (ii) in a nonarbitrary manner. I shall tentatively accept solution (c) in this paper. It fits well into a theory of scientific inference in which an a priori logical structure is supported and enriched by a posteriori considerations. And it illustrates one of Plato's central ideas: that the clarification of concepts is inseparable from the progress of knowledge. [Note added, July 1969: The acceptance of solution (c) now appears to me to be based upon wishful thinking. Consequently, I now prefer solution (b) but have not worked it out in detail.]

III. THE TEMPERED PERSONALIST CONCEPT
OF PROBABILITY

A

The discussion of section II has indicated that a rationalization of scientific inference should be sought in terms of a concept of probability which is endowed with a stronger normative character than that of the personalist concept, which derives this additional normative character from conditions for the progress of knowledge (rather than from internal relations among propositions as in the theory of Keynes), and which possesses greater openness to the contingencies of inquiry than Carnap's c-functions. A concept satisfying these requirements is to be found, a little beneath the surface, in the work of Harold Jeffreys. Although Jeffreys is generally recognized as one of the masters of Bayesian statistics (Edwards, Lindman, and Savage 1963: p. 194; Hacking 1965: pp. 201–02; Carnap 1950; pp. 245–46), he has not, in my opinion, been completely understood. The misunderstanding is largely due to the discrepancy between his philosophical statements on the foundations of probability theory and his treatment of probability evaluations in concrete contexts – the latter being, again in my opinion, by far the more profound. His characterization of probability resembles that of Keynes in having a Platonic ring, for he seems to regard the probability of h on e as objectively fixed by internal relations between h and e. Thus, he states,

> probability theory . . . is in fact the system of thought of an ideal man that entered the world knowing nothing, and always worked out his inferences completely, just as pure mathematics is part of the system of thought of an ideal man who always gets his arithmetic right. (1961: p. 38)

Also, "differences between individual assessments . . . can be admitted without reducing the importance of a unique standard of reference" (*ibid.:* p. 37). But it is by no means clear that these statements should be taken literally, especially in view of his recommendation that difficult problems of evaluating prior probabilities be settled by the decision of an "International Research Council" (*ibid.:* p. 37). After all, nothing concerned Plato more deeply than to refute the contention of the Sophists that norms are matters of convention! An implicit conception of probability as an instrument for extending knowledge, rather than logical considerations, underlies Jeffreys's actual proposals for evaluating prior probabilities. Thus, he says that

> philosophers often argue that induction has so often failed in the past that Laplace's estimate of the probability of a general law is too high, whereas the main

point of the present work is that scientific progress demands that it is far too low. (*ibid.:* p. 132)

His fundamental rule for assigning prior probabilities to scientific hypotheses (a rule which he misleadingly calls "the simplicity postulate") is the following:

Any clearly stated law has a finite prior probability, and therefore an appreciable posterior probability until there is definite evidence against it. (*Ibid.:* p. 129)

This rule makes no pretense at providing an ideal prior ordering and weighing of all possible laws as did an earlier proposal of Wrinch and Jeffreys (1921) (for which the name "simplicity postulate" was appropriate), but is rather a methodological prescription of open-mindedness. Because of the vagueness of this prescription, there is much arbitrariness, which he tends to gloss over, in his actual evaluation of probabilities (for example, his assignment of probability $1/2$ to the null hypothesis [1961: sec. 5.0]). However, he correctly points out that within wide limits the exact values of prior probabilities are often unimportant in comparing posterior probabilities, because the ratio of likelihoods becomes the dominant factor (*ibid.:* p. 194). My feeling at this point is that the basic insight of Jeffreys, which is the instrumental character of probability, would be better presented if all claims of approximating an objectively fixed value of the prior probability were dropped in favor of the acknowledgment that his prescription of open-mindedness leaves great latitude to subjective judgment.

A methodological principle of great importance, which permits probability to be a manageable instrument in scientific inference, seems to underlie tacitly much of Jeffreys's treatment of specific problems (especially in 1961: chap. VI, "Significance Tests: Various Complications"): it is that *the individual investigation delimits an area in which probabilities are calculated.* This principle in no way denies that investigations overlap and influence one another in intricate ways, but asserts only that the conditions of a single investigation establish a kind of "local" universe of discourse within which calculations strictly governed by the axioms of probability can be performed. The conditions of the investigation are (1) a set of hypotheses $h_1, \ldots, h_n$ (of which the last may be a "catchall" hypothesis equivalent to $\sim[h_1 \vee \cdots \vee h_{n-1}]$) which have been "suggested as worth investigating" (*ibid.:* pp. 268, 270); (2) a set of possible outcomes $e_1, \ldots, e_m$ of envisaged observations; and (3) the information i initially available. This information in actual circumstances is very heterogeneous, consisting partly of vague experience concerning the matter of interest, partly of experience which may be sharp but is of dubious relevance, partly of sharp evidence which does not seem to bear directly on the question at hand but is relevant to other questions in the same field, and partly of propositions which are regarded as established even though they go beyond the actual evidence and, therefore, have been accepted because of previous investigations.[18] According to the principle of letting the investigation delimit an area in which probabilities are calculated, $P(q|r)$ need not be taken as well defined in the present context unless the information i is included in (that is, entailed by) r; and, therefore, in particular, the prior probability of h can be taken to be $P(h|i)$. Without such a restriction on the domain of definition of P, one could formally apply Bayes's theorem to any decomposition of i into the conjunction of i_1 and i_2 to obtain $P(h|i) = P(h|i_2) \cdot P(i_1|h \& i_2)/P(i_1|i_2)$, thereby effectively assigning to probabilities upon i_2 the role of prior probabilities. The principle need not be taken as a *prohibition* against taking $P(q|i_2)$ to be well defined, and there may be instances of i and of decompositions of i into i_1 and i_2 in which it would be convenient to do so. But the *option* of relativizing prior probabilities to i should be preserved in order to permit the apparatus of probability theory to be applied *in medias res* (where actual investigations are always located), and also to avoid being driven to the ultimate decomposition in which i_1 is taken to be i itself and i_2 is taken to be the empty or tautological proposition t, so that the application of Bayes's theorem expresses $P(h|i)$ in terms of the *"tabula rasa"*[19] probabilities $P(h|t)$ and $P(i|t)$.

It should be noted that the methodological principle of letting the individual investigation delimit an area in which probabilities are calculated is nowhere stated explicitly by Jeffreys, but it is a natural complement to such dicta as the following:

The best way of testing differences from a systematic rule is always to arrange our work so as to ask and answer one question at a time. Thus William of Ockham's rule, 'Entities are not to be multiplied without necessity' achieves for scientific purposes a precise and practically applicable form: *Variation is random until the contrary is shown; and new parameters in laws, when they are suggested, must be tested one at a time unless there is specific reason to the contrary.* (1961: p. 342)

It should also be noted that Jeffreys often apologizes when he follows this principle: for example, "there is a limit to the amount of calculation

18. It is often convenient to exclude from i all sharp data bearing on the hypotheses of interest and to include these data as part of the evidence used to compute the posterior probabilities. Idealizing the structure of the investigation in this way permits a more uniform application of Jeffreys's prescription of open-mindedness: the prior probabilities of all the seriously proposed hypotheses h_i can then be evaluated without considering sharp data, whereas the latter are taken into account in likelihood calculations, on which there is generally much more intersubjective agreement than on prior probabilities.

19. This expression was used by L. Savage in a discussion.

that can be undertaken at all – another imperfection of the human mind" (*ibid.:* p. 366). In other words, he seems to treat the principle merely as an expedient; this fact is not surprising, since the principle entails a "local" view of probability that conflicts with his espousal (however superficial) of a logical concept of probability. However, if I am correct that an instrumental approach to probability is the profounder part of Jeffreys's work, then the principle of letting the individual investigation delimit an area for probability calculations can be regarded as partially constitutive of his working concept of probability.

B

The tempered personalist concept of probability, which I propose as a suitable explication of "rational degree of commitment," is very close to Jeffreys's working sense of "probability," except for the role explicitly assigned to subjective judgment. Alternatively, it can be regarded, like the personalist concept of probability, as an idealization of subjective probability, except that it is governed not only by the axioms of probability but by a prescription of open-mindedness. A definition of the proposed term will now be given, and the remainder of section III will be devoted to comments and clarifications. The definition assumes that "the investigation" is schematized as in section III.A.

> Suppose that at the beginning of an investigation a person X has a body i of information and assumptions, and suppose that S is a set of propositions closed under truth-functional operations and containing all seriously proposed hypotheses of which X is aware on the matter under investigation and all propositions regarding the possible outcomes of an envisaged set of observations. *Then a tempered personalist probability function over S for X with the body i of information and assumptions* is a map assigning to every ordered pair (q, r) of members of S (such that r is consistent with i) a real number which is approximately equal to X's subjective degree of commitment to q upon supposition that r is his total evidence supplementary to i, subject to two conditions: (1) the map satisfies the standard axioms of probability, and (2) the prior probability (that is, the number assigned when r is tautological) of each seriously proposed hypothesis must be sufficiently high to allow the possibility that it will be preferred to all rival, seriously proposed hypotheses as a result of the envisaged observations.

The crucial features of this concept of probability are its locality, its latitude toward subjective judgment, and its prescription of open-mindedness, the last being expressed by condition (2), which I call "the tempering condition."[20] I do not preclude that additional conditions may reasonably be imposed on methodological grounds in order to make the concept of probability a more sensitive instrument. One such possibility is the invariance rule of Jeffreys, discussed in section III.E. At present, however, I do not see that the incorporation of this rule would be an improvement.

It will be convenient to use a notation fitted to a somewhat idealized decomposition of i into a part a, which is both sharply formulable and prima facie relevant to the matter under investigation, and a residue b, which is either vague or of no apparent relevance and, therefore, classifiable as background information. If this decomposition is specified, then the tempered personalist probability assigned under the conditions given in the definition will be denoted by "$P_{X,b}(q \mid r \& a)$"; if it is clear from the context whose probability function is being considered and at what time (thereby implicitly fixing b as well as possible, since an explicit inventory of X's experience is out of the question), the abbreviated notation "$P(q \mid r \& a)$" can be used. There are two advantages in a notation which displays a explicitly: first, the logical relations between i and the propositions of S are essential to calculations which employ the axioms of probability, but only the part a of the total body i of information and assumptions has evident logical relations to the members of S. For example, assumptions about the design of instruments and about the physical laws governing their operation, in conjunction with a hypothesis about the matter at hand, often entail definite conclusions about what appearances may *normally* be anticipated (if there is no malfunctioning or mishandling of the equipment, no unexpected external disturbance, etc.), whereas most of the experience which is crystallized in a man's skill at using equipment is inaccessible for such sharp deductions, essential though it may be for his capacity to function as an experimentalist. Second, in the comparison of personal probability evaluations implicit in public discussions of the status of scientific hypotheses (which, a Bayesian must acknowledge, are often conducted in terms that obscure the probabilistic structure of scientific inference), it is important that all parties disclose to the others their information and assumptions, in order to determine whether discrepancies in probability evaluations may not be attributable at least in part to differences regarding these. But since it is hopeless to compare total bodies of experience, the process of diagnosing disagreement is expedited if the sharp and prima facie relevant part of each person's information and assumptions is set forth explicitly. In spite of these advantages

20. Among the recognized senses of the verb "to temper" which suggest my technical usage are the following: "to restrain within due limits, or within the bounds of moderation," "to bring (steel) to a suitable degree of hardness and elasticity or resiliency . . . ," "to bring into harmony, attune" (*The Oxford Universal Dictionary,* 3rd edition).

the decomposition of i into a and b is admittedly a psychological and methodological oversimplification. Occasionally, in fact, a crucial step in an inquiry (particularly concerning singular hypotheses, which are, however, of less importance in the natural sciences than in law and history and in practical concerns) is to retrieve and weigh more carefully some element of experience which had previously been relegated to the background. In most cases, however, the decomposition of i corresponds well to what happens in effective scientific procedure, and the advantages far outweigh the disadvantages.

Tempered personalism recognizes the possibility that X and X', with background information respectively b and b', may agree with respect to a and yet have unequal $P_{X,b}(h|a)$ and $P_{X',b'}(h|a)$; but the latitude toward subjective judgment in the tempered personalist concept of probability permits the question of whether this inequality is due to deep-lying differences between b and b' to be circumvented. The tempering condition, moreover, makes it possible for X and X' to move toward consensus without having to agree first upon a probability function and second upon the total body of evidence. For if X and X' are aware of the same set of seriously proposed hypotheses and are open-minded toward all of them, then their prior probabilities $P_{X,b}(h_i|a)$ and $P_{X',b'}(h_i|a)$ are not radically different in order of magnitude, and, consequently, a moderate amount of data e is capable of ensuring agreement as to which hypothesis is to be preferred. Thus, if for any two seriously proposed hypothesis h_i and h_j

$$P_{X,b}(h_i|a)/P_{X,b}(h_j|a) < 10^4 \quad \text{and} \quad P_{X',b'}(h_i|a)/P_{X',b'}(h_j|a) < 10^4$$

(thus allowing considerable free play to the subjective judgments of both X and X'), and if the likelihood of e upon h_1 is at least 10^6 times as great as upon h_i ($i \neq 1$) for both X and X' (not atypical with a moderate amount of data in sampling problems), then X and X' would agree that the posterior probability of h_1 upon e is at least one hundred times greater than that of any of its rivals. This is an example of what Jeffreys calls "swamping" the prior probabilities. If personal probabilities were used, unqualified by the tempering condition, then deep-seated prejudices on the part of either X or X' could produce immense (and possibly infinite) prior probability ratios; swamping would consequently not be achievable with a moderate amount of data, and consensus would be severely delayed. In a sense, therefore, my commitment to the tempered personalist concept of probability in preference to personalism is a reflection of Peirce's "social theory of logic": "He who would not sacrifice his own soul to save the whole world, is, as it seems to me, illogical in all his inferences, collectively. Logic is rooted in the social principle" (2.654). Or less dramatically, "the progress of science cannot go far except by collaboration; or to speak more accurately, no mind can take one step without the aid of other minds" (2.220). The tempering condition, which prescribes that seriously proposed hypotheses be given an opportunity to show their virtues under scrutiny, incorporates Peirce's social principle into a concept of probability.[21]

C

One of the points in the characterization of tempered personalism which requires clarification is the status of the axioms of probability. That $P_{X,b}(h \mid r \& a)$ satisfies these axioms is built into the definition of a tempered personalist probability function, but it is questionable whether this ought to be done in an instrumentalist explication of "rational degree of commitment." Because of the considerations in section II.D against interpreting "rational degree of commitment" in terms of the coherence of a set of betting quotients, the method of justifying the axioms due to Ramsey and DeFinetti – and accepted by most Bayesians – is inapplicable. A further question is whether the axioms of probability are even consistent with the tempering condition, since the latter prescribes that a person who learns of a new seriously proposed hypothesis on the matter under investigation must give it a nonnegligible prior probability, and it is not clear how this can be done without redistributing the probabilities assigned to the previously known hypotheses in a manner which violates the axioms.[22]

The second question is easily answered. The tempered personalist concept of probability is "local," in the sense that its application is circumscribed by the conditions of an individual investigation. When the set of seriously proposed hypotheses is augmented, the conditions are changed, and prior probabilities are reevaluated. The axioms of probability are specified, but cannot be used to infer the prior probabilities associated with the new conditions from those associated with the old. Indeed, if

21. This suggests that the theory of probability which I am proposing as a modification of personalism might properly be called "socialism," had this term not been preempted.
22. In effect this question is raised by Putnam in the following passage: "Consider a total betting system which includes the rule: if it is ever shown that a hypothesis S is not included in the simplicity ordering corresponding to the betting system at time t, where t is the time in question, then modify the betting system so as to 'insert' the hypothesis S at a place n corresponding to one's intuitive judgment of the 'complexity' of the hypothesis S. This rule violates two principles imposed by Carnap. First of all, it violates the rule that if one changes one's degree of confirmation in one's life, then this should be wholly accounted for by the change in E, that is the underlying c-function itself must not be changed. Secondly, it can easily be shown that even if one's bets at any one time are coherent, one's total betting strategy through time will not be coherent. But there is no doubt that this is a good rule nonetheless . . ." (1963a: p. 10).

one attempts to make such an inference formally, he will see that some very weird probability evaluations are required, such as $P_{X,b}(s\,|\,a)$, where s is the proposition that X *will soon become aware that* h_{n+1} *is seriously proposed by someone.* (In section II.B it was pointed out that a similar peculiar probability evaluation may be required in order to reconcile a purely logical set of confirmation rules with Carnap's concession that the laws actually proposed by scientists should perhaps be taken into account.) It is most unlikely that s would belong to the set S, explained in the definition of "tempered probability function," which is associated with the investigation at hand. Furthermore, an intellectual contortion would be required in order to evaluate subjectively the probability that a certain hitherto neglected hypothesis will be seriously proposed by someone, since posing this question ipso facto singles out the hypothesis for special consideration.[23]

The first question, whether a concept which explicates "rational degree of commitment" can be justifiably assumed to satisfy the axioms of probability, is more difficult. A promising approach, which derives the axioms from a relatively weak set of assumptions, is to use with some supplementation a remarkable argument invented independently by a number of writers, the earliest to my knowledge being Cox (1946, 1961) and Good (1950). An extensive bibliography and a rigorous presentation of the purely mathematical aspects of the argument are given by Aczél (1966: pp. 319–24). I shall begin by stating and sketching the proof of a modification of a theorem in Aczél's book, and then I shall discuss the application of this theorem to the present question. (Throughout the following discussion the notation $P(c\,|\,e)$ will be used without a subscript "X" even when a tempered personalist interpretation of the formalism is intended.)

Theorem. *Let* $P(c\,|\,e)$ *be a real-valued function of pairs of propositions* (c, e) *such that* $c \in S$ *(a nonempty set of propositions closed under truth-functional operations) and* $e \in T$ *(where* $T \subseteq S$, *no member of* T *is a contradiction, and for every* $e, e' \in T$ *there exists a finite set of propositions* $e_1, \ldots, e_n$ *such that* $e \,\&\, e_1 \in T$, $e_i \,\&\, e_{i+1} \in T$ $[i = 1, \ldots, n-1]$, *and* $e_n \,\&\, e' \in T$). *Suppose* P *satisfies the following conditions:*

23. The reasoning of this paragraph can be paralleled to show that the coherence of sets of betting quotients must also be considered as a "local" concept, applicable only when a definite set of outcomes is envisaged and bet upon. In order to use the condition of coherence for the purpose of relating a set of subjectively acceptable betting quotients when $h_1, \ldots, h_n$ are the only outcomes envisaged to a set when $h_1, \ldots, h_{n+1}$ are envisaged, it would be necessary to evaluate a betting quotient for bets upon the proposition *that the outcome* h_{n+1} *will be added to those already envisaged* – and this appears to be a psychological impossibility.

(1) *if c is logically equivalent to c' and e to e' then $P(c|e) = P(c'|e')$;*

(2) *for any $e \in T$ there is a real number u_0 (perhaps dependent on e) such that if o is a contradiction belonging to S then $u_0 = P(o|e) \leq P(c|e)$;*

(3) *there is a real number u_1 such that for all $e, f \in T$ $P(e|e) = P(f|f) = u_1 > u_0$;*

(4) *$P(c \& e|e) = P(c|e)$;*

(5) *for every $e \in T$ there is a function F_e such that $P(c \& d|e) = F_e[P(c|d \& e), P(d|e)]$; and*

(6) *for every $e \in T$ there is a function G_e which is continuous and monotonically increasing in both variables such that if e entails $\sim(c \& d)$ then $P(c \vee d|e) = G_e[P(c|e), P(d|e)]$.*

Then there exists a continuous and monotonically increasing function $h(t)$ such that $h(u_0) = 0$, $h(u_1) = 1$, and $P'(c|e) = h[P(c|e)]$ satisfies the standard axioms of probability.

This theorem can be proved by slightly modifying the demonstration on pages 321–24 of Aczél's book. Since P here is a function of propositions rather than of sets, as in the book, the propositional calculus together with conditions (1) through (6) must be used as follows in order to establish the functional equations which Aczél needs on page 322:

$$F_e(x, u_1) = F_e[P(c|e), P(e|e)] = F_e[P(c|e \& e), P(e|e)]$$
$$= P(c \& e|e) = P(c|e) = x,$$

$$F_e(u_1, x) = F_e[P(e|e), P(c|e)] = F_e[P(c \& e|c \& e), P(c|e)]$$
$$= P[(c \& e) \& c|e] = P(c \& e|e) = P(c|e) = x,$$

$$G_e(u_0, x) = G_e[P(o|e), P(c|e)] = P(o \vee c|e) = P(c|e) = x,$$

$$G_e[G_e(x, y), z] = P[(c \vee d) \vee f|e] = P[c \vee (d \vee f)|e] = G_e[x, G_e(y, z)],$$

$$F_e[G_{f \& e}(v, w), z] = F_e[P(c \vee d|f \& e), P(f|e)] = P[(c \vee d) \& f|e]$$
$$= P[(c \& f) \vee (d \& f)|e] = G_e[P(c \& f|e), P(d \& f|e)]$$
$$= G_e[F_e(v, z), F_e(w, z)],$$

where

$$x = P(c|e), \quad y = P(d|e), z = P(f|e), v = P(c|f \& e), w = P(d|f \& e).$$

Aczél manipulates these equations and the premisses of the theorem to show that $G_e(v, w) = G_{e \& f}(v, w)$, so that the last of these functional equations becomes a distributivity equation

$$F_e[G_e(x, y), z] = G_e[F_e(x, z), F_e(y, z)].$$

A previous theorem permits him to assert that the most general solution to this equation (with the assumptions regarding G_e) is

$$G_e(x,y) = H^{-1}[H(x)+H(y)],$$

$$F_e(x,y) = H^{-1}[H(x)C(y)],$$

where C is an arbitrary continuous function and H is an arbitrary monotonically increasing function such that $H(u_0) = 0$ and $H(u_1) = 1$. But from the second of the functional equations $x = F_e(u_1, x) = H^{-1}[H(u_1)C(x)] = H^{-1}[C(x)]$, so that $C(x) = H(x)$ for any x in the range of P. Hence the most general forms of G_e and F_e are

$$F_e(x,y) = H^{-1}[H(x)H(y)]$$

and

$$G_e(x,y) = H^{-1}[H(x)+H(y)].$$

If $P'(c|e)$ is defined as $H[P(c|e)]$, then substitution into conditions (1) through (6) yields the following:

(1') if c is logically equivalent to c' and e to e', then $P'(c|e) = P'(c'|e')$,
(2') $o = P'(o|e) \leq P'(c|e)$,
(3') $P'(e|e) = P'(f|f) = 1$,
(4') $P'(c\&e|e) = P'(c|e)$,
(5') $P'(c\&d|e) = P'(c|d\&e) \cdot P'(d|e)$,
(6') if e entails $\sim(c\&d)$, then $P'(c\vee d|e) = P'(c|e)+P'(d|e)$.

Aczél's argument ends at this point, but axioms (i) through (iv) trivially follow from conditions (1') through (6'). If $e \in T$ and e entails h, then $h\&e$ is logically equivalent to e, and therefore $1 = P'(e|e) = P'(h\&e|e) = P'(h|e)$, which is axiom (ii). Axiom (i) asserts that $0 \leq P'(c|e) \leq 1$; the first inequality is the same as condition (2'), while the second follows from $1 = P'(c\vee \sim c|e) = P'(c|e)+P'(\sim c|e)$ together with $0 \leq P'(\sim c|e)$. Axiom (iv) follows immediately from conditions (5') and (1').

The utility of this theorem for the question at hand obviously depends upon whether its premises are justifiable if $P(c|e)$ is taken as an explicatum of "rational degree of commitment," and especially if locality and instrumentalism (which are characteristics of the tempered personalist concept of probability) are maintained. The condition on S is evidently reasonable if one wishes to use the operations of deductive logic in the context of scientific inference. The first two conditions on T are trivial: $T \subseteq S$ is required if one wishes $P(e|e)$ to be defined for all $e \in T$, and a contradictory proposition cannot possibly serve as evidence. The third condition on T is a direct consequence of locality: for if e and f are admissible evidential propositions in the probability calculations associated with an investigation, then they are respectively of the form $e'\&i$ and $f'\&i$, where i is the initial body of information and assumptions. Conditions (1) through (4) are justifiable partly as straightforward rules for ensuring that inductive and deductive procedures should mesh and partly as conventions for ordering the rational degree of commitment to a contradiction relative to the rational degrees of commitment to other propositions. Conditions (5) and (6), by contrast, are nontrivial. The functional dependence in condition (5) seems very natural, since it merely asserts that one can evaluate the rational degree of commitment to a conjunction $c \& d$ relative to evidence e by proceeding in a stepwise fashion: by evaluating the rational degree of commitment to d on e, and to c on $d \& e$. Even more appealing intuitively is the functional dependence in condition (6), for it is hard to see what else $P(c \vee d \mid e)$ could depend upon than $P(c \mid e)$ and $P(d \mid e)$. However, there is an obscurity in calling upon intuition in this way, just because the concept of commitment is unclear. The naturalness of these assumptions may be derivative from the consideration that the addition and multiplication principles of probability – axioms (iii) and (iv) – are necessary conditions for the coherence of beliefs in a set of bettable propositions (that is, those for which definite payoff conditions exist). One may feel that in treating commitment, which is a modality of belief appropriate to nonbettable propositions, it is quite conservative to retain the bare assertion of functional dependence instead of the specific dependencies of axioms (iii) and (iv). But more would have to be known about the general character of commitment in order to make this reasoning forceful. Perhaps the best defense of the assumption of functional dependence in conditions (5) and (6) is just that they contribute to the clarification of the concept of commitment in a methodologically fruitful way. They permit the argument to proceed to the conclusion that the explicatum of "rational degree of commitment" satisfies the axioms of probability, thereby permitting a formulation of scientific inference in which there is a firm mathematical structure. On the other hand, the fact that these are quite weak assumptions makes it plausible that their denial will result in a loose formulation of scientific inference rather than in a formulation with a firm though non-Bayesian structure. In order to strengthen this plausibility argument, however, one should systematically investigate the alternatives to the functional dependencies of conditions (5) and (6) in the light of the fundamental purpose of learning about the world from experience. Once the existence of the function G_e of condition (6) is granted, its continuity is entirely natural. If I introspect about my subjective commitments, I certainly feel strongly that a very small change in my commitment to one of two exclusive propositions induces a very small change in my commitment to their disjunction, and I can see no reason why the normative concept of degree of rational commitment should differ from the concept of subjective commitment in this respect. (It is relevant to note that discontinuities in prior probability distributions are sometimes of great instrumental value, as will be argued in section III.E, since they permit the possibility of accepting a hypothesis asserting the exact value of a parameter which a priori can have any of a continuum of values; but the continuity of the function G_e is clearly compatible with this kind of discontinuity in the distribution function.)

The one premiss which prima facie seems unacceptable in view of locality is that $P(c\,|\,e)$ takes on all values between u_0 and u_1, for it is possible that the set of propositions S mentioned in the definition of "tempered personalist probability function" is denumerable or even finite, in which case the premiss would be false. However, if S is augmented by an appropriate set I of "ideal elements," which are propositions irrelevant to the investigation at hand but which are introduced in order to permit refined judgments of subjective commitment, then $P(c\,|\,e)$ will take on all values as a ranges over S', the closure of $S \cup I$ under truth-functional operations. A convenient choice of I is the set of all propositions c_r $(0 \le r \le 1)$ to the effect that a needlepoint constrained to come to rest somewhere on a scale one meter long will, because of an unspecified physical mechanism, come to rest at a point less than r meters from the 0-end. It is reasonable to suppose that the degree of subjective commitment to c_r (which in this case can surely be equated with belief) relative to information i is a continuous function of r, and because c_0 violates the assumed constraints while c_1 is required by them, the rational degree of commitment to c_0 and c_1 must respectively be u_0 and u_1. Then $P(c_r\,|\,i)$ takes on all values between u_0 and u_1. No other assumption need be made about the functional dependence of P upon r, and in particular there is no appeal to a principle of indifference regarding subintervals of $[0,1]$ of equal length. When a man reflects upon a proposition belonging to S, a comparison with the propositions of a set like I is a very useful method for arriving at a numerical assessment of his subjective commitments. Indeed, unless the propositions of S are bettable, it is hard to see how numerical assessments could be arrived at except by systematic comparisons of this kind (or by a variant procedure like Koopman's use of a sequence of "n-scales" [1940: pp. 290–91] instead of the continuous set I); and if they are all bettable, then considerations of coherence ensure that P satisfies the axioms of probability so that the present inquiry is unnecessary. The "ideal elements" are, to be sure, not mentioned in the definition of "tempered personalist probability function," but it is unnecessary to do so, since there a scale for measuring subjective commitments is presupposed. However, the present inquiry concerns the justification for conjoining the various conditions in this definition, and, therefore, it is legitimate to pay attention here to the ideal elements which are auxiliary to numerical assessments of subjective degree of commitment.

The argument up to this point is that for any function $P(c|e)$ which is suitable as an explicatum of "rational degree of commitment," there exists a continuous monotonic function $H(x)$ such that $P'(c|e) = H[P(c|e)]$ satisfies the standard axioms of probability. Because of the monotonicity of H, P' is suitable as a measure of degree of rational commitment if P is, and the choice between them is claimed by Cox (1961: p. 16) and by Good (1950: p. 106) to be merely a matter of the convenient selection of a scale. An improvement can be made upon their conventionalistic reasoning by considering I', the closure under truth-functional operations of the set I discussed in the preceding paragraph. All the members of I' are bettable, and furthermore they are such that the degree of subjective commitment to each is equal to the credence in it. DeFinetti showed that if credence is measured by the maximum acceptable betting quotient and if coherence is recognized as a necessary condition for rationality, then a rational credibility function must satisfy the axioms of probability. Hence, if $c, d \in I'$, then $P(c \& \sim c | i) = 0$, $P(c \vee \sim c | i) = 1$, and $P(c \vee d | i) = P(c|i) + P(d|i)$ if i entails $\sim(c \& d)$. It follows that $H(0) = 0$, $H(1) = 1$, and $H(x + y) = H(x) + H(y)$ (where $x = P(c|i)$ and $y = P(d|i)$). But by the construction of I' it is possible to choose c and d so as to let x and y be arbitrary real numbers in $[0,1]$ subject to the constraint $x + y \leq 1$. One easily sees then that $H(x) = x$ for all dyadic rationals (x equals the sum of a finite number of terms of the form 2^{-n}, where $n = 1, 2, 3, \ldots$), and hence by the continuity of H, $H(x) = x$ for all $x \in [0,1]$. Consequently, P is identical with P', so that P itself satisfies axioms (i) through (iv) for all $c \in S$ and $e \in T$.

The foregoing argument admittedly rests upon several idealizations, but they are just those which one must make in order to use the real number system freely in assessing subjective commitments. It should be emphasized that this argument does not depend upon construing all commitments in terms of dispositions to bet, but rather the fact that some commitments can be so construed is used in order to complete an argument which is based primarily upon considerations of orderliness in scientific inference.

D

The phrase "seriously proposed hypothesis" in the definition of "tempered personalist probability function" requires clarification. Although there is often general agreement in the context of a specific investigation as to which hypotheses are to be considered seriously proposed, it is very difficult to state a reasonable set of conditions upon the intrinsic characteristics of hypotheses and upon the circumstances of their proposal which would permit one to distinguish unambiguously those which are seriously proposed from those which are not. Probably the request for a sharp set of conditions ought not to be honored because of the danger of arbitrariness and of diminishing the flexibility of the scientific method. On the other hand, if decisions about classifying hypotheses as seriously proposed or not are left entirely to the subjective judgment of individual investigators, then the advantages which I have claimed for tempered personalism – as a "social theory of inductive logic" – over unqualified personalism are in danger of being lost. In order to navigate between these two dangers, I shall try to formulate some methodologically sensible guidelines for decisions on this question, without pretending to eliminate entirely the subjective judgment of the investigator. I shall also argue that under social conditions which are generally favorable to theoretical inquiry, the prospect of scientific progress is not imperiled by this informality, while social conditions which are adverse to theoretical inquiry are not likely to be ameliorated merely by formalizing scientific methodology.

A part of the problem is to determine whose conjectures are to be regarded as seriously proposed. Again there are opposing dangers of being too strict and too loose. If the only persons who are to be so respected belong to some delimited group, such as a definite profession, then there is a danger of "blocking the way of inquiry," for such groups are subject to stagnation, parochialism, and obeisance to authority. On the other hand, if the conjectures of everyone are to be taken seriously, won't a conscientious investigator find himself overwhelmed by capricious and crankish hypotheses? The dilemma is not very painful if one recognizes that there are different kinds of situations. Sometimes the mode of presentation of a hypothesis to an investigator conforms to professional standards: the hypothesis is clearly stated, the motivation for proposing it is explained, and the explanation indicates understanding (though not necessarily complete acceptance) of the recognized body of propositions regarding the subject, and it is not an arbitrary choice from a family of hypotheses which answer to the same motivation. (See the following paragraphs on this last point.) Whether the proposer belongs to the scientific establishment or not can reasonably be regarded as irrelevant in such cases. On the other hand, there are cases in which the presentation of a hypothesis fails to meet professional standards in one or more ways. For example, the formulation may be obscure, or it may seem obscure upon first reading so that time and energy evidently would be required in order to determine whether the apparent obscurity is due to the author's confusion or to profound originality. Since an investigator has limited time and energy, he may in good conscience decide not to study the proposal carefully unless the credentials of the proposer indicate competence. Respect of this kind for the scientific establishment does not imply the abandonment of one's own judgment, and it is reasonable if learning about nature is assumed to be generally progressive.[24] However, one should not rigidify a reasonable guideline. Even an obscurely presented theory by an unknown author may draw the reader in by exhibiting some intellectual freshness, but at this point the subjective judgment of the individual is evidently irreplaceable. Even if the hypothesis of an outsider is dismissed by the establishment, that particular avenue of inquiry is not thereby definitively blocked if the society as a whole is sufficiently tolerant and flexible. When there is freedom of research and communication and adequate provision of leisure, the outsider may exhibit his seriousness by mastering the subject sufficiently to make his presentation lucid by professional standards or by designing and performing his own experiments;[25] and he may be able to seek the advice of acknowledged experts, who are sometimes more sympathetic in face-to-face encounters than in reading a written page. Citadels of entrenched scientific opinions have been conquered often enough to indicate that receptivity to novel insight has been present in at least some members of the scientific establishment (and perhaps enough even to vindicate the Platonic assumption that the truth exercises a gentle but persuasive force upon the minds of men). At any rate, if the society becomes so inflexible, intolerant, and incurious as to prevent a dedicated outsider from receiving a hearing for his ideas, it is hard to see how a sharpened set of methodological rules would make the social climate more favorable to inquiry.

A more difficult part of the problem is to determine when "seriously proposed" should be applied to a single hypothesis and when it should rather be reserved for the disjunction of a family of hypotheses of which it is a member. Consider, for example, the family of hypotheses $\{h(\alpha)\}$ asserting that the electrostatic force between two point charges is of the form $F \sim r^{-\alpha}$, where r is the distance between the charges. The hypothesis

24. Although a posteriori considerations are out of place in the first stage of my treatment of scientific inference, I shall anticipate the second stage and assert a proposition which seems to be overwhelmingly supported by the history of science: that the sensitivity of scientific method to the truth is not diminished by a reluctance to classify a hypothesis as seriously proposed unless the proposer has taken pains to relate it clearly to the currently accepted body of knowledge. On the whole Burke's maxim is apropos: "People will not look forward to posterity, who never look back to their ancestors" (*Reflections on the Revolution in France*).

25. Feyerabend (1963) recommends a "principle of proliferation" as an antidote to intellectual sterility. But in scientific method, as in the civil law, there should be a "principle of paternity," according to which the man who engenders a hypothesis has some responsibility for supporting it; and, as in civil affairs, this principle would to some extent curb the principle of proliferation.

$\alpha = 2$ was proposed by Priestley and D. Bernoulli (Whittaker 1951: p. 53), though it is named for Coulomb because of his delicate experiments confirming it. Suppose that another eighteenth-century scientist had proposed $\alpha = 2.0001$. Were both of these hypotheses to be considered "seriously proposed," or were both to be considered as merely arbitrary specifications of the seriously proposed family $\{h(\alpha)\}$, or was the first to be given the status of "seriously proposed" and the second not? Scientific practice of the time surely favored the third course, unless some special and satisfactory motivation had been given for the proposal $\alpha = 2.0001$. Methodologists sometimes rationalize scientific practice in cases like this by appealing to simplicity, and there can be no doubt that the exponent -2 appears intuitively to be simpler than the exponent -2.0001. In this essay I shall not undertake to survey the various analyses which have been made of the concept of simplicity (cf. Kyburg 1964: pp. 19–20), although, in section IV, I shall criticize Kemeny's suggestions and, in section V, I shall argue that factual considerations are relevant to the concept. At this point I shall only say that there appears to be general agreement on the inadequacy (from the standpoint of inductive logic) of all explications which have so far been given of the concept of simplicity, and that it is, therefore, desirable to explain the favoritism shown to $\alpha = 2$ along other lines. Without appealing to simplicity, one can give several reasons for singling out the inverse square hypothesis from the continuum of possibilities. Historically the most important reason was undoubtedly the analogy to the law of gravitation, but this reason merely shifts the problem to that of justifying the *exact* value $\alpha = 2$ in the gravitational force law. Another reason, which was heuristically important in Newton's proposal and, therefore, indirectly in the proposal of Coulomb's law, is that $\alpha = 2$ is the only hypothesis of the family $\{h(\alpha)\}$ which implies that the flux across the surface of a sphere with the point source at its center is independent of the radius of the sphere. In fact, there is a more general consequence of this kind, which was not known to Newton or to the natural scientists of the eighteenth century: that the flux across any simply connected surface enclosing the point source is the same. These mathematical consequences made it plausible – even before the development of electromagnetic theory in the nineteenth century – that an inverse square law could be readily embedded in a comprehensive formulation of the principles of electricity, whereas the hypothesis $\alpha = 2 + \epsilon$, where ϵ is too small to be detected by known experimental methods, could not be readily embedded in this way. More speculatively, one could say that if a deep theory is to be found connecting the electrostatic force law with the dimensionality of space, as the results concerning the flux through surfaces suggest, then force laws with integral exponents appear much more promising than the other members of $\{h(\alpha)\}$. However, if the electrostatic force law was proposed without any envisagement of a more comprehensive theory, then there was no motivation for going beyond descriptive adequacy, and it would have been appropriate to interpret the proposal $\alpha = 2$ as shorthand for something like "α has a value which is indistinguishable from 2 by direct measurements of the electrostatic force between charges."

A general methodological guideline can be extracted from this example: if $\{h(\alpha_1, \ldots, \alpha_k)\}$ is a family of hypotheses with k parameters such that the point $(\alpha_1', \ldots, \alpha_k')$ in parameter space is associated with a comprehensive theory, which may be explicitly stated or only sketched, into which $h(\alpha_1', \ldots, \alpha_k')$ but none of the other members of the family would fit, then this hypothesis individually may be considered to be seriously proposed; the disjunction of all the other members of the family should be considered as seriously proposed but the individual components should not be. This guideline permits the machinery of the tempered personalist formulation of scientific inference to lead to the tentative acceptance of exact values of parameters only when the exact values are of theoretical interest. The vagueness of this guideline is evident, for there is nothing to prevent the proposer of a bizarre hypothesis from supplementing his proposal with a sketch of a bizarre comprehensive theory in which it might be embedded. Furthermore, the notion of a "sketch" of a comprehensive theory is extremely vague. At this point, however, I am dubious that anything of value can be accomplished by sharpening the methodological prescription; there is no substitute for an intelligent examination of the reasons for the special proposal $h(\alpha_1', \ldots, \alpha_k')$ and for a personal judgment about their plausibility.

It should be noted that when the predicate "seriously proposed" is applied to the disjunction of a family of hypotheses, the tempering condition can be amplified somewhat, so as to prescribe open-mindedness not only to the family as a whole but to its subfamilies. Thus, if the members of the family are $h(\alpha_1, \ldots, \alpha_k)$, then a reasonable amplification of the tempering condition would prescribe that the prior probability density $P_{X,b}[h(\alpha_1, \ldots, \alpha_k)|a]$ must be such as to permit a posterior probability which is strongly peaked about any preassigned point in the space of the parameters for some possible result r of the envisaged observations.[26] The peaking of the posterior probability distribution function in an unexpected way may stimulate new speculation, leading to the serious proposal of a point in parameter space which previously had not been classified as seriously proposed but which can be so considered in a new investigation.

---

26. Because $\int P_{X,b}[h(\alpha_1, \ldots, \alpha_k)|a] d\alpha_1 \cdots d\alpha_k$ must be ≤ 1, while some of the α_i may have infinite range, a closer parallel to the statement of the tempering condition in the discrete case – such as setting a lower bound to the integral over a fixed volume in parameter space – is in general not possible.

E

Since the tempered personalist concept of probability is derivative with rather small changes from Jeffreys's working sense of "probability," its utility in formulating scientific inference can be exhibited parasitically by referring to the wealth of applications in his *Theory of Probability*. I shall give one example, in order to show how and with what modifications Jeffreys's calculations can be borrowed by tempered personalism.

In the classical sampling problem of Laplace, the information a asserts that there is a population of N objects of which an unknown number r have a specified property α, and from this population objects are drawn without replacement by a method designed to favor neither α nor $\bar{\alpha}$. Let $h(x)$ be the hypothesis that $r = x$ and d_{nm} be the proposition that in a sample of $n + m$ members n have the property α and m do not. Laplace appealed to the principle of indifference (which was then called the "principle of insufficient reason") in assigning equal prior probabilities to all constitutions of the population, that is, $P^L[h(x)|a] = 1/N+1$ for $x = 0, \ldots, N$ (where the superscript L refers to Laplace's evaluation). From this assignment, together with the axioms of probability and the assumption that at each draw the probability of obtaining any object not yet removed is equal to the probability of obtaining any other, he demonstrated his "rule of succession"

$$P^L(s \mid d_{n0} \& a) = \frac{n+1}{n+2},$$

where s is the proposition that the next object to be drawn will have the property α. Laplace considered this result to provide a mathematical justification for inductive inference, since a large value of n yields a posterior probability close to 1 that the uniformity exhibited in the sample will extend to the new instance. However, the same set of premises implies that

$$P^L[h(N) \mid d_{n0} \& a] = \frac{n+1}{N+1},$$

and, therefore, the posterior probability of the generalization that all members of the population have the property α is small unless the sample is a large part of the population. Jeffreys correctly argues that scientific progress requires the possibility of tentatively accepting generalizations, and, therefore, he suggests (1961: p. 129) that Laplace's prior probability assignments be replaced by the nonuniform distribution

$$P^J[h(0)|a] = P^J[h(N)|a] = k,$$

$$P^J[h(x)|a] = \frac{1-2k}{N-1} \quad \text{for } x = 1, \ldots, N-1,$$

for some appropriate real number k (the superscript J referring to Jeffreys's evaluation). It follows that

$$\frac{P^J[h(N)\,|\,d_{n0}\,\&\,a]}{\sum_{x=0}^{N-1} P^J[h(x)\,|\,d_{n0}\,\&\,a]} = \frac{n+1}{N-n}\cdot\frac{k}{1-2k}(N-1).$$

"Hence if n is large, the ratio is greater than $(n+1)k/(1-2k)$ whatever N may be, and the posterior probability that $r = N$ will approach 1, almost irrespective of N, as soon as n has reached $1/k$" (*ibid.:* p. 130). This reasoning is instrumentalistic and makes no obvious appeal to objectively determined probabilities. Jeffreys recognizes a range of possible choices of k which are all methodologically reasonable, his preference being $k = 1/4 + 1/2(n+1)$, which he motivates by the following classification of possibilities:

(1) Population homogeneous on account of some general rule.
(2) No general rule, but extreme values to be treated on a level with others.
Alternative (1) would then be distributed equally between the two possible cases, and (2) between its $n+1$ possible cases. (*Ibid.:* pp. 130–31)

In evaluating the probabilities of $h(0)$ and $h(N)$, the procedure of tempered personalism is similar to that of Jeffreys. Since these two hypotheses can be expected to be seriously proposed if the sampling problem occurs within the context of a scientific investigation, $P_{X,b}[h(0)\,|\,a]$ and $P_{X,b}[h(N)\,|\,a]$ must be large enough to permit the possibility of high posterior probabilities of $h(0)$ and $h(N)$. The tempering condition also requires similar open-mindedness to other seriously proposed hypotheses of the family $\{h(x)\}$, but Jeffreys's procedure is obviously also adaptable to such complications. Tempered personalism differs from Jeffreys's theory primarily in the role which the former assigns to subjective judgment; and subjectively the prior probabilities of $h(0)$ and $h(N)$ may be unequal, and the distribution over $h(1),\ldots,h(N-1)$ need not be in uniform. How much latitude is permitted to subjective judgment is determined by the tempering condition and the circumstances of the investigation. Thus, if only $h(0)$ and $h(N)$ are singled out for special consideration from among the $h(x)$, then the only other seriously proposed hypothesis is the general hypothesis of heterogeneity $H = h(1)\vee\cdots\vee h(N-1)$, and the tempering condition requires the prior probabilities of $h(0)$, $h(N)$, and H to be large enough to permit each to have the largest posterior probability for some d_{nm}. If the experiment envisaged will yield a sample of size n_0, and if the prior probability distribution over the disjunctive components of H is roughly uniform (thus satisfying the amplified version of the tempering condition which was stated at the end of Section III.D), then Jeffreys's formula (1) above permits an approximate calculation of a lower bound k_0 on $P_{X,b}[h(0)\,|\,a]$ and $P_{X,b}[h(N)\,|\,a]$, namely,

$$k_0 = \left[2 + (n_0 + 1) \cdot \frac{N-1}{N-n_0} \right]^{-1}.$$

In the realistic case in which $N \gg n_0 \gg 1$, one obtains a simple approximate expression for k_0: $k_0 = 1/n_0$.

I must admit in conclusion my uncertainty about the relation between tempered personalism and one other important part of Jeffreys's treatment of prior probabilities: his use of invariance considerations to assign prior probabilities to hypotheses with adjustable parameters when no particular values are seriously proposed (*ibid.*: chap. III, especially secs. 3.1 and 3.10). There are technical difficulties in Jeffreys's analysis as it is now presented, such as the appearance of nonnormalizable distribution functions (cf. Hacking 1965: pp. 203–05), and it is possible that when these are removed, the residue will yield nothing that cannot be obtained by subjective probability evaluations modified by the tempering condition. If, on the other hand, these invariance considerations are fruitful, it may be desirable to incorporate them in some form into the prescription of tempered personalism, since the basic idea of a rule "that is applicable under any non-singular transformation of the parameters, and will lead to equivalent results" (Jeffreys 1961: p. 192) is reasonable.

F

Sections II and III have been largely concerned with the way in which probabilities should be evaluated if the concept of probability is to be a useful instrument in theoretical investigations. Little has been said, however, on the question of how correct probability evaluations can be used to serve the primary purpose of theoretical inquiry, which is (if one adheres to the etymon θεωρία) to obtain a *view* of the world.

For a Copernican epistemologist, who is skeptical of achieving certainty about the principles of nature, a "view" is properly to be construed not as a set of propositions about various aspects of the world, but rather as a set of modalities in which propositions about the world are entertained.[27] The view as a whole is tentative, since it is subject to modification in the light of further experience and of new proposals, and moreover

27. The meaning of "view" is, of course, a matter of convention. An alternative to the meaning adopted in the text is "a comprehensive set of propositions about various aspects of the world." If this sense is preferred, then a person who is in doubt about many things does not have a single view of the world, but rather he entertains a large number of views with varying weights. Were the two different senses of "view" spelled out with ideal care and detail (perhaps in terms of Leibnizian possible worlds), then one might be able to express the state of a man's theoretical knowledge equivalently in two different ways, and either would serve equally well in a rational reconstruction of scientific inference. However, the equivalence of the two modes of expression would be difficult

tentativeness is distributed in varying degrees among the parts, for usually a number of propositions (in addition to the ubiquitous "catchall" hypothesis *that something else is true*) are entertained simultaneously with different degrees of commitment. Thus the view which is actually attained by theoretical inquiry is characterized by intellectual tensions, rather than by the calmness which is the traditional connotation of θεωρία. Even the greatest achievements of natural science, which permit a large number of previously uncoordinated or anomalous facts to be seen as parts of a pattern, are not free from tentativeness – partly because of the remote possibility that apparent success has been due to a long series of coincidences, but more seriously because of the possibility that a highly successful theory may be displaced by a more general theory (perhaps with a radically different conceptual structure) which agrees approximately with the old theory over a limited range of circumstances but not in all of them (cf. Section II.D).

A general answer to the question about the use of probability in theoretical inquiry can be stated in terms of the foregoing characterization of a "view": it makes possible the systematization of a person's tentative entertainment of propositions about the world. Probability is useful for this general purpose in spite of the fact that scientists whose thinking about nature is judicious and orderly do not usually try to weigh their tentative commitments quantitatively, and also in spite of the fact that the most successful theories of natural science are so obviously better supported by the evidence than their rivals that detailed probability calculations are dispensable. In the first place, the axioms of probability are needed in order to give a unitary formulation of scientific inference, and the formulation in probabilistic terms of special procedures of inference, such as the hypothetico-deductive method, permits important qualifications and refinements (cf. Section II.A). Secondly, the evaluation of probabilities (and not merely the mathematical structure of probability theory) is valuable for systematizing the tentative entertainment of propositions whenever there is some delicacy in the relationship of evidence to hypotheses, for example, when considerations of possible experimental errors must be taken together with considerations about prior preferences among

to exhibit not only for practical reasons but in principle because of the obscurities in our concepts. Both senses of "view" are idealizations in that they neglect the presence of conceptual obscurities which are unavoidable as long as our knowledge is incomplete. When the analysis of the content of a person's theoretical knowledge is less than ideal, there is a strong reason for preferring the sense of "view" adopted in the text: subjective uncertainties are for the most part felt "locally" concerning relatively small sets of propositions. Unless the contents of views in the second sense were specified with great precision, the assignation of weights to them would not permit one to infer where the local uncertainties lay.

hypotheses, or when the evidence consists of complicated correlations. In the ordinary variety of research, as contrasted with the dramatic achievements of science which often preoccupy methodologists, this kind of delicacy of relationship is very common, and statistical analysis of data has proved indispensable. It is not surprising that the scientific work of Harold Jeffreys, whose *Theory of Probability* is by far the best treatise on the use of probability in scientific inquiry, has primarily been in the complicated and untidy field of geophysics. Finally, both in introspecting upon one's own commitments and in making comparisons with the commitments of others, it is important to determine as precisely as possible the loci of uncertainty and disagreement. The explicit use of the apparatus of probability theory, as contrasted with its tacit use in informal scientific inference, is valuable in this kind of analysis. Notably, if posterior probabilities are calculated by means of Bayes's theorem, then prior probabilities and likelihoods must be specified, and furthermore the sharp part of the body of assumptions and information must be made explicit. As explained in Section III.B, explicitness on these matters can be conducive to consensus among investigators by revealing differences among assumptions which are in need of supplementary investigations. The revelation of striking differences among prior probability evaluations is often symptomatic of prejudices or limitations of imagination on the part of one or more of the investigators. Even if scientific inference is formulated in terms of the (untempered) personalist concept of probability, the analysis of prior probabilities can be conducive to consensus by stimulating men to reexamine their own beliefs and introspect about possible prejudice or obtuseness. The formulation of scientific inference in terms of the tempered personalist concept of probability has, in this regard, the additional virtue of requiring open-mindedness toward the insights of other men.

This probabilistic account of a "view" of the world is incomplete without some remarks about the role of extremely well-confirmed hypotheses. Although scientific inference is based upon critical habits of thought, it nevertheless concludes on some occasions that one hypothesis is strikingly preeminent over its rivals. It often happens, for example, that the initial experiments which result in preferring a certain hypothesis are difficult to achieve and are not entirely convincing, but independent reconfirmations are abundant and relatively easy – and one has an impression of easy progress after emergence from tangled underbrush. There are various reasons for this kind of phenomenon: the initial experiments may show what precautions are necessary, as in Lavoisier's careful accounting for all reaction products in confirming the conservation of mass in chemical reactions; or the initial experiments may indicate the irrelevance of factors which previously were distracting, as in the demonstration of the etiology of malaria; or there may be conceptual clarification, as in Galileo's combination of experiment and analysis. Whatever the reason may be for the avalanche of independent confirmations of a hypothesis, grouping them all together as if they were parts of a single investigation would result in overwhelmingly large ratios of the posterior probability of the successful hypothesis to the posterior probabilities of all its rivals. The dispensability of numerical evaluations of probabilities for recognizing the most striking achievements of science – which anti-Bayesians correctly point out – is an obvious consequence of this effect.

Suppose not only that the ratio of the posterior probability of h_1 to that of h_i ($i = 2, ..., n$) is very large, but also that the number n of seriously proposed hypotheses is moderate (which is a reasonable assumption, in view of the discussion of seriously proposed hypotheses in section III.D). In that case $P_{X,b}(h_1 | e \& a) = 1 - \epsilon$, where

$$\epsilon = \sum_{i=2}^{n} \epsilon_i = \sum_{i=2}^{n} P_{X,b}(h_i | e \& a)$$

is a positive real number much less than 1. Then h_1 can be "accepted" as part of the body of assumptions and information which is relied upon in subsequent investigations.[28] In the notation of the "local" concept of probability introduced in section III.B, this acceptance consists of replacing the proposition a by a' (equivalent to $a \& h_1$) and working with a new local probability function $P_{X,b'}(q | r \& a')$, where b' differs from b by absorbing the data e of the previous investigation and possibly other new background information of negligible importance. (However, if e is directly relevant to the new investigation, it cannot be relegated in this way to background information; as pointed out in section III.B, the separation of the total body of information and assumptions into parts a and b is an idealization.) Suppose, for the purpose of closer analysis, that instead of accepting h_1 the residue of uncertainty of the preceding investigation were explicitly taken into account in the new investigation in the following way:

$$P_{X,b'}(q | r \& a) = P_{X,b'}([q \& (h_1 \vee \cdots \vee h_n)] | r \& a)$$

$$= \sum_{i=1}^{n} P_{X,b'}(q \& h_i | r \& a) =$$

28. A good discussion of the acceptance of hypotheses, which stresses different points from the ones made here, is found in Jeffreys's section entitled "Deduction as an Approximation" (1961: pp. 365–68). Kyburg (1964: pp. 29–30) surveys some of the extensive literature on the problem of acceptance. It should be noted, however, that most of this literature is concerned with induction in a practical context, and as Carnap points out (1963: p. 973), the problem of acceptance in the context of theory is quite different.

$$= \sum_{i=1}^{n} P_{X,b'}(q \mid r \& a \& h_i) \cdot P_{X,b'}(h_i \mid r \& a)$$

$$\cong \sum_{i=1}^{n} P_{X,b'}(q \mid r \& a \& h_i) \cdot P_{X,b'}(h_i \mid a \& e)$$

$$= (1 - \epsilon) P_{X,b'}(q \mid r \& a') + \sum_{i=2}^{n} \epsilon_i P_{X,b'}(q \mid r \& a \& h_i),$$

where the approximate equality in the next to the last step is contingent upon the irrelevance of the data r of the new investigation to the h_i. It is evident that the correction obtained in this way to the "local" probability function $P_{X,b'}(q \mid r \& a')$ is small, because of the magnitudes of ϵ and the ϵ_i. Furthermore, the correction term is hard to evaluate, since the overwhelming confirmation of h_1 implies that the posterior probabilities $\epsilon_2, \ldots, \epsilon_{n-1}$ are almost certain to be extremely small, and only the posterior probability ϵ_n of the catchall hypothesis h_n is likely to be nonnegligible; but since h_n is the proposition that *something other* than the definite proposals $h_1, \ldots, h_{n-1}$ is true, the subjective degree of commitment to a proposition q relative to $r \& a \& h_n$ (which is required in evaluating the tempered personalist probability $P_{X,b'}[q \mid r \& a \& h_n]$) is likely to be more than ordinarily indefinite. Thus, the comparison of a "local" probability evaluation with a treatment which takes into account the overlap between two investigations reveals a motivation for the local concept and also exhibits its approximate accuracy.

The acceptance of h_1 is not tantamount to an unqualified commitment to its truth. The neglected ϵ is preserved in a residual attitude of tentativeness toward h_1, which involves the willingness to reopen an investigation concerning h_1 if suitable motivation is provided, such as persistent failure to make good progress in subsequent investigations which take h_1 for granted. A reinvestigation of h_1, however, must be quite different from the initial investigations which confirmed and reconfirmed it, for h_1 cannot be simply dismissed without some explanation of its previous remarkable success; the explanation may take the form of a superseding theory of greater generality than h_1, or perhaps it will consist in the revelation of a deep-lying systematic error in the preceding investigations, but in any case the set of seriously proposed hypotheses in the reinvestigation must differ somewhat from the set $h_1, \ldots, h_n$ considered earlier. Even if ϵ were 0, a literal belief in the truth of h_1 would not be justified, for $P_{X,b}(h_1 \mid e \& a) = 1$ means that a rational degree of commitment for X to h_1, *relative to $a \& b$*, is 1. There will surely be some tentativeness in X's entertainment of a; and furthermore, as explained in section II.D, commitment to a specific hypothesis is somewhat weaker than belief in its literal truth (for indeed, if this were not the case, then it would be difficult to see how sufficient evidence could ever be accumulated to make the probability of the catchall hypothesis much smaller than 1).

A final remark about the role of highly confirmed hypotheses lies on the border line between the methodology and the psychology of science. There is no a priori reason why critical procedures of analysis, such as the deliberate design of experiments for testing the deductive consequences of theories, should not always lead to the rejection of all definite proposals in favor of the catchall hypothesis, or at best (by means of suitable appearance-saving assumptions) to suspense among many alternatives. However, it is extremely unlikely that if the outcomes of scientific investigations were uniformly negative or indecisive, critical habits of thought would ever have displaced the acceptance of loose explanations of natural phenomena provided by primitive religion and mythology.[29] The existence of highly confirmed hypotheses, which have been reconfirmed when reinvestigated and have been successfully used as assumptions underlying further investigations, demonstrates the fertility of procedures which are, as Popper especially has insisted, largely eliminative. The historical achievement of overwhelmingly successful hypotheses at various levels of generality has been essential for instilling confidence in critical thought concerning matters remote from mundane affairs (where a certain amount of critical thinking is a component of common sense), and also for counteracting the tendency of critical thought to lapse into sterile cynicism for lack of substantial results. A sophisticated scientific methodology can profit greatly from a detailed analysis by historians of science of this complex interplay of achievement and confidence.[30]

IV. THE SENSITIVITY OF SCIENTIFIC INFERENCE

A

In this section I shall consider to what extent an a priori justification can be given for the tempered personalist formulation of scientific inference. It is a delicate matter to make a priori claims in favor of a method of confirmation while disclaiming a priori knowledge about the world, and one invariably finds that excessive claims in any respect must be paid for by undesirable concessions and assumptions in other respects. An illuminating way to approach this problem is to examine the family

29. See Scheibe and Sarbin (1965) for a discussion of the momentum of a superstition in the absence of a countervailing explanation.
30. There is much to be salvaged from the Hegelian thesis that a dialectical process is at work in the development of intellectual disciplines, but painstaking research in the history of science and philosophy is required to determine what is salvageable. (See Feyerabend 1970.)

of arguments called "pragmatic justifications of induction," which has often been regarded as the most promising a priori defense of inductive inference. All versions of this argument purport to establish, with various qualifications, that by persisting indefinitely in the application of a properly formulated method of inductive inference, we can discriminate true from false hypotheses, however the world is constituted. A brief examination of three of the most interesting pragmatic justifications – those of Peirce, Reichenbach, and Kemeny – will indicate the difficulties of achieving a satisfactory defense along this line. Because Peirce is the most obscure of the three, but also in my opinion the most suggestive of an alternative analysis, I shall disregard historical order and discuss his argument last.

The a priori claims that I shall make in favor of the tempered personalist formulation of scientific inference are relatively modest. It is a method which can navigate systematically, though not infallibly, between those errors due to overskepticism and those due to credulousness; and it can do so in finite intervals of time without requiring the long run. It is designed to complement and support any powers which human beings may possess for making intelligent guesses about nature – though I do not suppose that tempered personalism is in any sense a logic of discovery. Finally, it is capable of assimilating and employing, without abandonment of its basic structure, any methodological device which analysis or experience indicates to be valuable. I feel that one may fairly sum up these claims by ascribing "sensitivity to the truth" to the tempered personalist formulation of scientific inference and also to the informal processes of confirmation used by scientists from which tempered personalism is extracted. There is likely to be disagreement on whether a justification of scientific inference is achieved in this way, but I doubt whether appreciably stronger claims can be established without taking into account the actual constitution of the world.

B

1. Reichenbach conceives the central problem of induction very narrowly: as consisting in the determination of the limits of relative frequencies in infinite empirical sequences. Letting a_n be the number of elements having a specified property among the first n members of a specified sequence of entities and letting $f^n = a_n/n$, he correctly reasons as follows:

If the sequence has a limit of the frequency, there must exist an n such that from there on the frequency f^i $(i > n)$ will remain within the interval $f^n \pm e$, where e is a quantity that we choose as small as we like, but that, once chosen, is kept constant. Now if we posit that the frequency f^i will remain within the interval $f^n \pm e$, and if we correct this posit for greater n by the same rule, we must finally come to the correct result. (1949: pp. 445–46)

He then supplements this conditional claim for the eventual success of the procedure of "positing" by an argument that nothing is lost by supposing the antecedent to be true:

Inductive positing in the sense of a trial-and-error method is justified as long as it is not known that the attempt is hopeless, that there is no limit of the frequency. Should we have no success, the positing was useless; but why not take our chances? (*Ibid.:* p. 363)

The suggestion that the long-run success of induction can be proved only conditionally, but that it is rational to accept the unconfirmable antecedent of the conditional argument, is reminiscent of earlier epistemological proposals, notably Pascal's wager and Kant's treatment of "regulative principles." It will also be seen that Peirce anticipated the employment of such *as if* argumentation for justifying scientific inference. But Reichenbach deserves great credit for bringing this mode of argumentation into prominence and making the explicit point that it provides a new approach to the problem of Hume (cf. 1938: pp. 356–57).

There are, however, a number of crucial defects in Reichenbach's theory of induction. No satisfactory suggestion has been made for treating the probability or the acceptability of theories in terms of the frequency concept of probability, although Reichenbach's intent is to provide a rationale for scientific inference. The ontological status of an infinite empirical sequence of future events of a given kind is dubious. The application of the frequency concept of probability to an individual case in which the outcome is uncertain requires that the case be embedded in a reference sequence, and this can always be done in infinitely many different ways. The mutual relevance of the limit of a sequence of frequencies and the structure of any specified finite initial segment of the sequence is dubious, unless a reason can be given for considering the segment to be a "good sample" of the entire sequence; but Reichenbach's theory prevents him from making any judgment about the goodness of a sample except in a state of advanced knowledge, when the structure of an infinite sequence of infinite sequences is known (1949: p. 443). Finally, the "straight rule" of positing that the limiting frequency is close to f^n is only one of an infinite set of methods which asymptotically lead to the correct value of the limiting frequency if it exists; and each of these methods has equal claim to being a guide to rational decisions about the future.[31] I shall not

31. Salmon (1961, 1963) attempts to show that the straight rule is the only one of this infinite set of methods which also satisfies the reasonable "principle of linguistic invariance"; but Hacking (1965) has proved that other members of this set also satisfy the principle.

discuss these defects in detail, however, since they have been thoroughly explored by Popper (1961), Russell (1948), Burks (1951), Lenz (1958), Katz (1962), and others.

2. Kemeny's work on simplicity (1953) has the great virtues of detaching the pragmatic argumentation from a preoccupation with limits of relative frequencies and of applying it to inductive inferences concerning quite general classes of scientific hypotheses. He assumes that a denumerable class of hypotheses h_i is considered concerning a certain question and that one and only one member h of the class is true. A sequence of experiments is envisaged such that the possible outcomes of the first n experiments are e_j^n, the actual outcome being e^n. Kemeny wishes to give a methodologically defensible rule for "selecting" a hypothesis on the basis of e^n. (The notion of "selecting" is not explained, but apparently he intends something like making a tentative commitment as to which hypothesis is true. He does not discuss other types of conclusions of inductive reasoning, such as ordering hypotheses or assigning to them degrees of credibility, though his remark in 1953, page 407, as well as his other works on induction, indicate his awareness of their importance.) The rule of selection which he proposes after rejecting several alternatives is:

Select the simplest hypothesis compatible with the observed values. (If there are several, select any one of them.) (*Ibid.:* p. 397)

Kemeny defines compatibility in terms of a measure $m(h_i, e_j^n)$ of the *deviation* between a hypothesis and the observed results, but he says little about the choice of m except to require that "the deviation between a given hypothesis and the observed results tends to 0 if and only if the hypothesis is the true one" (*ibid.:* p. 394). He adopts a convention, which is frequent in statistical practice, of taking h to be compatible with e_j^n if upon assumption of h there is at least a 1 percent probability that the outcome of the first n experiments will deviate from h_i by as much as e_j^n does. The crucial and ingenious innovation in Kemeny's work is the proposal of four conditions under which a set of hypotheses is said to be "ordered according to simplicity." None of the four depends upon factual assumptions about nature or upon subjective judgments. The essential condition is that for every hypothesis h_i there is an integer N_i such that if $n \geq N_i$, then the compatibility of h_i with e^n implies that any other hypothesis as simple as h_i, or simpler, is incompatible with e^n.[32] Kemeny's conditions for a simplicity ordering permit him to assert the theorem:

If the true hypothesis is one of the hypotheses under consideration, then – given enough experiments – we are 99 percent sure of selecting it. (*Ibid.:* p. 401)

32. This is a slight modification of condition (3) in Kemeny (1953: p. 403), made for the purpose of avoiding the unessential explanation of a technical term.

This theorem provides the justification for inductive procedures which incorporate Kemeny's selection rule (together with his explications of "compatibility" and "simplicity"). He claims that his procedure is similar to that of working scientists, who do not try to fit their hypotheses exactly to the data, but rather gradually try out more complex hypotheses – whatever that means to them – when the data are too much out of line with the simpler ones.

Kemeny's exposition is quite elliptical, and as a result some points are obscure which probably could be cleared up without difficulty. Thus, he should have stated that an ordering according to simplicity is relative to a sequence of experiments. Also, he does not explicitly state what concept of probability he is employing; apparently he does not regard this as a crucial issue because the only probabilities involved are likelihoods (of e_j^n conditional upon h_i), and there is more general agreement about evaluating these – whatever their significance – than in evaluating prior probabilities. Nevertheless, explicit assumptions need to be made about the likelihoods and also about the measure of deviation in order to justify the following statement, which is crucial to his argument:

As we know from statistics, as *n* increases, the deviations allowed by the compatibility requirement decrease. Hence for high *n* we can find an interval around the observed values (an interval that can be made as small as required by increasing *n*) such that all compatible hypotheses lie within this interval. (*Ibid.*: p. 400)

Indeed, until these points are clarified the criteria for constituting a simplicity ordering remain obscure.

A more important matter, however, is the extent to which Kemeny's theorem does justify his kind of inductive procedure. First of all, the theorem is noneffective: one cannot tell, without already knowing which is the true hypothesis, how many experiments are "enough experiments," since the integer N_i in the explication of "simplicity ordering" depends on *i* – as it obviously must if the hypotheses are permitted to come "closer" to one another as one proceeds in the ordering. It should be emphasized that "we are 99 percent sure of selecting it" does not mean that $P(h^n | e^n) = .99$, and in fact Kemeny does not claim to provide machinery for evaluating $P(h^n | e^n)$. In short, Kemeny's procedure shares with that of Reichenbach the disadvantage of requiring that one live to "the ripe old age of denumerable infinity"[33] in order to draw an inductive conclusion with confidence. Furthermore, as Kemeny recognizes, his criteria for a simplicity ordering do not determine the order uniquely, and evidently the use of different orderings would in general entail different selections of hypotheses at each stage of experimentation; and, as Katz emphasizes (1962:

33. Attributed to Carnap by Salmon (1965).

pp. 86ff), there seems to be little hope of choosing an optimum ordering by means of a condition that N_i should be made as small as possible.[34] Finally, Kemeny's theorem is conditional in form ("If the true hypothesis is one of the hypotheses under consideration . . ."), but the kind of *as if* argumentation which Reichenbach applied to the antecedent "If the sequence has a limit . . ." is not legitimate here. It may very well happen that the truth is a member of a different class of hypotheses, and it is not the case that nothing would be lost by pretending that the antecedent is true. Many of the dramatic episodes in the history of science consisted first in the suggestion of the plausibility of a previously unconsidered class of hypotheses and then in the exhibition that the truth probably is to be found in the new class.[35]

3. It is difficult to summarize and evaluate Peirce's theory of scientific inference, partly because he changed his opinions in important respects during his career without writing a definitive statement of his latest doctrine and partly because of obscurities of exposition. I suspect that these textual difficulties reflect the intellectual difficulties which Peirce experienced in attempting to develop to his own satisfaction the ingenious and attractive idea that one can be certain of asymptotic approach to the truth by means of the scientific method. However, the tension between his commitment to this idea and his self-criticism may have been responsible for a number of insights which are valuable even if his central idea is unworkable.

Peirce's broadest justification of the scientific method is that it approaches the truth because of its submissiveness to reality; it is a method

by which our beliefs are determined by nothing human, but by some external permanency – by something upon which our thinking has no effect. But which, on the other hand, unceasingly tends to influence thought; or in other words, by something Real. (5.384. The last sentence is a later addition to the passage by Peirce and is placed in a footnote by the editors.)

Again broadly speaking, the scientific method achieves this submissiveness by systematically and self-critically correcting beliefs in the light of experience (for example, 7.78).

34. A little progress along these lines is possible, however, by noting that good statistical discrimination between two hypotheses requires more data the "closer" the hypotheses are to each other. Consequently, it is desirable to choose orderings which keep very "close" hypotheses as far apart as possible. This desideratum ought to be formulable precisely so as to yield a *partial ordering* of the class of orderings.
35. One might attempt to answer this objection by taking the class of considered hypotheses to consist of all noncontradictory hypotheses formulable in a given language. But, as Putnam points out (1963: p. 775), in a nontrivial language this class cannot be effectively enumerated.

Peirce's characterization of the scientific method (7.80–88) and else-where is very rich, and it includes not only an analysis of modes of infer-ence but also fine considerations of heuristics, of the relation of science to metaphysics, and of the ethics of inquiry. He is evidently strongly drawn by the idea of an asymptotic approach to the truth by means of the scien-tific method *as a whole* (for example, 7.77), even though he recognizes the sporadic character of some elements in the method, especially the proposal of hypotheses (which he variously refers to as "abduction," "pre-sumption," "hypotheses," and "retroduction"). However, the only clear example of an infallible asymptotic approach which he offers is the simple one which is the heart of Reichenbach's treatment of scientific inference: the evaluation of the limit of relative frequencies in infinite sequences of events (for example, 2.650, 6.100, 7.77, 7.120). Since this kind of inference ("statistical" or "quantitative" induction) is only one of the three kinds of induction which he recognizes, and since induction taken generically is not the whole of the scientific method, even sympathetic commentators on Peirce have found that his demonstrations fall far short of realizing his general program (for example, Murphey 1961, Burks 1964, Lenz 1964, Madden 1964).

Peirce may have underestimated the gap in the realization of his pro-gram by overestimating the amount of information about the statistical structure of an infinite sequence that can be obtained from a finite seg-ment of it. He does not maintain, like Reichenbach, that the success of induction in dealing with a particular sequence depends upon the exis-tence of a limit of the sequence; nor does he resort to the argument that nothing is lost by acting as if the limit exists. Instead, he claims that

if experience in general is to fluctuate irregularly to and fro, in a manner to deprive the ratio sought of all definite value, we shall be able to find out approximately within what limits it fluctuates and if, after having one definite value, it changes and assumes another, we shall be able to find that out, and in short, whatever may be the variations of this ratio in experience, experience indefinitely extended will enable us to detect them, so as to predict rightly, at last, what its ultimate value may be, if it have any ultimate value, or what the ultimate law of succession of values may be, if there be any such ultimate law, or that it ultimately fluctuates irregularly within certain limits, if it do so ultimately fluctuate. (6.40)

Although he says disappointingly little about the means for extracting so much information in general circumstances, I find two hints about his thinking. One is that the structure of erratic sequences is to be investi-gated "with the aid of retroduction and of deductions from retroductive suggestions" (2.767), which indicates that he held no illusions about the existence of an algorithm for the purpose. The other is an appeal to a constructivist theory of the infinite, on the basis of which he seems to make excessive claims for the effectiveness of inductive procedures:

Whatever has no end can have no mode of being other than that of a law, and therefore whatever general character it may have must be describable, but the only way of describing an endless series is by stating explicitly or implicitly the law of the succession of one term upon another. But every such term has a finite ordinal place from the beginning and therefore, if it presents any regularity for all finite successions from the beginning, it presents the same regularity throughout. (5.170)

The first of these passages seems to me very sensible, but in view of the chance character of the proposal of hypotheses, it weakens rather than strengthens the program of establishing the long-range infallibility of the scientific method. The second passage does indeed seem to support his program, except that I see no way that it can be construed so as not to be fallacious: for whatever the true sequence $\{f^i\}$ of fractions may be, it is not the case that there exists an n (even an unknown n) such that $\{f^i\}$ is the only law-governed sequence with the initial segment $f^1, ..., f^n$.

I am inclined to believe that the statements of Peirce which throw the most light upon inductive inference are those which qualify his central idea or are tangential to it. For example, he presents an *as if* argument, though its locus is different from Reichenbach's, for it is a justification of the process of abduction rather than of the assumption that a sequence of relative frequencies converges:

I now proceed to consider what principles should guide us in abduction. . . . Underlying all such principles there is a fundamental and primary abduction, a hypothesis which we must embrace at the outset, however destitute of evidentiary support it may be. That hypothesis is that the facts in hand admit of rationalization, and of rationalization by us. That we must hope they do, for the same reason that a general who has to capture a position or see his country ruined, must go on the hypothesis that there is some way in which he can and shall capture it. (7.219. See also 5.145, 5.357, 6.529, 7.77)

It should also be noted that if statistical induction must be investigated "with the aid of retroduction," as he says in a passage cited earlier, then the justification of induction derivatively depends upon an *as if* argument.

One important qualification of Peirce's central argument for justifying induction is his requirement that the set of events upon which an estimate of probability (in the sense of relative frequency in the long run) is based should be randomly chosen from the population under investigation (for example, 2.726). The principle of random sampling permits Peirce to escape from one of the objections raised above against Reichenbach's theory – that the limit of a sequence of frequencies and the structure of a specified segment of the sequence are mutually irrelevant. Peirce is able to speak of the probable error at any finite stage of the process of investigation (cf. 2.770). However, the concept of a "random" or "fair" sample is implicitly probabilistic, and, therefore, its employment in the process of estimating a probability appears to be an inversion from the standpoint of the frequency theory of probability (as Reichenbach points out in 1949: p. 446). There is a further complication. According to Peirce, a sample is random if it is "taken according to a precept or method which, being applied over and over again indefinitely, would in the long run result in the drawing of any one set of instances as often as any other set of the same number" (2.726). But his doctrine of dispositions repudiates any identification of a "would-be" with what actually happens (for example, 2.664), and, therefore, it is difficult for him to provide a criterion for randomness of sampling without employing a nonfrequency concept of probability. The concept of "verisimilitude" (2.663) comes close to being such a concept, and he goes so far as to say (in a letter) that "all determinations of probability ultimately rest on such verisimilitudes" (8.224). This line of thought, which appears late in his career, is unfortunately not developed very fully.

The problem of establishing criteria for randomness led Peirce to insert ethical considerations into the inductive process itself. For example, "the drawing of objects at random is an act in which honesty is called for; and it is often hard enough to be sure that we have dealt honestly with ourselves in the matter, and still more hard to be satisfied of the honesty of another" (2.727). He also makes the following intriguing suggestion of a minimal presupposition regarding the reliability of data:

I am willing to concede, in order to concede as much as possible, that when a man draws instances at random, all that he knows is that he *tries* to follow a certain precept; so that the sampling process might be rendered generally fallacious by the existence of a mysterious and malign connection between the mind and the universe, such that the possession by an object of an *unperceived* character might influence the will toward choosing it or rejecting it. . . . I grant then, that even upon my theory some fact has to be supposed to make induction and hypothesis valid processes; namely, it is supposed that the supernal powers withhold their hands and let me alone, and that no mysterious uniformity or adaptation interferes with the action of chance. (2.749)

This passage suggests an *as if* argument to the effect that we have nothing to lose by assuming our experimental evidence not to be distorted by factors which are unperceivable by us.

To summarize, I find at least four methodological ideas of great value in Peirce's papers on scientific inference: that the scientific method achieves its successes by submission to reality, that a hopeful attitude toward hypotheses proposed by human beings is indispensable to rational investigation of the unknown, that a usable criterion of fair sampling involves subjective and ethical considerations, and that it is rational to make certain weak assumptions about the fairness of the data in order to permit inquiry to proceed. His suggestions on the instinctive basis of abduction go beyond methodology and will be discussed in section V.

C

Of the three treatments of inductive inference discussed above, Peirce's comes closest to exhibiting how the informal methods of confirmation actually used by scientists are sensitive to the truth, but even it requires corrections in various respects. Implicit in these informal processes is a wonderfully balanced and sinuous strategy with a strength that can be recognized a priori, despite the possibility that it will yield the truth only if the world is not too deceptively constituted. The primary virtue which I claim for tempered personalism is that it succeeds in catching much of the essence of this strategy: maintaining a balance between open-mindedness toward proposals and a critical attitude toward them; mediating between tenacity concerning plausible hypotheses, even in the face of a moderate amount of adverse evidence, and skepticism concerning artificial explanations to shore them up; utilizing the formally structured reasoning of probability theory in conjunction with informal and intuitive thinking; paying respect to the community of investigators and also to the intellectual conscience of the individual; disentangling questions for stepwise investigation while appreciating their interconnectedness.

Some of the strength which I attribute to tempered personalism can be conveniently exhibited by a comparison with Kemeny's proposals. His procedure partially catches the balanced strategy of informal scientific inference, for his rule of selection combines a tenaciousness regarding preferred hypotheses (by using a tolerant standard of compatibility between hypothesis and evidence) with a critical attitude toward them (by rejecting a more preferred hypothesis in favor of a less preferred one, if the former turns out to be incompatible with the data).[36] Although his

---

36. See also Putnam (1963: pp. 772, 775), who suggests that *corrigibility* and *tenacity* are essential characteristics of a good inductive method. He differs from Kemeny, however, by proposing inductive methods in which "the acceptance of a hypothesis depends on which hypotheses are actually proposed, and also on the *order* in which they are proposed" (*ibid.:* pp. 771, 775). The preferred status of actually proposed hypotheses is a crucial feature which both my treatment of scientific inference (derivative from Jeffreys) and Putnam's have in common. However, my formulation of scientific inference is Bayesian, whereas his is not. One consequence of this difference is that acceptance plays a less central role in my treatment than in his, for a Bayesian recognizes

procedure and tempered personalism are similar in this very important respect, there are obvious differences between them.

Tempered personalism has no rule for "selecting" a hypothesis at every stage of experimentation, nor even at the end of an investigation, which may very well conclude with the approximate equality of several seriously proposed hypotheses. The nearest approach to a rule of selection in tempered personalism is the process of "accepting" a hypothesis which has posterior probability close to 1 (discussed in section III.F), and the tentative nature of this acceptance has been pointed out. In Kemeny's procedure tentativeness is manifested only in the possibility that new evidence will lead to the selection of a different hypothesis, whereas in tempered personalism tentativeness is suffused throughout a person's theoretical view of the world.[37]

The most profound difference between the two procedures lies in their principles of preference among hypotheses antecedent to observation. Kemeny gives preference to the earlier members of a predesignated infinite sequence of hypotheses – the order of the sequence being arbitrary, subject to the limitations discussed in section IV.B. The principle of preference in tempered personalism, which is essentially contained in the tempering condition, is always relativized to a particular person and a particular time: he must assign nonnegligible prior probability to each seriously proposed hypothesis $h_1, \ldots, h_n$ of which he is aware concerning the matter under investigation, whereas all other hypotheses must be treated as disjunctive components of the catchall hypothesis and, hence, must be assigned prior probabilities which are generally many orders of magnitude smaller than those assigned to $h_1, \ldots, h_n$ (and indeed are infinitesimal in the case of a continuum of disjunctive components); the ordering and the relative weighting of $h_1, \ldots, h_n$ are subjective, within the limits set by the tempering condition. Although the principle of preference in tempered personalism seems unsystematic from the standpoint of any predesignated ordering of hypotheses, it has several great advantages: the most obvious is that tempered personalism dispenses with an assumption that the truth lies in a predesignated class of hypotheses. It was pointed out in section IV.B that historically assumptions of this kind have been false and methodologically one has no justification for acting

the possibility that a delimited investigation may terminate with approximately equal posterior probabilities assigned to rival hypotheses.
37. Although the problems of inductive inference in the context of practice are not within the scope of this paper, I think it is worth remarking that a theoretical view of the world which consists of varying degrees of commitment to rival hypotheses lends itself better to practical application than a view which always tentatively selects one hypothesis as true.

as if such an assumption were true. Tempered personalism permits an investigator to transcend any initially limited class of hypotheses simply by admitting as seriously proposed some hypothesis not belonging to that class. Even if a hypothesis seriously proposed by a very perceptive scientist does fall within Kemeny's predesignated class, a very long interval – perhaps longer than the duration of the human race – might have to elapse (at current rates of performing experiments) before the predesignated ordering of hypotheses would permit it a chance to be selected. The mechanical consideration of hypotheses in an order established a priori is, therefore, an abnegation of the clues which nature may provide unexpectedly as an investigation develops and of the powers which men immersed in a subject may possess of utilizing these clues. At the present stage in my treatment of scientific inference, I wish to avoid making factual assumptions about the world and, therefore, cannot say anything about the actual occurrence of reliable clues or about the capacity of investigators to utilize them fruitfully. I can say at this stage, however, that unless we act *as if* good approximations to the truth will occur among the hypotheses which will be seriously proposed within a reasonable interval, we are in effect despairing of attaining the objective of inquiry.

Kemeny's procedure disregards the proposals put forth by investigators on intuitive grounds because it is not only a method of confirmation but something of a "logic of discovery"; for once the class of admissible hypotheses has been chosen and ordered and a measure of compatibility has been defined, it provides an algorithm for selecting a hypothesis h as a function of the data e_j^n. But mechanizing the process of selection leaves no role for intelligent guessing about nature except in the initial choice of the sequence of hypotheses – a sweeping operation which requires much greater intuitive powers than do the individual conjectures of men immersed in specific problems. Tempered personalism, on the other hand, is in no way a "logic of discovery" but rather supports whatever powers human beings may have for making intelligent guesses. It is well adapted to the possibility that these powers are exhibited sporadically and in very different degrees by different people, and that they are refined and stimulated by the progress of knowledge. In contrast to a method which imposes an a priori ordering upon hypotheses, tempered personalism can take advantage of the possibility that a profound scientific discovery can have the effect of making hypotheses which previously would have been extremely remote in any plausible ordering seem, at least from a subjective standpoint, "natural" and simple. (The outstanding example is the discovery by Galileo and Newton of the relation between force and acceleration, whereby second-order differential equations became a familiar conceptual tool; the solutions to some of the commonest of these equations appear extremely complex from the standpoint of a naïve direct ordering of functional relationships.) Tempered personalism thus avoids the skepticism toward human abductive powers implicit in any formal scheme which treats on the same footing seriously proposed hypotheses, frivolously proposed hypotheses, and unsuggested hypotheses. By giving preferential treatment to seriously proposed hypotheses but insisting upon open-mindedness within this preferred class, the tempering condition provides a safeguard against one of the major types of error that could be committed by a method of confirmation: the error of rejecting, because of a priori commitments, a true hypothesis which some one has been fortunate enough to put forth. In this way tempered personalism incorporates Saki's great methodological maxim: "In baiting a mousetrap with cheese, always leave room for the mouse."

Because of the informality of its principle of preference among hypotheses, tempered personalism is in a sense on a metalevel relative to any formal procedure for prior ordering and weighting. If a formal ordering of hypotheses, such as one of Kemeny's or that of Jeffreys and Wrinch, is seriously proposed, and if the first n hypotheses in the ordering have been assigned extremely low posterior probabilities in previous investigations, the $(n+1)$th could be considered a seriously proposed hypothesis and, hence, would be assigned a nonnegligible prior probability in a new investigation. In this way the formal ordering would be interleafed with the order of intuitive proposals. The interleafing would permit the systematic exploration of a predesignated class of hypotheses, concurrently with excursions into possibilities which are very remote in the ordering or which lie outside it. The informal excursions would be fruitful if scientists are sufficiently imaginative and if the experimental data are sufficiently suggestive, whereas the plodding exploration of the ordered class of hypotheses would occupy idle equipment and would provide exercises in technique, with the possibility of unexpected striking confirmations, during intervals in which imagination is barren – though one may properly doubt whether the institution of scientific research would survive if these intervals became excessively long.

D

In order to complete the discussion of the sensitivity of the tempered personalist formulation of scientific inference to the truth, it is necessary to show that its receptivity toward seriously proposed hypotheses is adequately balanced by a capacity to evaluate them critically. Two elements of its apparatus are intended to serve this purpose: the catchall hypothesis, which says that none of the specific alternatives under consideration is true, is taken as seriously proposed and is, therefore, assigned a nonnegligible prior probability; and the posterior probability of each hypothesis is dependent, because of Bayes's theorem, upon the likelihood relative to it of the evidence actually gathered. The first of these has been discussed at several places in section III, but little has yet been said about likelihoods.

Prima facie the evaluation of the likelihoods $P_{X,b}(e\,|\,h_i\,\&\,a)$ is a subjective process, limited only by the axioms of probability, for the tempering condition refers explicitly only to assignments of prior probabilities. However, it will be seen that the tempering condition is relevant to the evaluation of likelihoods and restricts the subjectivity of these evaluations. Furthermore, the methodological arguments in favor of the tempering condition provide a partial justification for relying upon likelihoods in critically judging hypotheses.

Typically, the information and assumptions contained in a permit the evaluation of the likelihoods $P_{X,b}(e_m\,|\,h_i\,\&\,a)$ to be derivable from judgments of indifference concerning a set of possible experimental outcomes, once certain known or presumed differences are dismissed as irrelevant to their occurrence.[38] For example, in the Laplacean sampling problem (considered in section III.E, but without attention to likelihood evaluations),

38. The judgments of indifference that are made once all known and presumed differences among the possible outcomes are dismissed as irrelevant do not, as I see the matter, need to be based upon a "principle of indifference." The judgments $P_{X,b}(e_m\,|\,h_i\,\&\,a) = P_{X,b}(e_{m'}\,|\,h_i\,\&\,a)$ are immediate results of dismissing the differences between e_m and $e_{m'}$ as irrelevant, without the mediation of such a principle. Nor do I see what further justification can be given for this judgment other than the two grounds for dismissal of differences which were discussed above – namely, an investigation, with a negative conclusion, of the serious proposal that a certain difference is relevant, and the methodological argument against an insatiable suspiciousness of the presence of systematic errors. In other words, there seems to be no need for a principle of indifference which states the equiprobability of the propositions asserting various outcomes, once *all individuating differences* among these outcomes are suppressed; and, indeed, there are semantic difficulties even in *clearly formulating* a stringent principle of this kind. To be sure, there are some clearly formulated rules of equiprobability, notably Carnap's axioms of invariance. According to him, these axioms "represent the valid part of the principle of indifference, whose classical form, e.g., in the system of Laplace, was too general and too strong and was therefore correctly rejected by later authors" (1963: p. 975). These axioms are, in effect, systematic attempts to dismiss as irrelevant certain types of differences among propositions (though Carnap prefers to speak in terms of the sentences of a definite language rather than in terms of propositions). For example,

A7. The values of $c(h,e)$ remain unchanged under any finite permutation of individuals.

. . . .

A8. The value of $c(h,e)$ remains unchanged under any permutation of the predicates of any family. (*Ibid.:* p. 975)

It is, I believe, legitimate and important to systematize the dismissal of differences, but I can see no other kind of justification of these axioms of invariance than the two grounds mentioned above for dismissal of differences among cases. However, the *a* includes the information that the population of interest consists of N objects, of which r are selected successively by some process; and h_i asserts that i of the members of the population have the property α. Ordinarily the known and presumed differences among the $N-x$ objects remaining in the population after the selection of the first x members of the sample are dismissed as irrelevant to the selection of the $(x+1)$th. In particular, the individuating differences whereby the members are identifiable as "object 1," "object 2,"..., "object N" are dismissed as irrelevant, as is the difference between possession and nonpossession of the property of interest α. By dismissing these differences, a judgment of indifference is possible, that is, if s_x asserts that the first x objects selected were objects $j_1,...,j_x$, and if e_m and $e_{m'}$ respectively assert that the mth and m'th objects will be selected at the $(x+1)$th draw (where neither m nor m' equals any of the $j_1,...,j_x$), then

$$P_{X,b}(e_m \mid h_i \, \& \, a \, \& \, s_x) = P_{X,b}(e_{m'} \mid h_i \, \& \, a \, \& \, s_x).$$

The evaluation of $P_{X,b}(d_{mn} \mid h_i \, \& \, a)$ – which are the likelihoods involved in section III.E – follows straightforwardly by combinatorial analysis and the axioms of probability. A similar but more complex analysis can be carried through in more realistic problems, where the h_i are competing scientific hypotheses, *a* includes auxiliary assumptions about natural laws and boundary conditions and also about the statistical characteristics of the measuring instruments employed, and e_m and $e_{m'}$ are replaced by propositions concerning individual instrument readings. The evaluation of likelihoods in problems involving measurement depends upon a tacit judgment of indifference that the experimental error of each reading is drawn without bias from a population of errors associated with the instrument; and this judgment is possible only if the special circumstances of the investigation, which conceivably could cause a systematic error, are dismissed as irrelevant.

The tempering condition is relevant to the dismissal of differences among cases. Indeed, it permits a kind of social check upon personal judgments about the data. Someone may seriously make the proposal *g* that the method of selecting a sample is favorable or unfavorable in a certain manner to objects having the property α. For example, ecological experimentation requires sampling from all the members of a species in a given area by such means as trapping, and the susceptibility to being caught may be supposed to be correlated with the property α. The

process of dismissal of differences may not be amenable to the kind of generalization which Carnap is attempting. In particular, some of the criticisms which have been made against (A8) (e.g., Salmon 1961: p. 249) indicate that the irrelevance of differences among predicates of families must be investigated piecemeal, a family at a time.

tempering condition applies to g as to all serious proposals, for as Jeffreys remarks, "There is no epistemological difference between the Smith effect and Smith's systematic error; the difference is that Smith is pleased to find the former, while he may be annoyed at the latter" (1961: p. 300). The prescription that g be given nonnegligible prior probability could be followed by doubling the set of seriously proposed hypotheses in the investigation at hand – that is, by taking as seriously proposed not only $h_1, \ldots, h_n$ but also $h_1 \& g, \ldots, h_n \& g$. However, in view of the methodological advantages of disentangling problems (discussed in section III.A), it is generally preferable to conduct an auxiliary investigation concerning the bias of the sampling procedures.

Even after all serious proposals concerning possible bias have been checked and appropriate corrections in the sampling procedure (or in the analysis of sampling data) have been made, there always remain special circumstances concerning a sample which are unnoticed by investigators or which are habitually dismissed as trivial, and one might skeptically suspect that a systematic error is associated with them. The proper answer to this kind of skepticism is that a possible source of systematic error which no one has proposed for serious consideration is merely one of the infinite set of possible but unsuggested alternatives to the seriously proposed hypotheses in the investigation, and, hence, it should be treated as a component of the catchall hypothesis. It was argued previously that seriously proposed hypotheses should be treated differently from all others in order to give intelligent proposals a chance of acceptance into the corpus of scientific knowledge. In the case of hypotheses regarding systematic errors, an additional reason can be given for this differential treatment: that unless the unsuggested hypotheses are given very small prior probabilities, no evidence could be utilized as reliable grounds for critically judging the seriously proposed hypotheses. A certain amount of tentative trust is a prerequisite for critically probing and testing, whereas sweeping skepticism is methodologically sterile. Jeffreys gives a similar warning:

A separate statement of the possible range of the systematic error may be useful if there is any way of arriving at one, but it must be a separate statement and not used to increase the uncertainty provided by the consistency of the observations themselves, which has a value for the future in any case. In induction there is no harm in being occasionally wrong; it is inevitable that we shall be. But there is harm in stating results in such a form that they do not represent the evidence available at the time when they are stated, or make it impossible for future workers to make the best use of that evidence. (*Ibid.:* p. 302)

One qualification should be added at this point, even though it is quite obvious: that the recommendation to treat as statistically irrelevant those experimental circumstances which are not seriously proposed as relevant must not be construed as an excuse for slovenliness in experimentation or for relaxation of the vigilance of the investigator regarding his own prejudices. That samples may be skewed, perhaps unconsciously and in subtle ways, so as to favor the preferred hypothesis of the experimenter, should be a standing serious proposal of a possible source of systematic error. (See Peirce 2.727, quoted in section IV.B.)

E

There was no reference to the "long run" in the claims made above that the tempered personalist formulation of scientific inference has a kind of sensitivity to the truth. An examination of a situation which has been fully analyzed in the literature – the Laplacean sampling problem with replacement – will suffice to indicate that consideration of the long run does not lead to an essential strengthening of the foregoing claims. In the sampling problem with a finite population but with replacement, no apodictic statement can be made on the basis of a finite or infinite number of draws other than that h_0 is false if at least one object with property α is drawn, and h_N is false if at least one object lacking this property is drawn.

It is well known that if probability theory is formulated so as to permit the treatment of infinite sets of alternatives (by using the axiom of complete additivity – note 8), then there exist disjoint sets S_i of infinite binary sequences of α's and $\bar{\alpha}$'s such that $P_{X,b}(s \in S_j \mid h_i \& a)$ is 1 if $i = j$ and 0 if $i \neq j$, where s is the sequence actually obtained in an infinite sequence of draws (for example, Kac 1959: pp. 18–22). Bayes's theorem then yields $P_{X,b}(h_i \mid s \in S_j \& a) = 1$ if $i = j$ and 0 if $i \neq j$. But it is also well known that these results do not permit nontrivial inferences to be made with certainty about the composition of the population, since the probability 0 of a proposition does not entail its falsehood. The kind of conditional apodictic statement which Reichenbach's analysis permits is tangential to the point of interest here. He would say that if s has a definite statistical structure (which for him means that the sequence of relative frequencies of α's in finite segments converges), then for any $\epsilon > 0$ there exists an n such that the relative frequency of α's in every segment of length greater than n differs from the limit by less than ϵ. But the question under investigation concerns the composition of the *population,* and the sequence of draws from it is only a means for learning about that. To deny that this is so, and to say that the only question of human interest or of scientific interest is the structure of s, implies a commitment to a kind of phenomenalism, which is incompatible with the Copernican point of view. One more remark is too obvious to require elaboration: that nontrivial inductive problems in the natural sciences, in which the seriously proposed hypotheses imply propositions about the statistical distribution over possible observations in a potential infinity of situations, are closer in character to the sampling problem with replacement than without replacement.

The one point in the ascription of sensitivity to tempered personalism where reference to the long run may be required is its capability of assimilating other methodological devices. It was seen, for example, that any one of Kemeny's ordering of hypotheses according to simplicity can be interleafed with an ordering of seriously proposed hypotheses; but because of the intrinsically long-run objective of Kemeny's procedure, no advantage of this interleafing can be exhibited in any finite segment of the sequence.

The foregoing considerations indicate the futility of trying to establish without qualification that *the tempered personalist formulation of scientific inference is a method which, if persisted in indefinitely, can discriminate true from false hypotheses* (call the italicized statement "Ψ"). Nevertheless, it is tempting to take Reichenbach's pragmatic justification as a paradigm and to make a conditional apodictic claim – that is, *if conditions (1), (2), . . . are satisfied, then Ψ* – and to do so in such a way that a methodological argument can be given for acting as if conditions (1), (2), etc., are true, whether or not their truth can ever be checked. Indeed, a step in that direction was already taken in section IV.C, in reasoning that "unless we act *as if* good approximations to the truth will occur among the hypotheses which will be seriously proposed within a reasonable interval, we are in effect despairing of attaining the object of inquiry." This reasoning supplies a natural condition (1) for the desired conditional claim. Unfortunately, in spite of long reflection on the matter, I see no way of completing the list of conditions in a way that is neither hopelessly obscure nor hopelessly hobbled.

V. CIRCUMSTANCES FAVORABLE TO INDUCTION

A

The foregoing a priori analysis of the tempered personalist formulation of scientific inference showed several ways in which errors could be committed in spite of systematic apparatus to prevent them, and gave reasons for believing that any formulation of scientific inference would be fallible in the same way. Moreover, since scientific inference is not a method of discovery, it cannot exclude the possibility of a nonerroneous, but thoroughly demoralizing, rejection of every specific scientific hypothesis that will be proposed during an indefinitely long period of time concerning some aspect of nature. Consequently, if we wish to understand the fact that we are now in possession of a vast and detailed and apparently reliable system of scientific knowledge, the explanation cannot be entirely methodological but must refer to facts about the world.

What kind of explanation can reasonably be expected for the success of scientific inquiry is far from clear, for consider how many general and particular facts are relevant to any scientific discovery: the laws of the domain in which the discovery is made, the laws governing the instruments used in observations, the psychological principles of perception and concept formation, the facts of geology and meteorology that permit observations to be made, the sociological facts about the milieu in which the research was undertaken, and the crucial biographical facts about the discoverers. In a sense the only adequate explanation of the existence of the system of scientific knowledge would consist of an encyclopedia of the natural and social sciences, together with accounts of the contingencies in the history of science. In spite of this holistic consideration, it surely should be possible to give good partial explanations by intelligently following important threads in the total fabric; and grounds for optimism regarding such enterprises are provided by the existence of illuminating work in the history, sociology, and psychology of science.

The first purpose of this section is to propose a special line of partial explanation, which consists in *exhibiting circumstances complementary to and supporting the logical structure of scientific inference.* There are several crucial junctures in the formulation of scientific inference at which the methodological prescriptions are counsels of desperation unless circumstances are favorable to inquiry. At these junctures scientific inference proceeds as if human faculties and the natural environment are favorable in various ways to determining the truth: as if the background of information and assumptions (of which only a small part can be critically tested) is on the whole reliable, as if problems can be disentangled for stepwise treatment, and as if some good approximations to the truth are to be found among the hypotheses seriously proposed within some reasonable interval of time. The partial explanation of the success of scientific inquiry which I have in mind consists in showing that nature is indeed favorable to inquiry at these junctures. The explanation should be to some extent independent of a detailed characterization of the environment and of human faculties, because the strategy of scientific inference is designed to be sinuous and adaptable to a wide range of possible worlds. Within reasonable limits it should be possible to make a plausible case that if a certain fact (or even a law) had been otherwise, the truth about the situation could nevertheless have been discovered. An explanation along these lines has a unitary character even though it evidently involves a rather special selection of topics from psychology and other sciences and, therefore (like all partial explanations of the success of the natural sciences), must tacitly refer to a comprehensive body of knowledge in these fields.

The second purpose of the section is to examine the methodological consequences of the envisaged partial explanation of the success of scientific inquiry. The primary consequence is to provide an a posteriori justification for various characteristics of the tempered personalist formulation of scientific inference. Thus, the "localization" of problems is seen to be well suited to investigations in the actual world. Also the role of subjective judgment in evaluating probabilities and in other aspects of scientific inference is to some extent sanctioned, for the adaptation of human beings to the natural environment ensures that a crude rationality is operative in our mental activity even when we are unable to make it articulate. An entirely different consequence is that antecedent knowledge about the world can provide guidelines which supplement both subjective judgment and the a priori structure of scientific inference. Some examples will be given of guidelines for deciding whether or not to classify a hypothesis as seriously proposed, and for evaluating probabilities on the grounds of indifference and simplicity.

The problem of circularity is evidently raised by the fact that a body of natural knowledge, which was itself tentatively established by means of scientific inference, is methodologically significant. An analysis of the sense in which scientific inference is circular and an argument that the circularity is nonvicious will conclude the essay.

B

I shall begin the discussion of the circumstances which support the structure of scientific inference by considering the existence of reliable prescientific knowledge and then shall outline some of the reasons for believing that nature is also favorable to scientific inference at other junctures. It will be seen throughout that this favorableness is easily indicated in a sweeping and impressionistic manner but that a deep and precise analysis is difficult.

1. Although scientific and prescientific elements are inextricably mixed in the belief system of a scientist, it is possible to discern in this system a common sense picture of the everyday world which on the whole is independent of his professional training and research.[39] This picture is indispensable as a background to controlled scientific inquiry, since many

---

39. The common sense of our own culture is, of course, very much affected by the results of science, as Dewey points out (1938: p. 75).

of the terms occurring in the evidence, and some of those occurring in the assumptions a and in $h_1, \ldots, h_n$ as well, presuppose knowledge about the spatio-temporal structure of the everyday world, correlations among classes of appearances, object constancy under change of viewing conditions, etc.[40] The fact that we are at home in the everyday world, to the extent of coping with most of the practical problems of staying alive, suffices to show in a coarse way that this picture is reliable. But in order to achieve a fine understanding of its reliability, we need to know exactly what the content of this picture is, how much of it is culturally determined, by what processes it is acquired, and whether any part of it can be characterized as innate. These are deep questions in psychology (though overlapping somewhat with biology and with the social sciences), and all of them are at present largely unsettled.[41]

Nevertheless, I have the impression (based on a very limited knowledge of the literature) that an intricate but coherent set of answers to these questions is beginning to emerge. The evidence is accumulating that there is a common cross-cultural core to pictures of the everyday world. This thesis was challenged by Whorf (1941 and elsewhere), who emphasized

40. Even those philosophers who maintain that the observation reports of scientists could in principle be formulated in a phenomenalistic language usually recognize that a "thing" language is somehow much more convenient (for example, Carnap 1950: pp. 23–24). However, I think that a thorough psychological study of the ontogenesis of common sense knowledge will (and to a large extent already does) show the impossibility in principle of a purely phenomenalistic observation language, thus supporting the contentions of such antiphenomenalists as Sellars (1963: pp. 83–84 and elsewhere) and Strawson (1959).

41. Also, although there is no doubt that in some sense the pictures of the everyday world are reliable, the exact sense and the exact extent are far from clear. One finds great variation among philosophers of science in characterizations of the relationship between common sense and scientific pictures of the world – with Eddington, for example, emphasizing their discrepancies (1928: pp. xi–xiii and elsewhere), and Bohr, at the opposite extreme, insisting upon the continuity of the concepts of theoretical physics with common-sense concepts and indeed with the "forms of perception" (1934: p. 1). Clarification of these matters is difficult, since it requires an explicit characterization of the content of the common core of representations of the everyday world, a comparison with relevant scientific knowledge, and some sensible criterion of what constitutes reliability. I shall not attempt this clarification here but shall only mention a few considerations. (1) For the purpose of biological adaptation a "reality principle" is needed only in coping with events on a human scale, not with microscopic or with cosmic events. (2) Some of the features of everyday pictures of the world which are most profoundly erroneous from a scientific point of view, such as the attribution of sensed colors to physical objects, are practically advantageous (cf. Whitehead 1929, pt. II: chap. 8). (3) The common core of representations of the world is surely vague in view of its incorporability into very different conceptual systems. (4) The effect of this vagueness is mitigated in practical activity by the "inarticulate intelligence" of our sensorimotor behavior (Polanyi 1958: pp. 71ff; Piaget 1952: *passim*).

the great variation among languages in such fundamental respects as the syntactical treatment of time, action, and objects. However, the connection between language and world view is probably much looser than Whorf maintained (cf. Hockett 1954). Furthermore, the view that linguistic differences are revelatory of profound divergences in metaphysics is undermined by the mass of evidence in favor of the existence of formal linguistic universals (Chomsky 1965: pp. 29ff). A plausible proposal about the ontology of the common core was made by Strawson (1959): that objects, which are many-faceted, quasi-permanent things, have a more fundamental status than events and appearances, and that there exist persons, who are located in space and time like objects and who can interact with them, but are unlike objects in being the subjects of feelings, thoughts, and intentions. Yet this way of putting the matter may be excessively abstract and intellectual, and it should probably be supplemented by considering the sensorimotor and perceptual components of common sense (cf. Flavell 1963: pp. 129–50). Both the existence of a common core of representations of the world and the depth of its biological grounding are indicated by clinical findings that circumstances which seriously impede the internalization of basic common sense principles – for example, artificial handling which insulates an infant from persons or interference with the normal sequence of events in a way that affects the child's concept of objects – tend to induce psychoses (for example, Vernon 1962: pp. 182–83). We also have some insight into the way in which the common core of representations of the everyday world is ripe, so to speak, for incorporation into very diverse comprehensive conceptual systems: for example, objects are many-faceted and are universally recognized as being capable of deceptive appearances, and, furthermore, they probably are universally understood to be involved in causal relations, but these characteristics of objects are invitations to intellectual elaboration in the form of explanations, theories, myths, and cosmologies.[42]

The bare thesis of the existence of cultural universals is compatible with alternative theories of the genesis of concepts, specifically, both with suitable versions of the doctrine of innate ideas and with suitable theories of the operation of very general mechanisms of learning (for example, association) upon sensory input which is ordered only by the regularities

---

42. Although this remark is vague in content and conjectural concerning cognitive operations, I think that it is epistemologically important. It implies, for example, a very different relationship between observational and theoretical terms from that set forth by Carnap (1956), Hempel (1965), and Nagel (1961: chap. 6). However, systematic applications of current psychological knowledge about concept formation (and probably also some investigations beyond the present frontiers of the subject) are necessary for a thorough understanding of the status of theoretical terms in science.

in the environment. In language learning the additional consideration that the grammatical rules mastered by children are complicated whereas the sensory input is relatively meager strongly favors the thesis of innate ideas (Chomsky 1965 and elsewhere). It seems probable, however, either that special cognitive mechanisms (for example, for acquisition of grammar and for learning the geometry of the visual field) coexist and interact with more generalized learning apparatus, or else that the organism effectively exercises a general learning strategy by being able to switch from one specific mechanism to another and to collate the outputs of various mechanisms. Either hypothesis accounts in a general way for the conjunction of efficiency in achieving intricate skills with flexibility in solving problems under variable conditions. (A particularly good discussion of the second of these two hypotheses is in Sutherland 1959.)

Whatever the correct account of special and general cognitive mechanisms will be, it is to be expected that the maturation of reasoning powers and the growth of a picture of the everyday world will turn out to be interdependent processes. As a result, a thorough investigation of the derivation of the reliable body of information required for scientific inference will throw light upon the process of intellectual development which culminates in the capacity to perform controlled scientfic inferences (or more accurately, which culminates in capabilities which can be shaped to this end if cultural and individual circumstances are favorable). The most elaborate experimental and theoretical work on this interdependence has been done by Piaget, who divides the intellectual development of the child into three major periods (and various subperiods) and finds that the techniques and the tentative views of the world which are crystallized in one period are prerequisites for the development of intelligence in later periods (1952; also Inhelder and Piaget 1958, and Flavell 1963). The interdependence of the development of reasoning powers and the growth of a world picture is expressed in a sophisticated way by Bruner, who acknowledges (much more than Piaget) both the role of special cognitive mechanisms and the influence of culture (1966: chap. II and p. 321). It is evident that theories of cognition along the lines indicated by Piaget and Bruner are very remote from Hume's doctrine of "experimental reasoning," according to which a single principle of "custom" guides the inferences of animals, children, and philosophers,[43] and remote also

---

43. Hume anticipates an objection by considering the question: "Since all reasonings concerning facts or causes is derived from custom, it may be asked how it happens that men so much surpass animals in reasoning, and one man so much surpasses another" (*Inquiry:* sec. IX, n. 1). The answer which he gives is very fine within the limits of his method of investigation, but is symptomatic of the thoroughly nonexperimental character of his study of psychology.

from the varieties of learning theory which were dominant until quite recently.

2. The circumstances which are favorable to the disentanglement of problems are diverse – and certainly cannot be studied as part of a single discipline, in the way that the existence of a reliable body of prescientific knowledge can to a large extent be considered part of the subject matter of developmental psychology. Perhaps the most important of these circumstances is the deep-lying principle that a sharp distinction can be made between physical laws and boundary conditions (cf. Wigner 1964). Without the distinction between laws and boundary conditions, it is hard to see how physics could be studied systematically, since the art of experimentation consists largely of choosing boundary conditions in favorable ways, by achieving relative isolation from disturbing factors and by arranging spatial symmetries, so as to check hypotheses about regularities which are supposed to hold generally. Furthermore, the laws themselves are, at least to very good approximation, invariant under time translation – an invariance which provides the deepest factual basis for the conformity of the future to the past. The distinction between laws and boundary conditions is also essential in disentangling problems concerning the operation of the physical instruments used in biological research from problems concerning biological functions. I do not mean to prejudge the speculations of philosophers like Whitehead (1929: p. 162) that the intrinsic behavior of physical particles is modified by their incorporation into organisms. However, the evidence is overwhelming that if there is such a modification, it is extremely small and, therefore, of negligible importance, at least in the biological problems currently under investigation, in interpreting the biological data obtained by means of physical apparatus. Further preconditions for controlled experimentation in all the natural sciences are that the forces between physical systems fall off rapidly with their distance and that the laws of nature permit physical and biological configurations which are highly stable against small perturbations; otherwise, every experiment would have to take account of detailed astronomical, meteorological, and geological information.

An entirely different factor in the disentanglement of problems is psychological and is related to the general reliability of our common-sense picture of the world, namely, that to first approximation we have a good sense of the relevance and irrelevance of various factors to phenomena of interest. This crude sense of relevance is often wrong, and some of its errors, such as overestimating the influence of "wonders" in the heavens upon terrestrial events, have hampered the development of knowledge. Nevertheless, when this sense of relevance is controlled by critical intelligence, it makes crucial observations possible by enabling men to disregard the innumerable details that are potentially distracting in experimentation.

3. The last of the circumstances noted in section V.A which support the logical structure of scientific inference is the occurrence of good approximations to the truth among the hypotheses which men have seriously proposed.[44] The evidence of this occurrence is mostly contained in the history of science. It consists, first, of the record of striking confirmations and reconfirmations of many individual hypotheses which have been subjected to severe tests and, secondly, of the fact that on the whole this record has been cumulative and progressive. To be sure, there are historical examples of hypotheses which for long times were considered to be highly confirmed but which afterward were judged to be erroneous in the light of new evidence and further analysis. However, the usual pattern since the achievements of physical explanation in the seventeenth century has been that a well-confirmed hypothesis is not only reconfirmed by further tests but also fits fairly well into the general scientific world view – Kuhn's "normal science" (1962: chaps. II–IV). Even when discrepancies have developed between individual highly confirmed hypotheses and the prevailing world view, and have been resolved only by profound intellectual changes which deserve the name of "scientific revolution," the continuity of scientific knowledge is to some extent maintained by the existence of "correspondence" relations between old and new theories (cf. section II.D and note 17).

In order to obtain a deeper explanation of the existence of a vast system of scientific knowledge, it is necessary to understand the natural basis of the process which Peirce calls "abduction." Such an understanding would add little to the weight of evidence from the history of science in support of the proposition that good approximations to the truth have occurred among the hypotheses which have been seriously proposed. However, the question is evidently of great intrinsic interest, and it is also relevant to one of the methodological consequences to be discussed in section V.C (the sanctioning of subjective judgment).

Peirce is deeply impressed by the smallness of the number of fallacious guesses which men of genius – Kepler being his favorite example – have had to make concerning many phenomena before coming upon approximately correct ones. As a result, he finds the history of science to be

44. A proposition with similar content is Wigner's "empirical law of epistemology" which asserts "the appropriateness and accuracy of the mathematical formulation of the laws of nature in terms of concepts chosen for their manipulability, the 'laws of nature' being of almost fantastic accuracy but of strictly limited scope" (1967: p. 233).

incomprehensible unless the human mind possesses a *lume naturale* which results from the influence of the pervasive laws of nature (for example, 5.604). He appeals to the evolutionary history of the race in order to explain how this influence fostered human abductive powers; for example, he writes:

You cannot seriously think that every little chicken, that is hatched, has to rummage through all possible theories until it lights upon the good idea of picking up something and eating it. On the contrary, you think the chicken has an innate idea of doing this; that is to say, that it can think of this, but has no faculty of thinking of anything else. The chicken you say pecks by instinct. But if you are going to think every poor chicken endowed with an innate instinct toward a positive truth, why should you think that to man alone this gift is denied? If you carefully consider with an unbiassed mind all the circumstances of the early history of science and all the other facts specifically bearing on the question, . . . I am quite sure that you must be brought to acknowledge that man's mind has a natural adaptation to imagining correct theories of some kinds, and in particular to correct theories about forces, without some glimmer of which he could not form social ties and consequently could not reproduce his kind. In short, the instincts conducive to assimilation of food, and the instincts conducive to reproduction, must have involved from the beginning certain tendencies to think truly about physics, on the one hand, and about psychics, on the other. It is somehow more than a figure of speech to say that nature fecundates the mind of man with ideas which, when these ideas grow up, will resemble their father, Nature. (5.591)

I find Peirce's argument broadly convincing and inspiring but disappointing in its lack of details. To fill in the details, it would be necessary, for example, to investigate the question of the existence of innate ideas and, if they exist, to reconstruct their evolutionary development, presumably by studies of the comparative psychology of animals since paleontology is not informative about such matters. Some conjectures along these lines concerning the idea of causality have been made by Simpson (1963) and Barr (1964). Furthermore, since the intelligent proposal of hypotheses is one of the least mechanical of mental operations, it is important to understand not only the relatively fixed parts of our intellectual equipment but also the imagination and its play with possibilities. There is impressive evidence that the minds of gifted men who are immersed in their subjects survey and in some sense evaluate a wide range of possibilities on a subconscious level before consciously making serious proposals (Hadamard 1945), but little seems to be known about the machinery of this process and even less about its evolutionary background.

One respect in which the evolutionary account of human abductive powers requires supplementation is in understanding the formation of good hypotheses about matters which are remote from direct practical concerns. An evolutionary explanation of features of our cognitive faculties postulates a sequence of mutations and selections in the history of the species which had the effect of solving with partial success certain problems of surviving and propagating. One characteristic of the line of evolution leading to homo sapiens seems to be the development of a high degree of realism in ordinary judgments with respect to space, time, forces, objects, and persons (cf. Freud's "reality principle"). However, realism of this practical variety does not ensure the capability of making good guesses on theoretical questions remote from ordinary concerns, such as the microscopic constitution of matter; and it does not even ensure the ability to correct certain anthropocentric tendencies of common sense, such as the attribution of sensed qualities to physical things and of purpose to physical forces – attributions which may be biologically valuable in spite of being "unrealistic." In part, the transition from realism in practical affairs to realism in theorizing about nature can be accounted for by the impulse of curiosity and by the critical functions of intelligence, both of which fit into an evolutionary account of abductive powers since both are biologically useful in obvious ways. But this transition cannot be fully understood without reference to favorable characteristics of the laws of physics and of certain important contingent facts about the physical environment, for example, how the behavior of planets and of the approximately rigid stones available in our geological epoch are suggestive of the mechanics of point particles, how the range of temperatures we can control is sufficient to reveal many clues about the chemical composition of substances, and how the properties of visible light are a guide to the quantum properties of elementary particles.[45] Incidentally, it is an uncertain extrapolation to suppose that nature will continue to favor human abductive powers in domains increasingly remote from our immediate concerns. The formulation of good hypotheses in atomic physics required great flights of imagination, and the truth about the subatomic

---

45. An intriguing way to see what has to be done to account for human abductive powers is to compare human beings with Rothstein's "wiggleworms." He attempts "to sketch how science might develop for a race of blind, deaf, highly intelligent worms living in black, cold, sea-bottom muck, and possessing only senses of touch, temperature, and a kind of taste (i.e., a chemical sense)" (1962: p. 28). Rothstein empathizes with his worms impressively, and he concludes that if an objectively real world exists, then they would sooner or later discover the same laws governing its behavior that we do. However, his account of the discoveries of handless creatures, with extremely limited abilities to perform controlled experiments, strains our credulity. It forces us to reflect on what privileged children of nature we are, how lavish our environment is in clues and suggestions, and how often we are presented with situations of almost laboratory purity for exhibiting fundamental processes. The human condition may be miserable in many respects; but for the purpose of extracting natural principles from the tangle of appearances, it seems to be very good indeed.

domain may be orders of magnitude stranger and less accessible to human conjecture. (Nevertheless, the methodological arguments presented in sections III and IV for assigning nonnegligible prior probabilities to seriously proposed hypotheses, including those concerning remote domains, is unaffected by such doubts.)

Some remarks are necessary at this point to prevent a possible misapprehension. In discussing the factual circumstances which support the logical structure of scientific inference, the emphasis has been on physical, biological, and psychological factors favorable to scientific inquiry. However, I do not mean to disguise the cultural character of our system of scientific knowledge. Many cultures have had sophisticated technologies, governments, arts, etc., without coming close to an institutionalization of scientific research or the proliferation of scientific knowledge. One, therefore, cannot expect considerations of developmental psychology, evolutionary biology, and other natural sciences to explain more than the *potentiality* of sustained scientific thought. In particular, one should be skeptical of speculations that intellectual history parallels or is foreshadowed by the intellectual development of the individual (for example, Flavell 1963: pp. 252–56). A striking counterexample is the wonderfully abstruse and nonanthropocentric theorizing by Pythagoras at a very early stage in the history of science. Another example, which is damaging to many theories of scientific development, is the highly critical and sophisticated character of the cosmological argumentation of the Epicureans, as contrasted with the carelessness of their treatment of terrestrial phenomena (Cornford 1965: chap. 2). If we wish to understand how the potentiality for scientific knowledge has been realized, we cannot dispense with detailed sociological and historical studies.

The program which has been outlined is part of "the way down," that is, part of the enterprise of understanding how creatures like ourselves, capable of the kind of experience and endowed with the kind of faculties that make natural knowledge possible, can exist and function. To use a rough analogy, this understanding is comparable to understanding the functioning of optical instruments, such as lenses, in terms of the physical theories of light and of dielectric media. The roughness of the analogy is due in part to the fact that these physical theories are well established and precise, whereas genetic psychology and other sciences concerned with human behavior are still in their infancy. A more important reason for the roughness of the analogy, however, is that the acquisition of natural knowledge by human beings is an incomparably more complex phenomenon than the interaction of light with lenses. Indeed, because of the innumerable cultural and biographical factors which are relevant, the development of natural knowledge must be regarded as a matter of chance (entirely apart from the problem of determinism). Somewhat less roughly, our understanding of the acquisition of natural knowledge can be analogized to our understanding of the evolution of a specific anatomical feature of an animal. No evolutionary biologist claims to be able to predict from first principles that a species with a particular feature will evolve and survive for a long time. Nevertheless, when a biologist is presented with a particular feature of an animal, he can say illuminating things about its integration with other anatomical and physiological characteristics, about its role in enabling the species to fill a certain ecological niche, about its superiority in definite respects to the features of variant species which are rare or extinct, and also about the possibility that it is not biologically useful to the species but is present because of genetic linkage to useful features (cf. Huxley 1942 and Rensch 1960 for examples of the varied and intricate types of explanation which are fruitful in evolutionary theory). Similarly, the discussion above shows in outline that it is possible to understand the existence of a remarkably coherent body of natural knowledge by referring to general features of the environment, to the importance of realism in the life strategy of homo sapiens, to the ability of native human intelligence to achieve realism in important respects, to the psychological linkage between the biologically useful functions of intelligence and its capacity to acquire theoretical knowledge of no direct biological value, and finally to the remarkable potentiality of human intelligence for undergoing systematic refinement and increasing its sensitivity to the truth by the development of the scientific method. *Mutatis mutandis,* the patterns of explanation of specific anatomical features of a species and of the existence of natural knowledge are very similar.

C

Scientific methodology is part of "the way up," that is, part of the enterprise of showing that and how human beings can obtain an understanding of the universe beyond themselves. It has two distinct tasks: (i) to formulate the content of scientific method – to state as precisely as possible whatever can be formulated as rules and also to determine what part, if any, of scientific method cannot be formulated in this way, but must be embedded in intuition or habit; and (ii) to investigate the rationale or justification for scientific method and, in particular, for the part of scientific method which can be characterized as scientific inference. Thus, the Bayesian analysis of the hypothetico-deductive method in section II and the presentation of the tempered personalist formulation of scientific inference in section III belong to the first of these tasks, whereas the argument in section IV that the tempered personalist formulation of scientific inference has a kind of sensitivity to the truth belongs to the second.

Studying the acquisition of knowledge from the standpoint of "the way down," as in section V.B, has important consequences for "the way up" – primarily for the second task of methodology, though indirectly for the first task as well. Broadly speaking, one finds that there are other factors besides the bare logical structure of scientific inference which contribute to its sensitivity to the truth. There are, for example, certain characteristics of the tempered personalist formulation of scientific inference which can be justified on a priori grounds only as desperate expedients, but which a posteriori can be seen to be well suited to investigation of the actual world. Thus, the "local" use of the machinery of probability theory was prescribed in section III because subjective judgment becomes confused and indefinite without localization. But the second of the three circumstances considered in section V.B indicates that localization is appropriate to investigation of nature and will rarely cause irreversible errors.[46] Similarly, a body of common-sense knowledge must be utilized in order to permit consequences to be drawn from scientific hypotheses which can be confronted sharply with experience; and the rough reliability of our prescientific pictures of the world, which was the first of the three circumstances, ensures that this process will not usually be a source of errors. In short, these two circumstances sharpen the accuracy of the critical function performed by scientific inference. The prescription of assigning nonnegligible prior probabilities to all seriously proposed hypotheses is justifiable on a priori grounds as a necessary condition for accepting the truth about some phenomenon if it is ever proposed. But in view of the third circumstance discussed in section V.B, that human beings possess remarkable abductive powers, the hypotheses seriously proposed by men immersed in their subject are statistically much more likely to contain a good approximation to the truth than an equal number of hypotheses selected from among all possibilities by a random process.

The considerations of section V.B also show that the role of subjective judgment in the tempered personalist formulation of scientific inference is more than merely a desperate expedient. According to sections II and III, the logical structure of scientific inference is quite meager, and, therefore, a major role must be assigned at various points in the inferential process to subjective judgment, in order to ensure that definite conclusions can result from inquiry. It was also argued that the a priori principles of

46. The methodological prescription of localizing investigations provides a partial answer to Duhem's holistic conception of natural science (1954: pp. 187–90), but this answer is supplemented by the evidence that nature is favorable to the disentanglement of problems.

scientific inference exercise control over subjective judgment, so that large classes of errors can reliably be avoided and consensus can be eventually reached by investigators who differ greatly in their evaluation of prior probabilities. However, subjective judgment is not merely something to be kept in check, and its role in scientific inference is sanctioned (with reservations) by characteristics of human nature and the environment. How well our intelligence is adapted to the natural environment is exhibited by the rough reliability of our prescientific picture of the world, by our fairly good native sense of the relevance or irrelevance of various factors to phenomena of interest, and by the abductive powers of men who are immersed in a subject matter. Since intelligence is at work in our subjective judgments, even when its *modus operandi* cannot be adequately described, these judgments can enhance the sensitivity of scientific inference.[47] Incidentally, this helps explain why the achievement of scientific knowledge does not depend upon sophistication regarding scientific method. Locke's famous jibe, that "God has not been so sparing to men, to make them barely two-legged creatures, and left it to Aristotle to make them rational" (*Essay,* bk. IV, chap. xvii, sec. 4) applies to inductive inference as well as to deductive.

It must be acknowledged that the a posteriori justification of scientific inference is limited. There are no grounds for a conditional apodictic claim to the effect that if the propositions about human beings and the environment which were asserted in section V.B are correct, then eventually nonerroneous conclusions would be reached (or asymptotically approached) by means of scientific inference. The occurrence of large random sampling errors, regardless of the size of the sample, is compatible not only with logic but also with the known laws of nature. Furthermore, our knowledge of circumstances generally favorable to scientific inference is compatible with the occurrence of systematic errors which will remain undetected for an indefinitely long time in the investigation of special phenomena. However, the program of section V.B gives strong grounds for optimism that the persistence of such systematic errors, in spite of careful experimentation and analysis, is improbable. In particular, the examination of the acquisition of scientific knowledge from the standpoint of "the way down" provides evidence that there is no undetectable systematic error introduced by the uneliminable presence of the human subject in scientific inference.

47. I agree with much of Polanyi's position concerning the personal factor in scientific knowledge (Polanyi 1958: especially pt. 1), but I think he does not say enough about the circumstances which make the personal factor effective. Rescher (1961) makes some good comments on the reliability of subjective probability evaluations, though his treatment of the concept of probability (a kind of hybrid of the personalist and logical theories) seems to me obscure.

D

The foregoing considerations in defense of subjective judgment should not disguise the fact that subjective judgment is often distorted by proprietary interest in a theory, by prejudice, and by obtuseness; and, therefore, guidelines which provide an alternative to or a check upon subjective judgment are desirable. If the vein of a priori principles is exhausted, then these guidelines can only be obtained from knowledge of the actual world. The investigation of the acquisition of scientific knowledge from the standpoint of "the way down" provides a justification for a posteriori supplementation of scientific method. The manifest cumulative character of scientific knowledge has been seen to be almost certainly not the result of systematic error, and, therefore, the use of a tentative body of scientific knowledge to suggest guidelines in further investigations is unlikely to have the effect of propagating or compounding a systematic error. In this way the program of section V.B indirectly contributes to the first task of scientific methodology, which is to formulate the content of scientific method.

I shall give several examples of guidelines in deciding whether hypotheses are seriously proposed and in evaluating prior probabilities. A related problem, which I shall not discuss but which deserves attention, is the derivation of guidelines concerning the acceptance of hypotheses into the tentative body of scientific knowledge.

Two examples will show how guidelines for classifying hypotheses as seriously proposed or not are obtained from antecedent scientific knowledge. The first is the reevaluation of entire classes of hypotheses as a result of the triumph of classical mechanism. The conceptual clarification of dynamics and the accurate explanation of many celestial and terrestrial motions led physical scientists generally to discount hypotheses formulated qualitatively and in terms of sensuous properties of bodies and to pay serious attention to hypotheses formulated mathematically and in terms of nonsensuous properties. Less dramatic, but of comparable importance, was the discrediting of magic by the gradual extension of critical thinking into emotionally sensitive areas. A consequence was the devaluation of certain classes of hypotheses which are characteristic of magic, especially those motivated by surface analogies (for instance, that rain can be induced by bloodletting and that the waxing moon is favorable to conception).[48] In addition to such large-scale reevaluations of classes of

48. Analogy remains an essential part of scientific heuristics, but (to paraphrase a statement of Weyl's concerning simplicity) we must let nature train us to recognize good analogies. As Maxwell pointed out (1952: pp. 156–57), analogies of structure are especially fruitful, and his own proposal of the existence of the displacement current is a classical case in point.

hypotheses, there is usually an understanding among investigators, based upon their knowledge of the recent history of the subject, that certain types of hypotheses are likely to be fertile and others to be sterile. Consequently, specialized knowledge provides specialized guidelines for deciding whether a hypothesis is seriously proposed.

It may be objected that guidelines based upon a tentative body of scientific knowledge are likely to be excessively conservative in just the way that the tempering condition is intended to prevent. This possibility need not be a danger, however, if good sense is used. It is reasonable for an expert to refuse to take seriously a proposal in conflict with a theory which is well confirmed, especially when the proposer evidently is ignorant of it or does not fully understand it. Even if the proposal itself has never been directly tested, the expert may see that because of the established theory the testing would be routine and hence not worth his attention unless it would unexpectedly lead to anomalous results. On the other hand, it is important to take seriously a challenge to an established theory which is not careless about the evidence whereby the theory was established. Thus, conceptual analysis may show that a hypothesis which challenges one established theory is, nevertheless, compatible, contrary to appearances, with the deeper parts of the accepted body of knowledge; or the proposal of the hypothesis may be accompanied by an examination of weak points in the evidence upon which the established theories are based; or the approximate agreement within a limited domain between the predictions of established theories and those of the new hypothesis may be exhibited, thus accounting from a new point of view for the success of the former. The essential point is that without setting limits to the content of subsequent conjectures, the tentative body of scientific knowledge generally serves as an indispensable base of reference for them.

Within the limitations of the axioms of probability and the tempering condition, the evaluation of tempered personalist probabilities is left to subjective judgment. However, at least two guidelines for these evaluations are commonly recognized: (1) the principle of indifference and (2) the principle that the simpler of two hypotheses has higher prior probability than the less simple. Much of the discussion of these two principles seems to be based upon the assumption that if they are valid, then the grounds for their validity must be a priori; but my contention is that their valid content is largely derived from antecedent knowledge about the world.

1. Some comments on the principle of indifference are found in note 38. It is pointed out that a formulation of the principle which is weak enough to be a priori – namely, the equiprobability of propositions asserting various outcomes, once all individuating differences among the outcomes are suppressed – is superfluous in scientific inference, whereas more stringent forms of the principle depend upon the dismissal as irrelevant of classes of differences. Antecedent knowledge is used in several different ways in making judgments of irrelevance.

In evaluating "geometrical" probabilities, the differences among equal volumes in a parameter space with a properly defined measure are judged to be irrelevant, but antecedent knowledge is required to determine whether a measure is proper. The well-known paradox of Bertrand (1889: p. 5; Keynes 1921: pp. 41–42), which results from applying the principle of indifference to alternative measures on a certain continuum (namely, the class of chords of a given circle), was intended to show that there are no unambiguous objective prior probabilities in such cases. However, it may also happen that in these cases personal probability evaluations are confused or vacillating. If so, then the body of assumptions a must be augmented (thus essentially changing the character of the problem) in order to arrive at a situation in which personal probability evaluations become unambiguous, which in turn makes possible the evaluation of tempered personalist probabilities. The augmentation of a does not necessarily require that statistical data be gathered. In fact, the standard resolution of the Bertrand paradox (for example, Borel 1965: pp. 87–88) consists in specifying the physical means by which an element of the continuum of interest is selected. The probability distribution over this continuum is then determined – via the axiom of complete additivity and other probability axioms, together with physical principles which are contained in the total body of information $a \& b$ – by a distribution over a different continuum, namely, the points within the given circle. From the standpoint of personalism the problem is thereby solved if the personal probability distribution over the latter continuum is unambiguous. However, the considerations at the beginning of section V.C remain relevant. Our adaptation to the environment is reflected in our subjective probability evaluations and partly explains why practical decisions based upon these evaluations are fairly successful. In particular, a uniform personal probability distribution over the points within a circle results from the assumption of a Euclidean metric and the associated measures of area and volume, which are deeply embedded in the background information of every normal human being.

In more complicated problems it is often impossible to obtain a reduction to an unambiguous personal probability distribution without resorting to sophisticated antecedent knowledge. Notably, in classical statistical mechanics one wishes to find probability distributions of dynamical variables characteristic of systems of particles moving under given constraints and interactions. These distributions can always be determined in principle (that is, neglecting formidable mathematical difficulties) by assuming that equal volumes in the phase space of a closed system have equal prior probabilities (Tolman 1938: pp. 59–61). One of the deep problems of statistical physics is to derive this assumption from weaker ones (cf. Farquhar 1964), but at present its primary justification must still be said to be the great explanatory power of the statistical physical theories based upon it. The justification of the assumption is, therefore, typical of scientific inference concerning general theories. The important point for the considerations of the present section, however, is that the attempt to evaluate the personal probability of a physical system to be found in a specified region of phase space can be expected to result in *subjective confusion* unless something of the order of generality of this basic assumption of statistical mechanics has been confirmed, or unless at least clear alternative assumptions of this order of generality are under consideration and are all assigned personal probabilities. In short, it is naïve to expect a man to be able to form unambiguous subjective judgments (and a fortiori sensible ones) about such matters unless he entertains sophisticated statistical conjectures, which in turn presuppose an extensive knowledge of physical theory. It was argued in section II that probability evaluations are "local," and in the present context locality requires that the total body of information and assumptions $a \& b$ contain a certain amount of physical theory. That explicit antecedent knowledge is dispensable in making personal probability evaluations regarding more mundane questions indicates the richness of the background information of ordinary men concerning such questions.

In problems concerning gambling devices, which constitute the locus classicus for applications of the principle of indifference, the role of antecedent knowledge is entirely different. Most adults in our culture know, for example, that there is no correlation between the color of a playing card and its susceptibility to being drawn from a deck, unless the circumstances are abnormal (for example, the drawer is cheating). Hence, if information about the usual behavior of playing cards is contained in $a \& b$, if e is the proposition that a single drawing will occur from a standard deck of cards all turned face down, and if $r(j)$, $b(j)$, $d(j)$ are respectively the propositions that the jth card is red, is black, and is the one to be drawn, then

(1)
$$P_{X,b}[d(j)\,|\,a \& e \& r(j)] = P_{X,b}[d(j)\,|\,a \& e \& b(j)]$$
$$= P_{X,b}[d(j)\,|\,a \& e].$$

But how important is the antecedent information about playing cards for this judgment? Suppose, for example, that $a \& b$ is replaced by $a' \& b'$, in which no information about playing cards is contained. Unless X has some unusual beliefs about the efficacy of specific colors, his probability evaluations can be expected to satisfy equation (1') which results from (1) by replacing a and b by a' and b'. Or suppose that c is the proposition that the drawer has the facility and the intention to cheat, but it contains no information as to which color is advantageous to the drawer. Then past experience about the behavior of playing cards is of no obvious use to X, and, nevertheless, one can expect his judgment to be

$$(2) \qquad P_{X,b}[d(j)\,|\,a\,\&\,e\,\&\,c\,\&\,r(j)] = P_{X,b}[d(j)\,|\,a\,\&\,e\,\&\,c\,\&\,b(j)]$$
$$= P_{X,b}[d(j)\,|\,a\,\&\,e\,\&\,c].$$

The parallels among (1), (1'), and (2) indicate that antecedent knowledge is not needed for supplementing a subjective judgment concerning a *single* drawing. Where antecedent information about playing cards is important is in leading X to assign a low prior probability (that is, prior relative to a, in the local sense of section II) to the proposition that a correlating mechanism exists; and this assignment has the consequence that the judgment of irrelevance of black or red when no previous drawings have been made dominates the evaluation of the posterior probability even when the results of a small number of drawings are known. In other words, the effect of antecedent information in this case is to increase the applicability of a subjective judgment of irrelevance. This conclusion can be greatly generalized, since we have a large but diffuse body of knowledge that, in stochastic situations in which subclasses of events are differentiated from one another in physically minor ways, the frequencies of the various subclasses are approximately equal. Additional knowledge is required to judge reliably whether a characteristic is indeed minor, and, furthermore, the possibility always remains that a special mechanism favors the occurrence of a subclass which is distinguished by a prima facie negligible characteristic. In spite of these qualifications, however, our diffuse knowledge of such stochastic situations strongly reenforces the methodological prescription of section IV against extreme suspiciousness of the presence of unknown correlations.

2. After surveying the recent literature on simplicity, Kyburg remarked that, in spite of much effort, "the whole discussion of simplicity has been curiously inconclusive" (1964: pp. 19–20). I suspect that one reason for this situation is that methodologists have tended to ascribe too large a role to simplicity, namely, the role of being the primary criterion for comparing all hypotheses compatible with the experimental data. This role is too great a burden for the concept of simplicity to bear, and, indeed, one of the advantages of the tempered personalist formulation of scientific

inference is that it uses a different primary criterion for comparing hypotheses, namely, that of being or not being seriously proposed. Simplicity considerations are undoubtedly important in scientific investigations, but usually, if not always, within contexts that are circumscribed by assumptions or by antecedent knowledge. The circumscription often has the effect of suggesting, either unequivocally or with the possibility of small variations, the ordering of a favored family of hypotheses. Thus, if the motion of a planet is supposed to be a superposition of circular motions, then it is reasonable to identify an ordering according to decreasing simplicity with an ordering according to the number of epicycles. And if a theory of interacting fields is assumed to be Lorentz invariant, then the orders of the tensors entering into the interaction term are plausible indices of the complexity of the theory. Without the guidelines suggested by antecedent knowledge or by stringent assumptions, so many alternatives suggest themselves as candidates for "ordering according to simplicity" (as pointed out in the discussion of Kemeny in section IV.B) that the concept becomes methodologically valueless. Moreover, proposed orderings which purport to be a priori usually derive whatever intuitive appeal they possess from a tentative picture of the world. A case in point is the simplicity postulate proposed by Wrinch and Jeffreys (1921; also Jeffreys 1931: p. 45), but later disavowed by Jeffreys, according to which the highest prior probabilities are assigned to those laws which can be cast in the form of differential equations of low order and degree. This assignment is reasonable only if the basic natural regularities govern rates of change, as in Newtonian physics, rather than configurations, as in Kepler's astronomy, and only if continuity is a fundamental feature of nature. Far too much knowledge of physics is implicit in Wrinch and Jeffreys's proposal to admit the claim that "it would represent the initial knowledge of a perfect reasoner arriving in the world with no theoretical knowledge whatever" (Jeffreys 1961: p. 49).

Weyl's statement, "The required simplicity is not necessarily the obvious one, but we must let nature train us to recognize the true inner simplicity" (1949: p. 155) seems to me to approach the heart of the matter. It does justice to the history of science in a way that no formal theory of simplicity comes close to doing, and it expresses compactly the Platonic conceptions which I think are essential to an adequate naturalistic epistemology. However, one may be dubious about the methodological value of Weyl's statement; for if the secrets of nature must be known before simplicity can be understood, then it is hard to see how considerations of simplicity can sharpen a person's probability evaluations in the course of investigating nature. A partial answer is that evidence and analysis may indicate a general respect in which nature is simple without yielding the detailed laws. Nature has, in fact, "trained" us to recognize several kinds of inner simplicity: geometrical invariance principles such as Lorentz invariance, simplicity of the nature and behavior of elementary particles, and simplicity arising from the statistics of large numbers of particles. Nevertheless, this extensive body of knowledge leaves many physical questions open. For example, exactly which nongeometrical invariance principles govern the elementary particles? At the same time, it often suggests an obvious ordering of a family of hypotheses, for example, an ordering according to the magnitude of n of the hypotheses that the appropriate group for describing the nongeometrical invariances of strongly interacting particles is $SU(n)$ (see Carruthers 1966). There is another methodological consequence of Weyl's statement, which concerns heuristics as much as confirmation. Suppose that a hypothesis of great generality is highly confirmed, even though it is not manifestly simpler than its disconfirmed rivals. Then a point of view should be sought from which the simplicity of the successful hypothesis is evident. Such points of view have indeed been found upon analysis and have provided some of the most striking advances in scientific understanding. An outstanding example was the exhibition of the simplicity of Maxwell's electrodynamics within the theory of special relativity. The fertility of deliberate quests for points of view from which established principles appear simple is evidence that Weyl is pointing to an objective feature of the world and is not merely using poetic diction to describe our habituation to novel theories.

On a more mundane level Goodman makes an important and sensible contribution to the analysis of simplicity:

Formulation of general standards for comparing the simplicity of hypothesis is a difficult and neglected task. Here brevity is no reliable test; for since we can always, by a calculated selection of vocabulary, translate any hypothesis into one of minimal length, the simplicity of the vocabulary must also be appraised. I am inclined to think that the standards of simplicity for hypotheses derive from our classificatory habits as disclosed in our language, and that the relative entrenchment of predicates underlies our judgment of relative simplicity. (1961: p. 151; see also 1955: p. 119)

The argument which he gives for rejecting brevity as a reliable test for simplicity can evidently also be directed against other formal criteria, such as number of parameters or number of quantifiers, unless these criteria are supplemented by appraisal of the basic vocabulary. However, I think that Goodman's analysis can be deepened by considering the concept of entrenchment from a naturalistic standpoint. Ordinary language habits reflect an extensive common-sense knowledge of the world, which is roughly reliable in that it permits normal human beings to survive under normal circumstances (cf. Jeffrey 1965: pp. 176–77). Furthermore, as pointed out in section V.B, studies in biology and genetic psychology have provided at least some insight into the basis of this reliability, for example, the role of innate ideas which presumably are the product of natural selection and the fact that the learning processes of children are critical in crude but important ways. When a man judges the hypothesis that all emeralds are green to be simpler than the hypothesis that all emeralds are grue (where "grue" is a predicate which "applies to all things examined before *t* just in case they are green but to other things just in case they are blue" [Goodman 1955: p. 74]) and assigns far higher prior probability to the former in spite of the fact that from a logical point of view the "green" – "blue" and the "grue" – "bleen" vocabularies are coordinate (*ibid.:* p. 79), he is relying upon his common-sense picture of the world. He is not hopelessly conservative in so doing, for there are anomalies within common-sense pictures of the world which have had the social consequence of stimulating the scientific theorizing whereby common sense is transcended. However, most philosophers will agree that the route whereby common sense is transcended – and Weyl's view of nature is achieved – does not pass via grue-like concepts, even though further analysis would be required to demonstrate this decisively.

E

I shall now briefly summarize the theory of scientific inference which has been proposed and make some comments about its rather complicated structure, in particular about the circularity which arises because of a posteriori contributions to methodology.

The first part of the theory is a formulation of scientific inference in terms of the tempered personalist concept of probability. The formulation has a meager logical structure, consisting only of the axioms of probability and the tempering condition (with the possibility of some supplementation). An a priori methodological justification can be given for this logical structure. The axioms of probability are necessary conditions for orderly thinking about propositions with unknown truth values, whereas the tempering condition is only a way of prescribing open-mindedness within a probabilistic scheme of inference; and open-mindedness is a necessary condition for the acceptance of true hypotheses in case they should ever be proposed. Despite these a priori considerations, the logical structure of this formulation is inseparable from the context of actual inferential processes – in contrast to the situation in deductive logic. This is shown by the preferred status of hypotheses which are actually proposed and by the crucial role which had to be assigned to subjective judgment in order to ensure that a person can use the apparatus of probability theory in arriving at a view of the world. Although the logical structure is meager, it suffices to provide systematic (but not infallible) protection against major classes of possible errors. In this way a sensitivity to the truth can be claimed for scientific inference on a priori grounds, though the exact content of this sensitivity was not determined. For example, no attempt was made to survey models of the universe with the intention of making a dichotomy between those in which errors of acceptance and rejection of hypotheses could be eliminated in the long run and those in which they could not; and in view of the role of subjective judgment in scientific inference, it is hard to see how this would be a well-posed problem for statisticians, even if the set of models were accurately described.

The second part of the theory complements the first part by investigating the context of actual inferential processes which the tempered personalist formulation refers to but does not characterize. The relevant scientific knowledge that we now possess indicates that this context is favorable to scientific inference: that even without special cultivation human intelligence is sufficiently well adapted to the environment to be quite apt at discriminating, on the basis of experience, between those hypotheses which are good approximations to the truth and those which are not, and that the informal methods of confirmation used by scientists (which I claim are made explicit in the tempered personalist formulation of scientific inference) improve this native adaptation and extend it beyond the area of direct biological utility. In this way an examination of the acquisition of scientific knowledge from the standpoint of "the way down" greatly strengthens the claim that scientific inference has a sensitivity to the truth and thereby contributes to "the way up." Furthermore, this general knowledge of the favorableness of nature to scientific inquiry provides a rationale for relying upon special bodies of scientific knowledge to augment the scientific method with a posteriori principles, for example, with guidelines for the evaluation of probabilities.

The epistemological status of the propositions discussed in section V.B, to the effect that nature is favorable to scientific inference, is complex. In a sense they are undoubtedly presupposed in the operations of native intelligence. Any nontrivial process of induction makes use of a background of common-sense knowledge and, therefore, implicitly assumes the rough reliability of that knowledge; any actual inquiry localizes a problem and, therefore, assumes the propriety of localization; and any actual inquiry (as contrasted with bloodlessly going through the motions of inquiry) is optimistic about the ability of human beings to propose approximately true hypotheses. However, the explicit formulation of these propositions requires reflection upon inquiry; and their confirmation, as is evident from the sketches of section V.B, requires extensive physical, biological, and psychological investigations. Furthermore, such investigations are essential for eliminating the vagueness of the naïve formulations of these propositions: to determine *how* reliable common-sense knowledge is, *how* amenable nature is to the disentanglement of problems, and *how* good the abductive powers of human beings are. Scientific inference is, therefore, required to clarify and confirm these propositions; on the other hand, the truth of these propositions is presupposed by a strong justification of scientific inference, that is, a justification which goes beyond the counsel of desperation which a priori considerations provide. This is the circularity which arises in the theory of scientific inference that has been proposed.[49]

I claim, however, that the circularity is nonvicious in the following sense: *the theory as a whole is open to critical evaluation in the light of experience, for the reciprocal support of a methodology and a scientific world picture does not render it impregnable to criticism.*[50] This openness is partly due to the open-mindedness which is incorporated into the logical structure of the formulation of scientific inference. The negation of any of the propositions of section V.B could be seriously proposed and then, in accordance with the tempering condition, would have to be assigned a sufficiently high prior probability to permit its survival under experimental scrutiny. It is logically possible then that experimental evidence would lead to a high posterior probability of the negation of one of the propositions upon which the a posteriori contributions to methodology were based. Whether it is psychologically possible is a different matter. Experience which would lead to the rejection of the proposition that common sense is roughly reliable would be intolerable to creatures like us and would almost certainly induce disillusionment with critical thinking at the least, and more likely insanity or death. Similar consequences could be expected from experience which leads to the thorough rejection, even in matters which are vital to our biological functioning, of the propositions that problems can be disentangled and that human beings have good abductive powers. However, the judgments that these would be the probable outcomes are a posteriori, derived from our knowledge of human nature, and, therefore, do not contradict the logical openness of the theory of scientific inference. Furthermore, it does seem to

49. This circular argument is quite different from that proposed by Black (1966). However, I strongly concur with his idea that the effectiveness of a circular justification of induction depends upon a global point of view: "Perhaps the place to find a connection between induction and intelligible human interests in arriving at the truth is in the entire *practice* and not in some artificially dissected component of it" (*ibid.:* p. 199).

50. I have also discussed nonvicious circularity in a review (1954: pp. 657–58) and in greater detail in an unpublished thesis (1953: pp. 119–27).

be psychologically possible for creatures like ourselves to undergo experience which disconfirms the propositions that in matters remote from direct biological concern, problems can be disentangled and human beings have good abductive powers.

There is also an openness implicit in the derivation of guidelines from specific bodies of scientific knowledge. The reciprocal support of methodology and the scientific picture of the world is compatible with a gradual correction of both. This dialectic would undoubtedly terminate in confusion (or worse, as indicated in the preceding paragraph) if certain basic propositions assumed in the early stages of the dialectic were radically false. But correction is possible and fruitful precisely because these basic propositions are true only with qualifications: for example, common sense knowledge is not completely reliable, and the loci of its unreliability become evident with the growth of scientific knowledge. It is even possible that the dialectic would lead to the adoption of a scientific method radically different from that of contemporary scientists. Suppose, for example, that psychological evidence would come to support Aristotle's contention that the properly prepared intellect is able to grasp the universal form of any species (*On the Soul,* bk. III, chaps. 4–8); then Aristotle's method of obtaining major premises by intuitive induction (*Posterior Analytics,* bk. II, chap. 19) would be justified, and the current scientific method would become a stepping-stone to a more direct one. One of the virtues claimed in section IV for the tempered personalist formulation of scientific inference is an ability to subsume methodological devices which experience or analysis indicate to be powerful. A corollary is the possibility of its being dominated by that which it subsumes, as the Mongolians by the conquered Chinese.

The possibility remains that the dialectical interplay of science and methodology will come to an end. In this unlikely event it is to be hoped that the reason will be the attainment of essentially complete knowledge of the place of human beings in nature. There would then be an exact meshing of methodology and of the scientific picture of the world, not because embarrassing questions will have been evaded and critical experiments left unperformed, but because the perspective will be complete and all pieces will fit into place. Conceivably, however, such a meshing could occur in spite of deep-lying errors in the scientific picture, which failed to be exposed in spite of strenuous experimental probing. There would, so to speak, be a probabilistic version of the Cartesian deceiving demon. Despite this conceivable eventuality, a full commitment to a methodology and a scientific picture which mesh in this way would be rational, for no stronger justification is compatible with the status of human beings in the universe. After all, in spite of the advances which have been made upon

Hume's treatment of inductive inference, his basic lesson remains valid: that the conditions of human existence which force us to resort to inductive procedures also impose limitations upon our ability to justify them.

REFERENCES

Aczél, J. (1966) *Lectures on Functional Equations and Their Applications*. New York, Academic Press.

Aristotle (1941) "On the Soul," in *The Basic Works of Aristotle*, ed. R. McKeon. New York, Random House.

Aristotle (1941a) "Posterior Analytics," in *The Basic Works of Aristotle*, ed. R. McKeon. New York, Random House.

Barr, H. J. (1964) "The Epistemology of Causality from the Point of View of Evolutionary Biology," *Philosophy of Science*, 31: 286–88.

Bertrand, J. (1889) *Calcul des probabilités*. Paris, Gauthier-Villars.

Black, Max (1966) "The Raison d'Être of Inductive Argument," *The British Journal for the Philosophy of Science*, 17: 177–204.

Bohr, Niels (1934) *Atomic Theory and the Description of Nature*. Cambridge, University Press.

Borel, Emile (1965) *Elements of the Theory of Probability*. Englewood Cliffs, N.J., Prentice-Hall.

Bruner, Jerome (1966) *Cognitive Growth*. New York, Wiley.

Burks, Arthur (1951) "Reichenbach's Theory of Probability and Induction," *Review of Metaphysics*, 4: 377–93.

Burks, Arthur (1964) "Peirce's Two Theories of Probability," in *Studies in the Philosophy of Charles Sanders Peirce* (2d ser.), eds. Edward Moore and Richard Robin. Amherst, University of Massachusetts Press, pp. 141–50.

Carnap, Rudolf (1950) "Empiricism, Semantics, and Ontology," *Revue Internationale de Philosophie*, 11: 20–40.

Carnap, Rudolph (1952) *The Continuum of Inductive Methods*. Chicago, University of Chicago Press.

Carnap, Rudolph (1956) "The Methodological Character of Theoretical Concepts," in *Minnesota Studies in the Philosophy of Science*, I, eds. H. Feigl and M. Scriven. Minneapolis, University of Minnesota Press, pp. 38–76.

Carnap, Rudolph (1963) "Replies and Systematic Expositions," *The Philosophy of Rudolf Carnap*, ed. P. A. Schilpp. La Salle, Open Court.

Carnap, Rudolph (1968) "A Basic System of Inductive Logic." Dittographed. Los Angeles, University of California. Published in revised form in *Studies in Inductive Logic and Probability*, vol. I, eds. R. Carnap and R. Jeffrey, vol. II, ed. R. Jeffrey. Berkeley and Los Angeles, University of California Press, 1971 and 1980.

Carruthers, P. (1966) *Introduction to Unitary Symmetry*. New York, Wiley.

Chomsky, Noam (1965) *Aspects of the Theory of Syntax*. Cambridge, M.I.T. Press.

Cornford, Frances (1965) *Principium Sapientiae*. New York, Harper and Row.

Cox, Richard (1946) "Probability, Frequency, and Reasonable Expectation," *American Journal of Physics,* **14**: 1–13.

Cox, Richard (1961) *The Algebra of Probable Inference.* Baltimore, Johns Hopkins Press.

DeFinetti, Bruno (1937) "La prévision: ses lois logiques, ses sources subjectives," *Annales de l'Institut Henri Poincaré,* **7**: 1–68. English translation (1964) in *Studies in Subjective Probability,* eds. Henry Kyburg and Howard Smokler. New York, Wiley.

Dewey, John (1929) *Experience and Nature* (2d ed.). La Salle, Open Court.

Dewey, John (1938) *Logic: The Theory of Inquiry.* New York, Holt, Rinehart and Winston.

Duhem, Pierre (1954) *The Aim and Structure of Physical Theory.* Princeton, Princeton University Press.

Eddington, Arthur (1928) *The Nature of the Physical World.* New York, Macmillan.

Edwards, W., Lindman, H., and Savage, L. J. (1963) "Bayesian Statistical Inference for Psychological Research," *Psychological Review,* **70**: 193–242.

Farquhar, I. E. (1964) *Ergodic Theory in Statistical Mechanics.* New York, Interscience.

Feigl, Herbert (1950) "Existential Hypotheses: Realistic vs. Phenomenalistic Interpretations," *Philosophy of Science,* **17**: 35–62.

Feyerabend, Paul (1963) "How to Be a Good Empiricist," in *Philosophy of Science: The Delaware Seminar,* II, ed. B. Baumrin. New York, Interscience, pp. 3–39.

Feyerabend, Paul (1970) "Problems of Empiricism, Part II," in *The Nature and Function of Scientific Theories,* ed. R. G. Colodny. Pittsburgh, University of Pittsburgh Press, pp. 275–353.

Flavell, J. H. (1963) *The Developmental Psychology of Jean Piaget.* Princeton, N.J., Van Nostrand.

Fodor, Jerry (1966) "Could There Be a Theory of Perception?" *Journal of Philosophy,* **43**: 369–80.

Gödel, Kurt (1946) "Russell's Mathematical Logic," in *The Philosophy of Bertrand Russell,* ed. P. A. Schilpp. Menasha, Wis., George Banta Press, pp. 123–53.

Good, I. J. (1950) *Probability and the Weighing of Evidence.* London, C. Griffin.

Goodman, Nelson (1955) *Fact, Fiction, and Forecast.* Cambridge, Harvard University Press.

Goodman, Nelson (1961) "Safety, Strength, Simplicity," *Philosophy of Science,* **28**: 150–51.

Hacking, Ian (1965) *Logic of Statistical Inference.* Cambridge, University Press.

Hacking, Ian (1965a) "Salmon's Vindication of Induction," *Journal of Philosophy,* **62**: 269–71.

Hadamard, Jacques (1945) *Psychology of Invention in the Mathematical Field.* New York, Dover.

Harris, E. E. (1965) *Foundations of Metaphysics in Science.* New York, Humanities Press.

Hempel, C. G. (1965) *Aspects of Scientific Explanation and Other Essays in the Philosophy of Science.* New York, Free Press.

Hirst, R. J. (1959) *The Problems of Perception.* New York, Macmillan.

Hockett, Charles (1954) "Chinese Versus English: An Exploration of the Whorfian Theses," in *Language in Culture,* ed. Harry Hoijer. American Anthropological Association, Memoir no. 79: 106–23. Reprinted (1959) in *Readings in Anthropology,* I, ed. Morton Fried. New York, Crowell.

Hume, David (1955) *An Inquiry Concerning Human Understanding,* ed. Charles Hendel. New York, Liberal Arts Press.

Huxley, Julian (1942) *Evolution: The Modern Synthesis.* New York, Harper.

Inhelder, B., and Piaget, J. (1958) *The Growth of Logical Thinking from Childhood to Adolescence.* New York, Basic Books.

Jeffrey, Richard (1965) *The Logic of Decision.* New York, McGraw-Hill.

Jeffreys, Harold (1931, 2d ed., 1937) *Scientific Inference.* Cambridge, University Press.

Jeffreys, Harold (1961) *Theory of Probability* (3rd ed.). Oxford, Clarendon Press.

Kac, Mark (1959) *Probability and Related Topics in Physical Sciences.* New York, Interscience.

Kant, Immanuel (1929) *Critique of Pure Reason,* trans. Norman Kemp Smith. New York, Macmillan.

Katz, Jerrold (1962) *The Problem of Induction and Its Solution.* Chicago, University of Chicago Press.

Kemeny, John (1953) "The Use of Simplicity in Induction," *Philosophical Review,* **62**: 391–408.

Keynes, John M. (1921) *A Treatise on Probability.* London, Macmillan.

Koopman, B. O. (1940) "The Axioms and Algebra of Intuitive Probability," *Annals of Mathematics,* ser. 2, **41**: 269–92.

Kuhn, Thomas (1962) *The Structure of Scientific Revolutions.* Chicago, University of Chicago Press.

Kyburg, Henry (1964) "Recent Work in Inductive Logic," *American Philosophical Quarterly,* I, no. 4: 1–39.

Lenz, John W. (1956) "Carnap on Defining 'Degree of Confirmation,'" *Philosophy of Science,* **23**: 230–36.

Lenz, John W. (1958) "Problems for the Practicalists' Justification of Induction," *Philosophical Studies,* **9**: 4–8.

Lenz, John W. (1964) "Induction as Self-Corrective," in *Studies in the Philosophy of Charles Sanders Peirce* (2d ser.), eds. Edward Moore and Richard Robin. Amherst, University of Massachusetts Press, pp. 151–62.

Locke, John (1894) *Essay Concerning Human Understanding,* ed. A. C. Fraser. 2 vols. Oxford, Clarendon Press.

McKeon, Richard (1951) "Philosophy and Method," *Journal of Philosophy,* **48**: 653–82.

Madden, E. H. (1964) "Peirce on Probability," in *Studies in the Philosophy of Charles Sanders Peirce* (2d ser.), eds. Edward Moore and Richard Robin. Amherst, University of Massachusetts Press, pp. 122–40.

Mandelbaum, Maurice (1965) *Philosophy, Science, and Sense Perception.* Baltimore, Johns Hopkins Press.

Maxwell, James C. (1952) *Scientific Papers.* New York, Dover.

Murphey, Murray (1961) *The Development of Peirce's Philosophy.* Cambridge, Harvard University Press.

Nagel, Ernest (1961) *The Structure of Science.* New York, Harcourt, Brace and World.

Nagel, Ernest (1963) "Carnap's Theory of Induction," in *The Philosophy of Rudolf Carnap,* ed. P. A. Schilpp. La Salle, Open Court, pp. 785–826.

Passmore, John (1957) *A Hundred Years of Philosophy.* London, Duckworth.

Peirce, Charles Sanders. *Collected Papers.* Vol. I–VI (1931–35), eds. Charles Hartshorne and Paul Weiss. Cambridge, Harvard University Press. Vol. VII–VIII (1958), ed. Arthur Burks. Cambridge, Harvard University Press. (Note: In the Peirce references I have followed the standard practice of indicating the volume and paragraph number of *Collected Papers.*)

Piaget, Jean (1950) *Introduction à l'épistémologie génétique.* 3 vols. Paris, Presses Universitaires de France.

Piaget, Jean (1952) *The Origins of Intelligence in Children.* New York, Norton.

Polanyi, Michael (1958) *Personal Knowledge.* Chicago, University of Chicago Press.

Popper, Karl (1961) *The Logic of Scientific Discovery.* New York, Science Editions.

Popper, Karl (1962) *Conjectures and Refutations.* New York, Basic Books.

Putnam, Hilary (1963) "'Degree of Confirmation' and Inductive Logic," *The Philosophy of Rudolf Carnap,* ed. P. A. Schilpp. La Salle, Open Court, pp. 761–84.

Putnam, Hilary (1963a) "Probability and Confirmation," The Voice of America Forum Lectures. Washington D.C., U.S. Information Agency. Reprinted (1967) in *Philosophy of Science Today,* ed. S. Morgenbesser. New York, Basic Books.

Quine, W. V. (1957) "The Scope and Language of Science," *The British Journal for the Philosophy of Science,* **8**: 1–17. Reprinted (1966) in *The Ways of Paradox.* New York, Random House.

Raiffa, H., and Schlaifer, R. (1961) *Applied Statistical Decision Theory.* Boston, Harvard University Graduate School of Business Administration.

Ramsey, Frank (1931) "Truth and Probability," in *The Foundations of Mathematics and Other Logical Essays,* ed. Frank Ramsey. London, Rutledge & Kegan Paul. Reprinted (1964) in *Studies in Subjective Probability,* eds. Henry Kyburg and Howard Smokler. New York, Wiley.

Reichenbach, Hans (1938) *Experience and Prediction.* Chicago, University of Chicago Press.

Reichenbach, Hans (1949) *The Theory of Probability.* Berkeley and Los Angeles, University of California Press.

Rensch, Bernhard (1960) *Evolution Above the Species Level.* New York, Columbia University Press.

Rescher, Nicholas (1961) "On the Probability of Nonrecurring Events," in *Current Issues in the Philosophy of Science,* eds. Herbert Feigl and Grover Maxwell. New York, Holt, Rinehart, and Winston, pp. 228–37.

Rothstein, Jerome (1962) "Wiggleworm Physics," *Physics Today,* September: 28–38.

Russell, Bertrand (1948) *Human Knowledge: Its Scope and Limits.* New York, Simon & Schuster.

Salmon, Wesley (1961) "Vindication of Induction," in *Current Issues in the Philosophy of Science,* eds. Herbert Feigl and Grover Maxwell. New York, Holt, Rinehart, and Winston, pp. 245–56.

Salmon, Wesley (1963) "Inductive Inference," in *Philosophy of Science: The Delaware Seminar,* II, ed. B. Baumrin. New York, Interscience, pp. 341–70.

Salmon, Wesley (1965) "What Happens in the Long Run?" *Philosophical Review,* **84**: 373–78.

Salmon, Wesley (1966) "The Foundations of Scientific Inference," in *Mind and Cosmos: Essays in Contemporary Science and Philosophy,* ed. Robert Colodny. Pittsburgh, University of Pittsburgh Press, pp. 135–275.

Savage, Leonard J. (1954) *Foundations of Statistics.* New York, Wiley.

Savage, Leonard J. (1961) "The Foundations of Statistics Reconsidered," in *Proceedings of the Fourth Berkeley Symposium on Mathematical Statistics and Probability.* Berkeley, University of California Press, pp. 575–85. Reprinted (1964) in *Studies in Subjective Probability,* eds. Henry Kyburg and Howard Smokler. New York, Wiley.

Scheibe, K., and Sarbin, T. (1965) "Towards a Theoretical Conceptualization of Superstition," *The British Journal for the Philosophy of Science,* **16**: 143–58.

Sellars, Wilfrid (1963) *Science, Perception and Reality.* New York, Humanities Press.

Shimony, Abner (1953) *A Theory of Confirmation.* Ph.D. dissertation, Yale University.

Shimony, Abner (1954) "Braithwaite on Scientific Method," *Review of Metaphysics,* **7**: 644–50.

Shimony, Abner (1955) "Coherence and the Axioms of Confirmation," *Journal of Symbolic Logic,* **20**: 1–28.

Shimony, Abner (1965) "Quantum Physics and the Philosophy of Whitehead," in *Philosophy in America,* ed. Max Black. London, Allen and Unwin, pp. 240–61. Reprinted (1965) in *Boston Studies in the Philosophy of Science,* vol. 2, eds. Robert Cohen and Marx Wartofsky. New York, Humanities Press.

Simpson, George G. (1963) "Biology and the Nature of Science," *Science,* **139**: 81–88.

Smart, J. J. C. (1963) *Philosophy and Scientific Realism.* New York, Humanities Press.

Stein, Howard (1967) "Newtonian Space-Time," *The Texas Quarterly,* Autumn: 174–200.

Strawson, Peter (1959) *Individuals.* London, Methuen.

Sutherland, N.S. (1959) "Stimulus Analyzing Mechanisms," *Mechanization of Thought Processes,* vol. 2. London, National Physical Laboratory Symposium No. 10.

Tolman, Richard (1938) *The Principles of Statistical Mechanics.* London, Oxford University Press.

Vernon, M. D. (1962) *The Psychology of Perception.* Baltimore, Penguin Books.

Weiss, Paul (1938) *Reality.* Princeton, Princeton University Press.

Weyl, Hermann (1949) *Philosophy of Mathematics and Natural Science.* Princeton, Princeton University Press.

Whitehead, A. N. (1929) *Process and Reality.* New York, Macmillan.

Whittaker, Edmund (1951) *History of the Theories of Aether and Electricity: The Classical Theories.* London, Thomas Nelson.

Whorf, Benjamin (1941) "The Relation of Habitual Thought and Behavior to Language," in *Language, Personality, and Culture: Essays in Memory of Edward Sapir,* ed. Leslie Spier. Menasha, Wis., Sapir Memorial Publication Fund, pp. 75–93. Reprinted (1956) in *Language, Thought, and Reality: Selected Writings of Benjamin Lee Whorf,* ed. John Carroll. Cambridge, M.I.T. Press.

Wigner, Eugene (1964) "Events, Laws of Nature, and Invariance Principles," in *The Nobel Prize Lectures.* Amsterdam, N.Y., Elsevier.

Wigner, Eugene (1967) *Symmetries and Reflections.* Bloomington, Indiana University Press. (A reprint of the preceding reference is included.)

Wrinch, D., and Jeffreys, H. (1921) "On Certain Fundamental Principles of Scientific Inquiry," *Philosophical Magazine* (6th ser.), **42**: 369–90.

10

Reconsiderations on inductive inference

I . BAYESIAN GENERALITIES ON INDUCTIVE INFERENCE

My most systematic discussion of inductive inference was "Scientific Inference," published in 1970. The title indicates its primary concern with reasoning in the sciences, though many of its considerations also apply elsewhere. I aimed at articulating principles of inductive reasoning actually used by working scientists, sometimes explicitly and sometimes tacitly. And I proposed justifications of these principles, in spite of the powerful tradition from Hume onward that this exercise is futile. Of course, the proposals for justification had to be made in tandem with a dialectical examination of the criteria that "justification" should satisfy.

My proposals fell under the denomination of "Bayesian," with some reservations. The central concept of inductive inference was taken to be epistemic probability: a quantitative measure of reasonable degree of belief in a hypothesis on specified evidence. The standard epistemic probability notation "$P(h/e) = r$" is to be read, "the reasonable degree of belief in h, given e as the total body of evidence, is (the real number) r." This rough explication is vague, for the constraints on "reasonable" are not specified, and indeed the lines of cleavage of Bayesianism are largely determined by different specifications of these constraints. But all the Bayesians agree (in contrast with frequency theorists of probability and with statisticians of the schools of Fisher and Neyman-Pearson) that the domain of pairs of propositions over which the function P is well defined is very wide, permitting one *inter alia* to speak of the "prior probability" of a hypothesis h, the evidential proposition e being taken to be a tautology t or possibly the general background information b. It is assumed generally that e is a noncontradictory proposition.

All Bayesians agree that the epistemic concept of probability satisfies the so-called "Axioms of Probability," despite different ways of arriving at them and variant ways of expressing them:

(i) $0 \leq P(h/e) \leq 1$;
(ii) if e logically implies h, then $P(h/e) = 1$;

(iii) if e logically implies the falsity of $h_1 \& h_2$, then

$$P(h_1 \lor h_2/e) = P(h_1/e) + P(h_2/e);$$

(iv) $P(h_1 \& h_2/e) = P(h_1/e) P(h_2/h_1 \& e) = P(h_2/e) P(h_1/e \& h_2)$.

From these axioms one easily derives Bayes's theorem, which is the fundamental instrument of inductive inference according to all Bayesians: let h and o be two propositions whose conjunction is consistent (the choice of notation suggesting that b is background information and o is an observational proposition); then

(1)
$$P(h/b \& o) = \frac{P(h/b) P(o/h \& b)}{P(o/b)}.$$

The term on the left-hand side of this equation is called "the posterior probability of h," the first factor in the numerator on the right-hand side is "the prior probability of h," the second is "the likelihood of o given h," and the term in the denominator is "the prior probability of o." An immediate corollary of Bayes's theorem is an expression for the ratio of the posterior probabilities of two hypotheses:

(2)
$$\frac{P(h_1/b \& o)}{P(h_2/b \& o)} = \frac{P(h_1/b) P(o/h_1 \& b)}{P(h_2/b) P(o/h_2 \& b)}.$$

Obvious advantages of using the corollary are the absence of $P(o/b)$ due to cancellation and the occurrence on the right-hand side of the prior probability ratio $P(h_1/b)/P(h_2/b)$ and the likelihood ratio $P(o/h_1 \& b)/P(o/h_2 \& b)$, both of which are usually easier to evaluate than their numerators and denominators separately.

Personal probability theory is a variety of Bayesianism which holds that there is no objective value of $P(h/e)$, and in fact that this expression is elliptical without some additional notation referring to a person X, whose evaluation of the probability is simply an expression of X's degree of personal belief in the hypothesis given the evidential proposition; a suitable notation is "$P_X(h/e)$." It usually happens that $P_X(h/e)$ and $P_Y(h/e)$ are unequal if X and Y are two different persons, and nevertheless according to personalists it need not be the case that either is irrational. The logical probabilists, by contrast, would hold that at least one of X and Y has made an irrational evaluation. More details are given about the logical probabilist and the personalist versions of Bayesianism in Sections II.B and II.C of "Scientific Inference."

All Bayesians agree that the various types of inductive inference that have been worked out in the past – induction by simple enumeration, eliminative induction, the hypothetico-deductive method, maximum likelihood inference, significance testing, et cetera – can be formulated in terms of the concept of epistemic probability and usually refined thereby, so that manifestly invalid modes of inference can be winnowed out and those that are roughly valid can be fine-tuned.

"Scientific Inference" proposed a theory of inductive inference inspired by H. Jeffreys (1961), called "tempered personalism," that is intermediate between logical probability theory and personalism. It takes the probability function P_X to be a somewhat idealized version of X's actual belief function, but the idealization makes no reference to a putative objective credibility function as postulated by logical probabilists. Instead, the idealization results from a kind of strategic and pragmatic constraint, called "the tempering condition":

In any investigation, assign to each seriously proposed hypothesis sufficiently high prior probability to allow the possibility that it will achieve a higher posterior probability than any rival hypothesis as a result of the envisaged observations.

Tempered personalism is a "local" strategy for inductive inference, in the sense that the tempering condition concerns a specified investigation, and the hypotheses governed by the prescription are those that are seriously proposed for that investigation. The tempering condition treats the set of seriously proposed hypotheses entirely differently from those that are frivolously proposed or not proposed at all (and hence are buried in a "catch-all" alternative to the explicitly articulated hypotheses); but within the set of seriously proposed hypotheses the tempering condition imposes a rough parity, allowing great differences in prior probability assignments but not by so many orders of magnitude as to preclude a reversal of preference as a result of empirical evidence.

Tempered personalism has some virtues that, in my opinion, ought to be preserved in a revised theory of inductive inference.

1. Without postulating objective epistemic probabilities (which, as the personalists argue, have no factual status either in the actual world or in the world of Platonic formal relations), tempered personalism nevertheless provides a corrective to personal prejudice. It is a way of realizing Peirce's "social theory of logic" (1932, p. 398).

2. It provides a strategy for navigating between credulity and excessive skepticism. Credulity is avoided by prescribing a kind of critical open-mindedness toward all seriously proposed hypotheses in an investigation, not just toward those which are antecedently favored by the subject; and this step, in turn, permits the likelihoods, which depend upon empirical

results, to determine the final preference. Excessive skepticism is avoided by abstaining from giving roughly the same prior probabilities to the nondenumerably many hypotheses that have never been proposed as to those relatively few that have been seriously proposed. The set of unproposed hypotheses so dominates the space of logically possible hypotheses that without the tempering condition it is hard to see how the seriously proposed hypotheses could have sufficient prior probability for eventual a posteriori acceptance.

3. As a special case of protection against excessive skepticism, tempered personalism opens the possibility of assigning high posterior probabilities to the strong general propositions of the natural sciences, contrary to the claim of Popper (1961, p. 405) and others that by any Bayesian inductive method these propositions must be assigned zero probability relative to any finite body of evidence.

4. Tempered personalism avoids the arbitrariness and rigidity of a priori orderings of hypotheses (like the simplicity orderings of Jeffreys and Wrinch and of Kemeny) - orderings which might fail to include hypotheses that are good approximations to the truth, or might place them so late in the ordering that they would never be reached during the likely lifetime of the human race.

5. By giving preferential weight to seriously proposed hypotheses, tempered personalism makes good use of any abductive powers that human beings might happen to possess.

6. Without conflating the context of justification with the context of discovery, tempered personalism brings these contexts into closer relation than do the usual formulations of inductive inference.

There are, however, some serious weaknesses in tempered personalism.

A. The tempering condition is obscure, because no criterion is given for a hypothesis to be considered "seriously proposed" and there is a danger that any prescribed criterion will be restrictive, authoritarian, or arbitrary.

B. The tempering condition transforms a relatively straightforward concept - namely, actual degree of belief - into a modified, distorted, idealized modality of belief, which some Bayesians have considered a psychological monstrosity (e.g., Dorling 1972, pp. 184-85). Nor is it clear that the tempered personalist concept of probability will satisfy the standard Axioms (i)-(iv), for neither the usual coherence arguments nor the Adamite derivation (Shimony 1988) are applicable, and the Cox-Good-Aczél argument recommended in "Scientific Inference" has a byzantine complexity.

C. The status of the catch-all hypothesis is troublesome. If $h_1, \ldots, h_{n-1}$ are explicitly formulated hypotheses (e.g., proposed laws of nature governing the phenomenon under investigation), then the catch-all, which is the negation of $h_1 \vee h_2 \vee \cdots \vee h_{n-1}$, surely merits the denomination "seriously proposed," because any undogmatic investigator is aware of the possibility of an ingenious theory not yet proposed by any one, and the wisdom of a fallibilistic attitude toward the array of hypotheses so far proposed is supported both by logical considerations (i.e., the domination of the space of logical possibilities by unproposed hypotheses) and by reflections on the history of science. It may even seem rational to assign a prior probability close to unity to the catch-all hypothesis. How does one balance these fallibilistic considerations against the desideratum of avoiding excessive skepticism (virtue 2 of tempered personalism, mentioned previously)?

I intend in the remainder of this paper to make certain modifications of my theory of inductive inference that will preserve the virtues of tempered personalism and ameliorate its weaknesses. The proposals will move closer toward ordinary (untempered) personalism in one respect, and away from it in others.

Once an investigation is prepared for the application of the Bayesian machinery, with a set of explicitly articulated rival hypotheses, I now see no harm in construing the probability function P_X as X's actual degree of belief. Tempering is not needed in order to make the Bayesian machinery sensitive to the truth if the virtues of the tempering condition can be achieved in other ways. But a posteriori principles of induction may be indispensable for the preparatory stage of inductive inference. Furthermore, non-Bayesian procedures are often needed subsequent to the evaluation of posterior probabilities of the rival hypotheses, in the decision to accept a hypothesis tentatively or in a decision to accept nothing and instead to widen the field of search.

As in "Scientific Inference" I still maintain that inductive logic has a complex structure, partly a priori and partly a posteriori. It is the indispensability of a posteriori principles that most sharply sets inductive logic apart from deductive logic and gives it a dialectical character. We must not only learn by experience, but we must learn by experience how to learn by experience. In these reconsiderations I shall attempt to show more clearly than in "Scientific Inference" the interdependence of the a priori and the a posteriori elements of inductive logic, instead of proceeding in two stages, the first entirely a priori and the second having recourse to factual information distilled from scientific investigations of the past.

III. TEMPERING WITHOUT TEMPERING

The tempering condition prescribes critical open-mindedness toward all seriously proposed hypotheses, by prohibiting the prior probability of each from being so small that no possible likelihood ratios emerging from the envisaged observations could overcome its initial handicap. Typically, there is a small number of specific seriously proposed hypotheses $h_1, \ldots,$ h_{n-1}, and the catch-all h_n is the abbreviation for $\sim (h_1 \vee h_2 \vee \cdots \vee h_{n-1})$. For example, in the chapter on the experimental basis of special relativity in Panofsky and Phillips (1955, p. 240) seven specific hypotheses are listed: stationary ether with no contraction, stationary ether with Lorentz contraction, ether attached to ponderable bodies, three variants of emission theories, and finally special relativity. The catch-all is not listed, but it can be considered as an eighth hypothesis. No prior probabilities are evaluated, even roughly, by the authors. Nevertheless, it seems fair to make the conditional assertion that if Panofsky and Phillips themselves were to use Bayesian procedures explicitly and attach prior probabilities to each of the seven hypotheses in their list, these would surely conform to the tempering condition. (I set the catch-all aside, because it requires special attention.) My confidence in making this assertion is based upon the fact that they dismiss none of the seven out of hand, but take the trouble to compile how well each of thirteen observational results (aberration, Fizeau convection coefficient, Michelson–Morley, Kennedy–Thorndike, de Sitter spectroscopic binaries, Michelson–Morley using sunlight, variation of mass with velocity, general mass–energy equivalence, radiation from moving charges, muon decay at high velocities, Trouton–Noble, and unipolar induction using a permanent magnet) agrees with each of the seven hypotheses. Their table (1955, p. 240) only states A (agrees), D (disagrees), and N (not applicable), but the discussion and the references provide finer information. As a matter of fact, the only hypothesis with which all thirteen observational results agree is the theory of relativity; but had all the results agreed with more than one of the hypotheses, there is no indication in the discussion that one of these would nevertheless have been overwhelmingly preferred to the others. Thus the tacit prior probability assignments of Panofsky and Phillips are effectively tempered without any overt use of the tempering condition.

There is so much individual and social variation in open-mindedness toward sets of competing scientific hypotheses in various investigations that no generalization can be made from the example just given, especially since a textbook was cited rather than documents from the period when special relativity was at the frontier of research. Nevertheless, one can make some plausible reconstructions of the methodological thinking implicit in the textbook discussion of special relativity and its rivals. I have two suggestions. The first is that all seven hypotheses offer a kind of "understanding," especially by making an analogy with other physically well-explored phenomena, such as propagation of sound in material media; and an offer of understanding that is not subject to immediate destructive criticism in the light of known scientific results is remarkable enough in the advanced sciences that it deserves a fair trial. The second suggestion is that the background b of anyone currently immersed in scientific research is permeated with much vague but massive historical lore: anecdotes about the narrowness of imagination of the scientific establishment at crucial junctures, moralizing stories concerning the courage of innovators and the ultimate vindication of their visions, and accounts of the occurrence and recurrence of scientific revolutions that convert the innovations of one epoch into the orthodoxies of the next. Hence the prior probability $P_X(h/b)$ is only relatively prior, in the sense that the explicit observational data are not included in the evidential proposition, but it is posterior to a considerable body of diffuse experience about investigations of the past, and this experience supports an attitude of critical open-mindedness toward hypotheses lying beyond the strict limits of orthodoxy.

I have delayed discussing the catch-all hypothesis, because it presents special problems. Up to this point one could regard the tempering condition solely as a provision to counteract a person's bias in favor of one specific hypothesis against others. But the tempering condition also serves as a methodological shield for the entire set of specific hypotheses $h_1, \ldots, h_{n-1}$ against the catch-all h_n. The reason is that in actual scientific theorizing, the specific hypotheses are strong and definite, and even their disjunction $h_1 \vee \cdots \vee h_{n-1}$ is quite strong, so that its negation h_n is quite weak; consequently, h_n occupies the preponderance of the relevant logical space (to speak in a way that has precise meaning if one is dealing with a formalized language like those of Carnap 1980, Hintikka and Niiniluoto 1980, and Kuipers 1978, but only metaphorical meaning otherwise). Consequently, the untempered judgment of person X may very well make $P_X(h_i/b)$ have the order of magnitude of a very small ϵ if i is less than n, and make $P_X(h_n/b)$ be close to unity. If so, it may be the case that no achievable body o of observational data would yield the posterior probability relation

$$P_X(h_i/b \,\&\, o) > P_X(h_n/b \,\&\, o) \quad \text{for any } i < n.$$

The conclusion of any scientific investigation would then be the frustrating judgment that "something else" is true! And, literally speaking, may that not be reasonable? Consider the comparison of special relativity theory with all its rivals. Is it not the case that special relativity theory has been superseded by general relativity theory, because the former cannot account for gravitational phenomena? And general relativity is likely to be superseded by something deeper – quantized general relativity theory or something more radical. Hence, it is the catch-all hypothesis, unlisted by Panofsky and Philips, that seems to have the highest posterior probability, and should have been accorded the highest prior probability. Is this not what one should have expected, in view of the immense scope of the catch-all?

The difficulties of the catch-all were discussed in Section II.D of "Scientific Inference," and some of the suggestions made there should be preserved, with appropriate modifications. It was suggested that a new modality of belief, called "commitment," is suitable when dealing with the strong specific hypothesis $h_1, \ldots, h_{n-1}$. To say that X has degree of commitment r to hypothesis h, conditional upon evidence e, is to attribute to X the degree of belief r (in the normal sense) in the approximate truth of h, where "approximate truth" is explicated by the following three conditions:

(i) within the domain of current experimentation, h yields almost the same observational predictions as the true theory;
(ii) the concepts of the true theory are generalizations or more complete realizations of those of h;
(iii) among the currently formulated theories competing with h, there is none which better satisfies conditions (i) and (ii).

But if $P_X(h/e)$ is interpreted in terms of degree of commitment, it is by no means clear that the function P_X will satisfy the standard axioms of probability. One can, however, retain the usual personalist sense of $P_X(h/e)$, keeping the normal modality of belief, and nevertheless achieve the desirable features of commitment by reinterpreting the specific hypotheses h_i, construing each (if $i < n$) as asserting conditions (i), (ii), and (iii). Some notational distinction is needed for clarity: h_i could still be taken to be the strictly stated specific hypothesis, taken literally (asserting, for example, that a certain general law holds throughout the universe and for all time), while $\bar{h}_i$ asserts that h is approximately true in the sense of satisfying conditions (i)–(iii). The catch-all hypothesis $\bar{h}_n$ is taken to be $\sim(\bar{h}_1 \vee \cdots \vee \bar{h}_{n-1})$. If these notational conventions are accepted and internalized, there would be no harm in dropping the bar on $\bar{h}_i$ and understanding the unbarred letter to refer to the assertion of approximate truth; any resulting systematic ambiguity would be resolved by context. But I shall keep the bars.

A brief remark is in order at this point about a very deep matter, the relation between physical law and exact mathematics. Questions of approximation arise not just because of the fallibility of the best theories so far proposed and confirmed, but because of the intrinsic character of physical systems themselves. At the beginning of Galileo's *Dialogue*

Concerning the Two Chief World Systems (1967, p. 14), Simplicio asserts, "But I still say, with Aristotle, that in physical (*naturali*) matters one need not always require a mathematical demonstration." And Sagredo answers, "Granted, where none is to be had; but when there is one at hand, why do you not wish to use it?" The deliberately understated optimism of Sagredo about the prospects of mathematical physics has, of course, been gloriously vindicated by the history of the science. But rigorous mathematical arguments in physics must always be carried out with a tacit codicil of caution regarding such matters as the neglected effect of the remote environment and the possible microscopic breakdown of the continuous structure of space–time. I do not wish to be drawn away from the main purpose of this paper by this difficult problem, but shall merely give one reference where it is handled very judiciously. In Tisza (1966, pp. 119–25), a section entitled "Discussion of the Postulational Basis" points out difficulties in applying the principles of the thermodynamics of equilibrium: the difficulty of knowing when thermodynamic equilibrium is achieved, the breakdown in scale invariance due to long-range forces, the occurrence of surface effects, and so forth. It is a remarkable fact about nature and human beings that the thermodynamic principles can nevertheless be used with control and fertility.

Since the hypotheses $\bar{h}_i$ ($i = 1, \ldots, n-1$) are much weaker than the corresponding h_i, the newly defined catch-all $\bar{h}_n$ is much stronger than h_n and hence occupies much less of the logical space. Consequently, without resorting to the tempering condition but simply by an honest representation of X's beliefs (following the nearly tautological rule of *crede quod credis*), it need not be the case that $P_X(\bar{h}_n/b)$ is many orders of magnitude greater than $P_X(\bar{h}_i/b)$ for $i < n$. However, a Bayesian treatment of scientific inference must now face another difficulty: the vagueness of the conditions (i), (ii), and (iii) implies that each of the $\bar{h}_i$ is vague even if h_i is clear; therefore, over and above the usual problem of evaluating one's own subjective degree of belief in a definite proposition on definite evidence – by introspection, by the Device of Imaginary Results (Good 1983, p. 126), or otherwise – there is here an additional problem of assessing one's attitude toward an indefinite proposition. Three relevant suggestions were made in Section II.D of "Scientific Inference," of which the second (with slight modifications) seems to me the most promising. The difficulty is manageable because the hypotheses $\bar{h}_1, \ldots, \bar{h}_{n-1}$ are *all vague in the same way,* for each asserts that a specific sharp hypothesis (one among $h_1, \ldots, h_{n-1}$) is an approximation to the truth in the sense of conditions (i), (ii), and (iii). Consequently, the probability ratios

$$\frac{P_X(\bar{h}_i/b)}{P_X(\bar{h}_j/b)}, \; \frac{P_X(o/\bar{h}_i \& b)}{P_X(o/\bar{h}_j \& b)}, \; \frac{P_X(\bar{h}_i/b \& o)}{P_X(\bar{h}_j/b \& o)} \quad (i < n, \; j < n, \; i \neq j)$$

are insensitive to the vagueness of conditions (i), (ii), and (iii) and presumably would be invariant under any reasonable sharpening of these conditions. If one uses the corollary to Bayes's theorem, Eq. (2), in order to make comparisons between $\bar{h}_i$ and $\bar{h}_j$ (i and j both less than n), only these insensitive ratios enter on both sides of the equation. On the other hand, if i is less than n but j equals n, then the three ratios just displayed (of prior probabilities, of likelihoods, and of posterior probabilities) clearly *are* sensitive to the vagueness of conditions (i), (ii), and (iii). It is therefore difficult to use the Bayesian machinery in order to adjudicate between $\bar{h}_i$ ($i < n$) and the catch-all. We have reached a juncture in scientific methodology that is beyond the resources of personalism and, I believe, beyond any version of Bayesianism. It does not follow that there is no rational adjudication concerning the catch-all, but none may be possible without some appeal to a posteriori considerations. Therefore I shall return to this question after a general discussion of a posteriori principles. (See also Mirabelli 1978.)

IV. DESIDERATA FOR A POSTERIORI PRINCIPLES OF METHODOLOGY

Inductive logic differs from deductive by requiring some principles that are not analytic. If, contrary to Kant, human beings are incapable of making synthetic a priori judgments whose truth is assured by the nature of our faculties, then the principles required for induction must be a posteriori. And since they are general, they transcend any finite evidential base and must therefore be obtained inductively. Two philosophical problems clearly must be faced: (1) what are the a posteriori principles that induction requires – principles that are strong enough to be methodologically useful, yet modest enough to be justified on the basis of collective human experience? (2) How is the obvious circularity of the foundations of inductive logic to be treated – can it be exorcized, or is it unavoidable and nevertheless not vicious?

The second of these two problems seems to have been more intensively discussed in the literature than the first. I also think that it is the easier of the two, and I have little to add to my discussion of it in Section V.E of "Scientific Inference." Most of my attention here will be devoted to the first problem, which is insufficiently discussed in the literature. Some of the difficulties of this problem become apparent from an examination of two important proposed solutions, a classical proposal of Mill and a recent one of Michael Friedman.

According to Mill, the four Methods of Experimental Inquiry (Agreement, Difference, Joint Method of Agreement and Difference, and Concomitant Variation) depend on the law of causation, that is, on the assumption that every event, or the beginning of every phenomenon, must have some cause, some antecedent, on the existence of which it is invariably and unconditionally consequent." (1949, Bk. III, Ch. XXI, Sect. 1)

Mill's justification for the law of causation is notoriously weak, for it is inferred by induction *per enumerationem simplicem* (*ibid.*, Sect. 1), which elsewhere he criticizes as loose and fallible. My main objection to using the law of causation as a principle of induction is its inappropriateness for the task which Mill assigns to it: it is too weak in one way and too strong in another. Its weakness consists in the fact that even if it were true there is no way of applying it in an inquiry without an exhaustive catalogue of all factors that might possibly bear upon the phenomenon of interest, a criticism presented in detail by Cohen and Nagel (1934, pp. 249–72). It is too strong, because it asserts a universal determinism in nature that is strongly disconfirmed by the evidence of contemporary physics. I suggest that Mill's impasse is due to his neglect of Aristotle's wise distinction between the order of knowing and the order of being (*Physics* I, 1, 84a, 16–21; *Metaphysics* VII, 3, 1029 a33–b12; discussed by McKeon 1947, pp. 27–31). If the law of causality were true, it would be a fundamental metaphysical principle, with priority in the order of being; and such principles are arrived at subsequent to the more accessible knowledge of nature. But Mill uses the law of causality as a prerequisite for applying his methods of experimental inquiry, even though it does not have priority in the order of knowing (as indicated by his reliance upon induction *per enumerationem simplicem* to justify it). I conjecture that any attempt to use a high-level metaphysical proposition as the basis for induction will engender one or both of Mill's troubles: it will be stronger than we can justify and very likely false, and it will be too remote from the arena of inquiry and hence too weak to apply in actual inductions.

Friedman (1985) proposes schematically an a posteriori treatment of methodology that avoids high-level metaphysics and aims at deploying empirical findings about human psychological mechanisms. He takes confirmation theory to seek the derivation of reliability statements about methods. The only method that Friedman discusses in detail is that of generalization based upon instances:

Under what conditions is a sentence of the form $\ulcorner \forall v (Fv \rightarrow Gv) \urcorner$ confirmed by the observed truth of a conjunction of instances of the form $\ulcorner (Fc \rightarrow Gc) \urcorner$? From the present point of view, the problem is to find a relation between sentences $C(S, R)$ such that if S is a generalization and R is a conjunction of its instances, we can derive a statistical law of the form . . . [f]or all S and R, the probability that S is true, given that R is observed to be true and $C(S, R)$, is p. (1985, p. 163)

The word "probability" here is to be understood in the sense of "objective or physical probability, not epistemic probability" (*op. cit.*, p. 154). After some discussion of candidates for the relation $C(S, R)$, centering around Goodman's concept of projectability, he concludes programmatically:

we specify the projectible predicates in terms of the history of their acquisition, for example. We then use a causal theory of reference – a general theory of how physical and psychological mechanisms related predicates with such-and-such a history to instances of the environment of such-and-such a kind – together with the rest of our theory of the world, to show that inductive inferences to generalizations whose vocabulary is limited to projectible predicates are in fact reliable. (*op. cit.*, pp. 164–65)

Friedman is right in trying to bring scientific information to bear upon the assessment of the reliability of a method, but his reliability statement is a will-o'-the-wisp: it is not likely to be achieved, and it is not needed for actual scientific inferences. My skepticism about achieving the assessment of a reliability statement along the lines that Friedman suggests is largely due to a doubt that a well-defined reference class exists when interesting methods of human inquiry are to be assessed. The reliability of a method used by a pigeon for deciding when to peck may possibly be assessed, because the reference class is designed by the experimenter, who also knows how to keep score on the experimental animal's successes and failures. But the reference class for assessing the reliability of the hypothetico-deductive method or induction *per enumerationem simplicem* seems to depend on so many contingencies of history, sociology, and individual biography that objective probabilities are ill defined; nothing like the objective probability of radioactive decay of a radium atom in a specified interval of time can be envisaged, contrary to Friedman's comparison (1985, p. 154). And even if the reliability of human methods of inference were somehow evaluated, it is hard to see how the resulting objective probability p would apply to a concrete case of scientific inference, with its own unique constellation of evidence, background information, competing theories, and conceptual considerations. Any objective theory of probability faces a problem of the relevance of a probability characterizing an ensemble to an individual case with all its peculiarities. This problem arises in a concrete case of scientific inference as much as in a political election or a war.

An alternative to Friedman's procedure is to look for a posteriori principles of modest strength, principles which are well suited to expediting the operation of Bayesian (epistemic) probability theory and which permit it to work in a manner that avoids the possible perversities and biases of personalism. The matter can be put metaphorically: Bayesian formalism is really intended to admit the light of the external world into the arcana of our subjective beliefs by using likelihoods, which are evidence dependent, to shape the posterior probabilities. But in order to admit the light the shutters must be opened, and this turns out to be a nontrivial operation, due to the vastness of the space of logically possible hypotheses and to the possibility (pointed out by Duhem) of skeptically and obstructively challenging all auxiliary assumptions in an inquiry. The a posteriori principles that are needed are those that provide instructions on opening the shutters. When the opening is properly performed, without the inadvertent insertion of distorting transducers (which Bacon warned us against in his discourse on the "Idols"), the submission of subjective belief to objective influences is expedited. This submission, according to a famous article of Peirce (1934, pp. 242–43), is the essence of the scientific method. There is no certainty that objectivity can be achieved in this way, nor even that this method will yield approximately true assertions with high frequency, as required by Friedman's definition of reliability. But the accumulative coherence, depth, predictive power, and extent of the scientific world view provides qualitative evidence for reliability even without statistics (cf. Hesse 1974, pp. 295–302 and K. Friedman 1990, pp. 92–109).

A list of four a posteriori principles will be stated in the following section, chosen in order to expedite the machinery of Bayesian probability theory. All are obtained by wide-ranging reflection on the enterprise of human knowledge and especially on the history of science. They may seem at first inspection to be banal, but they are far from being empty, and indeed each of them fails in some readily constructed model world. But there is abundant evidence that each is true in our actual world. Of course, it is not desirable that any of the a posteriori principles used in inductive inference should be tailored to the actual world in its particularity, for then one would require a quite complete knowledge of the world in order to design an effective methodology for investigating the world – an epistemological analogue of Kafka's *The Castle*. What is needed, and what indeed seems to be the case for the following principles, is that they fail for some possible worlds, but hold in a wide class of worlds, including the actual one. They have enough strength, but in no obvious way more than enough, to permit Bayesian inferences to proceed on the basis of empirical data for the purpose of progressively eliminating subclasses of the vast set of possible worlds, thereby bringing the actual world into increasingly sharp cognitive focus.

V. A SUGGESTED LIST OF A POSTERIORI PRINCIPLES

Principle 1: The class of hypotheses which at a given time offer "understanding" is statistically more successful in predicting subsequent empirical data than the complement of this class, where the word "understanding" is to be interpreted primarily in terms of the prevalent scientific standards of the time, but with deliberate openness to alternative well-articulated criteria.

Principle 2: A hypothesis that leads to strikingly successful empirical predictions is usually assimilable within a moderate time interval to the class of hypotheses that offer "understanding," possibly by an extension of the concept of understanding beyond earlier prevalent standards.

Principle 3: Nature has a pluralistic structure, with the parts enjoying a considerable degree of autonomy, and the causal impingement of one part upon another (whereby the autonomy of the parts is qualified) roughly falls off with their spatial separation.

Principle 4: Anomalies from the standpoint of generally successful theories are to a remarkable degree amenable to "trouble-shooting," in the sense that upon sufficiently serious and sustained investigation (the standards for which are established by reflection upon the relevant part of the history of science), an anomaly can usually either be explained away by noting appropriate contingencies or else converted into a non-anomaly in the light of an appropriate successor theory.

These four principles require some commentary, particularly since they all contain some vague expressions. The vagueness, I believe, is unavoidable in formulating a methodology which emphasizes the open and exploratory character of inquiry, but of course there is an obligation to show that the vague expressions do not drain the principles of content.

It was pointed out in Section II (weakness A) that the tempering condition is vague unless definite criteria are laid down for judging whether a hypothesis is "seriously proposed." But the specification of criteria for this purpose can hardly avoid a choice between a Scylla and a Charybdis: either the criteria are quite strict, in which case an authoritarianism is built into scientific methodology, and the way of inquiry may be blocked; or the criteria are quite loose, and any crackpot hypothesis that is advanced earnestly by its genitor would have to be considered "seriously proposed" and endowed with a nonnegligible prior probability. Section III.D of "Scientific Inference" struggled with this problem without a definitive resolution. Since I am retreating from the tempered personalist concept of probability in the present reconsiderations, I need not engage directly in the quest for a reasonable explication of "seriously proposed." However, I am trying to achieve "tempering without tempering" within a personalist version of Bayesianism, and therefore the problem of formulating criteria for "seriously proposed" emerges in a new guise. In an actual inquiry one must recognize a set of explicitly formulated hypotheses $h_1, \ldots, h_{n-1}$ which are set off from the catch-all hypothesis h_n and hence from the unarticulated hypotheses that are buried disjunctively in the catch-all. After the inquiry is structured in this way, the machinery of

Bayesian probability theory can be applied. An individual investigator who accepts Principles 1 and 2 (as can hardly fail to be the case if the investigator is immersed in some scientific research and is aware of the history of the discipline) will use them, either consciously or unconsciously, to identify the $h_1, \ldots, h_{n-1}$. And somehow, by complicated psychological and sociological mechanisms that are not fully understood, the scientific community comes to recognize a set of "live options," partially agreeing and partially disagreeing with the listing made by individual members of the community.

Principles 1 and 2 influence the selection of the explicit hypotheses $h_1, \ldots, h_{n-1}$ in radically different ways, each supplementing and effectively controlling the other. Principle 1 states that as a matter of fact, in the actual world, *rationality is empirically efficacious*. And Principle 2 states that as a matter of fact, in the actual world, *experience instructs rationality*. The interplay of Principles 1 and 2 constitutes a kind of dialectic of experience and reason, which is a recurrent motif in the history of science. These elliptical assertions require amplification and illustration.

If Principle 1 is read narrowly, without attention to the caution stated regarding the extension of the concept of understanding, it would appear to entail methodological conservatism. It is widely recognized, and stressed particularly by Kuhn (1970, pp. 23–51), that criteria for understanding are to a large extent determined in a given epoch by the paradigms of the dominant scientific theories. To the extent that this is the case, an investigator following Principle 1 would take the explicitly formulated hypotheses to fall within the general framework of the dominant theory. A multiplicity of hypotheses could be entertained because of the latitude of the framework; but any hypothesis lying outside that framework would fail to offer understanding (in the sense of conformity to the paradigm) and hence would be relegated to the catch-all. The outstanding example of the determination of criteria of understanding by a dominant theory (at least subsequent to the scientific revolution of the sixteenth and seventeenth centuries) is provided by the triumph of mechanism. "Understanding" in physics and to some extent in other sciences as well was widely taken to mean "understanding within the framework of mechanism." For instance, Helmholtz wrote, "the object of the natural sciences is to find the motions upon which all other changes are based, and their corresponding motive forces – to resolve themselves, therefore, into mechanics" (quoted by Nagel 1961, p. 155, along with a number of other similar expressions of opinion).

However, Principle 1 is not to be understood merely as an endorsement of the status quo. In fact, if that were all that was intended, a Bayesian could properly wonder whether it deserves the status of a principle of inductive inference. If X takes the dominant (and presumably highly successful) theory of the field of inquiry to be part of the background b, then it would automatically follow that $P_X(h/b)$ would be many orders of magnitude greater than $P_X(h'/b)$ if h conforms to the framework of the dominant theory and h' does not, ceteris paribus. But Principle 1 is indeed an a posteriori principle of inductive inference, and not an entirely conservative one, because of the clause that "understanding" may be construed not only in accordance with the dominant theory but according to alternative well-articulated criteria. A hypothesis that offers understanding in a novel way will obviously and automatically be given nonnegligible prior probability by the originator of the novel way of looking at things or by someone whose imagination has been captivated by the novelty; again, there is no need to legislate *crede quod credis*. But even someone whose imagination has not been captured may be open-minded about prior probability assignments because of Principle 1, which encapsulates the wide historical evidence for the empirical fruitfulness of novel modes of understanding; and the open-mindedness achieved in this way would constitute "tempering without tempering."

But does not the clause "with deliberate openness to alternative well-articulated criteria" open the floodgates to crank hypotheses, justified by crank criteria of understanding? I think not, for reasons that are inescapable when one descends from the high level of discussion thus far to the level of actual inquiry. First of all, it is possible to question a clearly articulated assertion and to demand reasons, and reasons for reasons. There are various typical patterns in the ensuing dialectic.

(i) It may disintegrate into inarticulateness.
(ii) It may eventuate in sharp predictions, which are susceptible of experimental test.
(iii) It may slide into equivocations, which have the effect of hedging assertions against empirical disconfirmation (cf. Giere 1991, pp. 103–8, on astrologers and psychics).
(iv) The dialectic may be inconclusive because of the subtlety of the concepts involved and the difficulty of formulating well-posed questions (which seems to me to occur in some of the debates on cosmology and psychophysics, perhaps indicating an unripeness of those disciplines).

Each pattern needs to be treated distinctively. Patterns (ii) and (iii) predominate in the sample of crank hypotheses that have come to me by unsolicited mail. And the majority of those of type (ii) could be refuted by citing a body of experimental findings that the authors were unaware of. In a mature science, like atomic physics, it is extremely difficult for an unorthodox theory that is sharply enough formulated to make definite predictions to evade a rapid refutation. An exception is an unorthodox theory that has developed a strategy for incorporating (cleverly or parasitically, according as it is assessed sympathetically or unsympathetically) the detailed modes of calculation of the orthodox theory; but when this happens, there are typically problems of conceptual interpretation, characteristic of pattern (iv) (e.g., the hidden-variables theory of Bohm 1952, with the device of "quantum potentials").

Although dialectical examination is often unavoidable concerning a criterion for the subtle and fundamental concept of "understanding," there are many examples both in actual scientific investigations and in methodological studies which present no obvious difficulty. Each of the seven hypotheses mentioned in Section III concerning propagation of light unequivocally offers understanding in some identifiable sense – mainly in the sense of providing an analogy to some domain of well-studied phenomena, such as the propagation of sound waves in fluid media and the propulsion of projectiles from sources. By contrast, the "grue" hypothesis of Goodman (1965, p. 74), which has occupied the attention of numerous methodologists, offers no promise of understanding by any articulated criterion. Define the predicate "grue$_t$," as applying to anything that is green and first examined before t, or blue and not examined before t. (I have slightly altered Goodman's definition and have inserted the temporal subscript in order to expedite the discussion.) If t is chosen to be subsequent to the time at which the inductive inference on the color of emeralds is being made, then the evidential proposition that all emeralds so far examined are green is equivalent to the proposition that all of them are grue. Hence, if in an inquiry an investigator assigns prior probabilities of the same order of magnitude to h and h', where h asserts that all emeralds are green and h' that all are grue$_t$, then the posterior probabilities of h and h' on the evidence e are likewise of the same order of magnitude, since the likelihoods of e on h and h' are exactly the same. But why should an investigator assign a nonnegligible prior probability to h'? Or more to the point of the present consideration, how should an investigator who is initially neutral and open-minded on the whole issue, but recognizes the historical correctness of Principle 1, articulate a criterion of "understanding" in such a way that h' offers a promise of understanding according to that criterion? I do not know whether this question is addressed anywhere in the literature, but I shall try to play devil's advocate by giving the following answer. Suppose that I take seriously the proposal of Peirce (1935, pp. 15–16, p. 84) and Wheeler (1983) that the laws of nature are not fixed for all time but are subject to cosmic evolution; and suppose that we charitably construe this metaphysical speculation as permitting secular changes of the laws concerning the colors of minerals. If so, one could say that the hypothesis H, "all emeralds have color depending in a variable way on t," offers "understanding" in accordance with a criterion of understanding based upon Peirce's evolutionary metaphysics. Even with this charitable interpretation, however, one could still not assign a nonnegligible prior probability to h', for this hypothesis, with a definite value of t, is one of a nondenumerable set of hypotheses disjunctively equivalent to H. In short, it is hard to see how any one could single out the hypothesis that all emeralds are grue$_t$ on the ground that it promises a new mode of understanding.

I conclude that Principle 1 serves in spite of its vagueness to delimit a subspace of hypotheses of small measure compared to that of the entire space of logically possible hypotheses. In this way it satisfies the desideratum stated in Section IV of preparing for the application of the Bayesian apparatus in scientific inference.

Principle 2 provides another antidote to the conservative side of Principle 1, for it recognizes a specific way in which the concept of "understanding" is extended: by the guidance and suggestiveness of phenomenological regularities. Illustrations are provided by two famous episodes in the history of physics.

The first is the transition from Newton's law of gravitation to the Newtonian natural philosophy. An extension of the concept of "understanding" did occur, as stated in Principle 2, in a "moderate time interval" – somewhat less than the time between Newton's initial work on gravitation during the plague year 1665–66 and the completion of the *Principia* in 1686 – and the transition was accomplished by Newton himself (see Stein 1990, pp. 37–38). At the beginning of his investigations Newton accepted that the laws of motion are laws of impact of absolutely hard bodies and that force is "the pressure or crowding of one body upon another," as formulated by the mechanistic philosophers of his time, most authoritatively by Huygens. In the course of establishing that gravitational phenomena, governed by the phenomenological law which he had discovered, could not be accounted for by contact forces, Newton arrived at a new metaphysics. "Its central notion is that of a *vis naturae* or *potentia naturalis* – a 'force of nature' or 'natural power'" (*op. cit.*, p. 38), of which the gravitational attraction between two bodies, however remote, was the first exemplification. Newton's concept of understanding was a corollary of his metaphysical innovation. The grip of his new concept of understanding upon the epistemology of working physicists for the following two and a half centuries is indicated by the quotation from Helmholtz in Section IV.

The second illustration is the success of Bohr's quantum theory of the atom of 1913 in accounting for the Balmer series in the spectra of hydrogen-like atoms, including quantitative derivations of the Rydberg constant and of the systematic difference between the spectral lines of hydrogen and those of singly-ionized helium. Bohr's strange combination of classical and quantum postulates was baffling, but his phenomenological derivation was indeed striking. Einstein publicly commented, "Very remarkable! There must be something behind it. I do not believe that the derivation of the absolute value of the Rydberg constant is purely fortuitous" (Jammer 1966, p. 86). The "moderate time interval" until Bohr's result was assimilated (with rectifications) into the new conception of physical understanding provided by the new quantum mechanics was about one and a half decades.

It is impossible to do justice to the concept of understanding without a serious epistemological study going far beyond the considerations of inductive inference to which this essay is devoted. I shall therefore only make some brief concluding remarks about the content of Principles 1 and 2 and the significance of the two illustrations just given. Both illustrations show that the new mode of understanding is recognizably much more than a feeling of habituation to a phenomenological regularity. The frameworks of both classical mechanics and quantum mechanics give criteria for understanding by providing a mathematical formalism, metaphysical concepts, linkages to common sense and experience, and paradigms of solved problems. Neither completely satisfies Leibniz's principle of sufficient reason, because each lays down postulates for which no *ratio essendi* is given, but only an indirectly empirical *ratio cognoscendi*. In view of the failure of the ambitious rationalist programs of Leibniz and Hegel, one can hardly escape the conclusion that an ultimate unexplained level is an inescapable part of any explanatory framework. From a human point of view it is experience that has the last word in the dialectic of experience and reason.

The primary methodological role of Principles 3 and 4 is to provide guidelines for assessing and relying upon auxiliary assumptions in inductive inference. It was stressed by Duhem (1954, chap. 6) that typically a hypothesis of interest must be supplemented by assumptions $a_1, \ldots, a_k$ in order to make possible the derivation of empirically testable consequences; accordingly, the empirical refutation of one or more of these consequences does not constitute a refutation of h, but only of the conjunction of h and the auxiliaries. (An extensive critique of Duhem's theses is given by Grünbaum 1973, pp. 585–629.) When the hypothetico-deductive method is assimilated into Bayesian probability theory, the first of Duhem's theses is transformed into a comment on the likelihood factor on the right-hand side of Bayes's theorem, Eq. (1): that $P_X(o/h \& b)$ is typically neither very close to 0 nor very close to 1, so that in comparisons of h with rival hypotheses the likelihood ratios are not overwhelmingly large or small; but if X computes the likelihood $P_X(o/h \& a \& b)$, which may well be very close to 0 or 1 if a is the conjunction of appropriate auxiliary assumptions, and uses this likelihood in order to evaluate the posterior probability $P_X(h/o \& a \& b)$, the result is of no value for theoretical or practical decisions unless a is well supported (an echo of Duhem's second thesis). Any one of the auxiliaries a_i can be checked by a separate investigation, but a pervasive skepticism regarding inductive inference would question the auxiliary assumptions in *that* investigation, ... In actual scientific work and in application of scientific results, the sequence of auxiliary investigations must be curtailed when a point is reached where no further reasons are given for further checking. The methodological value of Principles 3 and 4 is to provide rational support for what might otherwise be an excessively optimistic strategy of curtailing auxiliary investigations.

Principle 3 sanctions auxiliary assumptions discarding as negligible the effects of events remote from the locus of the phenomenon under investigation. Cautious phrasing of the Principle ("considerable degree of autonomy," "roughly falls off") is intended to permit exceptions to the general strategy of localizing the domain of investigation. The intention is not to foreclose in advance the possibility of transatlantic cable communication, well-collimated laser beams, radar signals, or astrological influences. But Principle 3 does assert that we live in a non-Parmenidean universe, in which it is *not* the case that everything is equally relevant to everything, and it provides a rough guideline for judging relevance in the absence of evidence or heuristics to the contrary.

The main methodological use of Principle 4 is to provide guidelines for the questioning of auxiliary assumptions: when to persist in probing them, when to accept them tentatively without further probing. Troubleshooting is a well-known procedure when a device fails to operate according to specifications. The operating instructions for my stereo receiver include a table with three columns, headed "Symptom", "Cause", and "Remedy". For example, on one line the symptom is "Noise is produced during reception of AM broadcasts when the power switch of other components is turned ON or OFF"; the cause is "AM loop antenna picks up electrical noise created by switch contacts"; and the remedy is "Put AM loop antenna as far away as possible from this and other components." When an apparatus is used in experimentation, trouble-shooting has the same pattern as for the devices serving our domestic needs and pleasures, and of course it is a standardized procedure just because the apparatus is a product of engineering, whose principles of construction and functioning are well understood by some of the personnel who produced it. But there is also trouble-shooting of a different sort, when the experimental results do not conform to expectations that are based upon well-entrenched assumptions. Checking for perturbations, human failures, instrument failures, systematic errors, and random errors has a logic generically like that of trouble-shooting the malfunction of an engineering product, but with one great difference: the space of logical possibilities is much greater. This is the reason why it is methodologically valuable to rely upon Principle 4, which summarizes a vast body of experience concerning scientific investigation. The principle asserts in effect that human beings (at least those that are sufficiently well trained) are able to sample the logical space of possible troubles with sufficient fairness that in general the cause of the nonconforming result is located, if indeed it is due to an error or a trouble – that is, to an uncontrolled contingency rather than to the falsity of the assumptions on which the expectations were based. It follows that the failure of trouble-shooting to locate an error or trouble constitutes prima facie evidence that the nonconforming experimental result is significant: that the assumptions upon which the expectations were based are incorrect or incomplete in some respect. The latter class of cases constitutes much of the drama of experimental and observational science – for example, Becquerel's discovery of radioactivity, the discovery of pulsars by Bell and Hewish, the discovery of cosmic background radiation by Penzias and Wilson, in each case after futile attempts at trouble-shooting. In general, if sustained efforts at trouble-shooting have failed to detect the falsity of one of the auxiliary assumptions a_i, where "sustained" is judged by the standards of the discipline, then it is rational to look for an alternative hypothesis in order to explain the anomaly. In other words, Principle 4, together with reflections upon the success of trouble-shooting in the field, provides some rational guidance at a time of crisis, to use Kuhn's language (1970, chap. VII), in deciding whether to seek a new hypothesis.

It may be recalled that at the end of Section III the problem of adjudicating between the catch-all and the entire set of explicitly formulated hypotheses was raised, but then postponed. The foregoing discussion of the use of Principle 4 is essentially a proposal to solve this problem. The problem becomes acute when the posterior probability ratio $P_X(\bar{h}_i/b \& a \& o)/P(\bar{h}_j/b \& a \& o)$ is very large for some $i \neq n$ and all j different from both i and n, and nevertheless the likelihood $P_X(o/\bar{h}_i \& b \& a)$ is disappointingly small. In other words, $\bar{h}_i$ is the best of all the explicitly formulated hypotheses, and nevertheless it is not very good at accounting for the empirical data. Trouble-shooting for weakness among the auxiliary assumptions is the entirely reasonable standard procedure, but if no weaknesses are found, then a decision has to be made between continuing the search, or else keeping the auxiliary assumptions and accepting the consequence of preferring the catch-all $\bar{h}_n$ to $\bar{h}_i$. Some resort to an a posteriori principle seems to be essential at this juncture, and Principle 4, conjoined with knowledge of the standards of success at trouble-shooting in the discipline, seems to provide the requisite guidance.

VI. SOME GENERAL REFLECTIONS ON THE A POSTERIORI PRINCIPLES

The methodological use of each of the four recommended a posteriori principles was discussed singly in the preceding section, but a few supplementary remarks about their collective character will help to give a perspective on the enterprise of inductive logic.

First, it should be noted that none of the four principles belongs to a well-recognized scientific discipline, such as physics, biology, or psychology. The first, second, and fourth share the peculiarity of slicing vertically through the hierarchical structure of nature, since they say something about the world which is the object of inquiry and the human beings who are investigating the world. The third principle does not explicitly refer to human investigators, but it still cannot be placed in any one of the natural sciences; it is rather a proposition of empirical metaphysics, serving all the special sciences, but with much less rigidity than the law of causation of Mill. The unusual character of the four a posteriori principles is inseparable from their role of expediting the application of the Bayesian machinery.

Second, it is possible to reconstruct to some extent the history of all four of the principles and to see how they were arrived at on the basis of a cruder methodology. The history of Principle 4 is probably the easiest to reconstruct, since much of the art of trouble-shooting is connected with error theory and statistical analysis, which are relatively late methodological achievements; but one comes upon some wonderful and surprising earlier examples of sophistication concerning trouble-shooting, for instance in the experimental treatises of Robert Boyle (see Sargent 199-, chap. 8). The history of the first and second principles is a continuation of classical controversies on the roles of reason and experience in human knowledge, going back to Plato and Aristotle and some of their predecessors. The history of Principle 3 is inseparable from classical philosophical and scientific debates between monists and pluralists.

Third, the most important consequence of the historical character of the four principles concerns the future. Just as those principles were arrived at by reflection upon earlier achievements of human knowledge, which themselves were arrived at by means of a crude or inarticulate methodology, so one can expect the four principles to be refined and supplemented by reflection upon more recent discoveries. The continued refinement of the a posteriori principles of inductive inference accords with the dictum stated at the end of Section II, that we learn by experience how to learn by experience. But the dictum needs to be enriched so as to allow a place for a rationalistic element – the search for understanding – in the dynamical development of methodology. One domain where a new methodological principle can be expected is fundamental physics. In an epoch when the dominant philosophies of science are empiricistic it is striking to hear recurrently the proclamation of the heuristic power of mathematics in deepening physics. The most eloquent of these to my knowledge is Hermann Weyl's pronouncement, "Ein paar Grundakkorde jener Harmonie der Sphären sind in unser Ohr gefallen, von der Pythagoras und Kepler träumten" ("A few fundamental chords of that harmony of the spheres, of which Pythagoras and Kepler dreamed, have entered into our ear") (Weyl 1970, p. 317). Lest Weyl's authority be impugned because he was primarily a mathematician, it must be emphasized that similar statements were made by Einstein, Dirac, Heisenberg, Wigner, and Wheeler, all of whom were very attentive to the relation between theoretical and experimental physics. The fact of the matter is that mathematical beauty and depth, not unconstrained by experience but very lightly constrained, has deepened and sharpened our characterization of the physical world. A good example of the commitment of fundamental physicists to this thesis is the following conclusion to an encyclopedia article on supersymmetry and supergravity:

Up to the present, supersymmetry remains a fascinating theoretical tool whose ultimate role, if any, within our understanding of physical laws eludes us. Its beautiful features suggest that this role is to be found, if not now then perhaps sometime in the future. (Burgess 1990, p. 1210)

It should be emphasized, however, that the heuristic power of mathematics for physics is not seen only in esoteric developments in twentieth-century physics. There is no better illustration of the power of mathematics to deepen physics than the history of classical mechanics, in the successive versions of Newton, Lagrange, Euler, Hamilton, Jacobi, and Cartan et al. (cf. Lanczos 1966, pp. vii–ix, and Saunders 199–). I am not prepared to extract from these remarks a methodologically usable supplement to Principles 1–4. Nevertheless, I believe on empirical grounds that Platonism, lightly monitored by experience, is physically efficacious, and herein lies a methodological principle waiting to be articulated.

Fourth, the imprecise wording of the four a posteriori principles needs to be controlled by a systematic series of case histories. There have recently been some excellent illustrations of Bayesian probability theory with scientific case histories, especially Franklin (1986) and Howson and Urbach (1989). A great virtue of Franklin's book is the attention paid to auxiliary assumptions, particularly in Chapter 2 ("The Discovery of Parity Nonconservation") and Chapter 3 ("CP or not CP"). The historical material in both of these chapters could be profitably used to illustrate the importance of Principles 3 and 4 in actual scientific inference, or else to challenge and correct some of the claims that I made about their utility. Further case histories are needed for this purpose, however, and should be taken from diverse areas of scientific investigation. Neither of the two works cited says much that is valuable concerning the initial structuring of a problem of scientific inference, when the explicitly formulated hypotheses are distinguished from the catch-all. Case histories are needed to illustrate Principles 1 and 2, which I recommend for this purpose, and possibly to refine and correct them.

Finally, something should be said about the relation between the a posteriori principles of induction and the method of argumentation introduced by Reichenbach (1938, chap. 5) and usually called "pragmatic justification" and "vindication" (Salmon 1966, p. 52 and p. 135). Such an argument in favor of a proposition ϕ has the form: "ϕ can neither be proved nor disproved, but nothing will be lost and something may be gained by tentatively accepting ϕ." In "Scientific Inference" I criticized Reichenbach's own version of this argument but used another version to defend the tempering condition (see especially Section IV.C therein), without invoking any a posteriori principles of induction. In fact, my justification of inductive inference consisted of two stages: the first being a priori and consisting of the axioms of probability and a pragmatic argument; the second relying upon certain a posteriori propositions. Teller (1975) presented an interesting criticism, claiming that pragmatic arguments must always rely implicitly upon some factual assumptions, and in fact he wishes to dispense entirely with pragmatic arguments in justifying inductive inference. In my reply to him (Shimony 1975) I accepted part of his criticism and said that it probably was an error to make a sharp separation between a priori and a posteriori stages in the justification of induction. But I did not, and still do not, agree with his dismissal of the entire genre of pragmatic arguments. Any a posteriori propositions used in induction, such as my Principles 1–4 or any of Teller's principles, are extrapolations beyond the evidence that supports them. To be sure, the extrapolation is tentative, and any of these principles is vulnerable to disconfirmation in the light of further evidence, even though the principles be employed in the inferences that eventuate in their disconfirmation. In fact, the circularity of using a posteriori principles as instruments of induction is rendered nonvicious by this very possibility of disconfirmation. But suppose, as I anticipate, that the a posteriori principles of induction stand up under empirical scrutiny. What should be the proper philosophical assessment of the entire theory of induction that relies upon them? At this juncture, I believe, a pragmatic argument is reasonable and indispensable, and I propose the following (which is similar to the last paragraph of "Scientific Inference").

The purpose of the a posteriori principles (in the metaphorical language at the end of Section IV) is to open the shutters, so that the Bayesian machinery can assess hypotheses in the light of evidence from the outside world. The efficacy of these principles for opening the shutters is not simply assumed, but is subject to empirical scrutiny. If the principles stand up under sustained scrutiny, then it is reasonable to accept them and to use them for expediting the application of Bayesian probability theory, for what better means are available to human beings for assuring that our beliefs are shaped by objective fact? It may indeed be the case that the world and we ourselves are so made that undetected sources of error will thwart our aspirations for good approximations to the truth, perhaps because of a Cartesian deceiving demon. But nothing that is salvageable is lost, and something may be gained, by using a method that is designed to be sensitive to objective fact and that assiduously monitors its own deviations from objectivity.

Acknowledgment. This work was supported by the National Endowment for the Humanities.

REFERENCES

Aristotle (1941), *The Basic Works of Aristotle,* New York: Random House.

Bohm, D. (1952), "A Suggested Interpretation of Quantum Theory in Terms of 'Hidden' Variables." *Physical Review* 85: 166–93.

Burgess, C. P. (1990), "Supersymmetry and Supergravity." In R. G. Lerner and G. L. Trigg (eds.), *Encyclopedia of Physics.* New York: VCH Publishers.

Carnap, R. (1980), "A Basic System of Inductive Logic," Part 2. In R. C. Jeffrey (ed.), *Studies in Inductive Logic and Probability,* vol. II. Berkeley: University of California Press.

Cohen, M., and Nagel, E. (1934), *An Introduction to Logic and Scientific Method.* New York: Harcourt, Brace.

Dorling, J. (1972), "Bayesianism and the Rationality of Scientific Inference." *British Journal for Philosophy of Science* 23: 181–90.

Duhem, P. (1954), *The Aim and Structure of Physical Theory,* transl. P. Wiener. Princeton, NJ: Princeton University Press.

Franklin, A. (1986), *The Neglect of Experiment.* Cambridge: Cambridge University Press.

Friedman, K. (1990), *Predictive Simplicity.* Oxford: Pergamon.

Friedman, M. (1985), "Truth and Confirmation." In H. Kornblith (ed.), *Naturalizing Epistemology.* Cambridge, MA: MIT Press.

Galilei, G. (1967), *Dialogue Concerning the Two Chief World Systems,* trans. S. Drake. Berkeley: University of California Press.

Giere, R. (1991), *Understanding Scientific Reasoning,* 3rd ed. Fort Worth, TX: Holt, Rinehart, and Winston.

Good, I. J. (1983), *Good Thinking.* Minneapolis: University of Minnesota Press.

Goodman, N. (1965), *Fact, Fiction, and Forecast,* 2nd ed. Indianapolis, IN: Bobbs-Merrill.

Grünbaum, A. (1973), *Philosophical Problems of Space and Time,* 2nd ed. Dordrecht: Reidel.

Hesse, M. (1974), *The Structure of Scientific Inference.* Berkeley: University of California.

Hintikka, J., and Niiniluoto, I. (1980), "An Axiomatic Foundation for the Logic of Inductive Generalization." In R. C. Jeffrey (ed.), *Studies in Inductive Logic and Probability,* vol. II. Berkeley: University of California Press.

Howson, C., and Urbach, P. M. (1989), *Scientific Reasoning: The Bayesian Approach.* La Salle, IL: Open Court.

Jammer, M. (1966), *The Conceptual Development of Quantum Mechanics.* New York: McGraw-Hill.

Jeffreys, H. (1961), *Theory of Probability,* 3rd ed. Oxford: Clarendon.

Kuhn, T. (1970), *The Structure of Scientific Revolutions,* 2nd ed. Chicago: University of Chicago Press.

Kuipers, T. (1978), *Studies in Inductive Probability and Rational Expectation.* Dordrecht: Reidel.

Lanczos, C. (1966), *The Variational Principles of Mechanics,* 3rd ed. Toronto: University of Toronto Press.

McKeon, R. (1947), "Aristotle's Conception of the Development and the Nature of Scientific Method." *The Journal of the History of Ideas* 8: 3–44.

Mill, J. S. (1949), *A System of Logic,* 8th ed. London: Longmans, Green.

Mirabelli, A. (1978), "Belief and the Incremental Confirmation of One Hypothesis Relative to Another." In P. D. Asquith and I. Hacking (eds.), *PSA 1978,* vol. I. East Lansing, MI: Philosophy of Science Association.

Nagel, E. (1961), *The Structure of Science.* New York: Harcourt, Brace.

Panofsky, W., and Phillips, M. (1955), *Classical Electricity and Magnetism.* Cambridge, MA: Addison-Wesley.

Peirce, C. S. (1932), *Collected Papers,* vol. II, C. Hartshorne and P. Weiss (eds.). Cambridge, MA: Harvard University Press.

Peirce, C. S. (1934), *Collected Papers,* vol. V, C. Hartshorne and P. Weiss (eds.). Cambridge, MA: Harvard University Press.

Peirce, C. S. (1935), *Collected Papers,* vol. VI, C. Hartshorne and P. Weiss (eds.). Cambridge, MA: Harvard University Press.

Popper, K. (1961), *The Logic of Scientific Discovery.* New York: Science Editions.

Reichenbach, H. (1938), *Experience and Prediction.* Chicago: University of Chicago Press.

Salmon, W. (1966), *The Foundations of Scientific Inference.* Pittsburgh: University of Pittsburgh Press.

Sargent, R. M. (199–), "The Diffident Naturalist." Unpublished manuscript.

Saunders, S. (199–), "Physics in Mathematics." Unpublished manuscript, Philosophy Department, Harvard University.

Shimony, A. (1970), "Scientific Inference." In R. G. Colodny (ed.), *The Nature and Function of Scientific Theories.* Pittsburgh: Pittsburgh University Press. (Reprinted as Chapter 9 of this volume.)

Shimony, A. (1975), "Vindication: A Reply to Paul Teller." In G. Maxwell and R. M. Anderson (eds.), *Induction, Probability, and Confirmation.* Minneapolis: University of Minnesota Press.

Shimony, A. (1988), "An Adamite Derivation of the Principles of the Calculus of Probability." In J. H. Fetzer (ed.), *Probability and Causality.* Dordrecht: Reidel. (Reprinted as Chapter 7 of this volume.)

Stein, H. (1990), "On Locke, 'the Great Huygenius, and the Incomparable Mr. Newton'." In P. Bricker and R. I. G. Hughes (eds.), *Philosophical Perspectives on Newtonian Science.* Cambridge, MA: MIT Press.

Teller, P. (1975), "Shimony's A Priori Arguments for Tempered Personalism." In G. Maxwell and R. M. Anderson (eds.), *Induction, Probability, and Confirmation.* Minneapolis: University of Minnesota Press.

Tisza, L. (1966), *Generalized Thermodynamics.* Cambridge, MA: MIT Press.

Weyl, H. (1970), *Raum-Zeit-Materie,* 6th ed. Berlin: Springer.

Wheeler, J. A. (1983), "On Recognizing Law without Law." *American Journal of Physics* 51: 398–404.

11

Comments on two epistemological theses of Thomas Kuhn

I

The Structure of Scientific Revolutions[1] and Kuhn's afterthoughts (1970a, 1974) have been subjected to so much penetrating criticism that any further examination might be supposed redundant.[2] There are, however, two epistemological theses in his work which have not, in my opinion, received adequate analysis. They are not essentially dependent upon considerations of Gestalt switches, theory-ladenness of observation, incommensurability, or the ambiguity of the word 'paradigm', which have received much critical attention.

The first thesis is that *the progress of science ought not be construed as the approach to a fixed goal which is the truth about nature.*

> But need there be any such goal? Can we not account for both science's existence and its success in terms of evolution from the community's state of knowledge at any given time? Does it really help to imagine that there is some one full, objective, true account of nature and that the proper measure of scientific achievement is the extent to which it brings us closer to that ultimate goal? (1970, p. 171)

Kuhn maintained this thesis in his replies to critics and even strengthened it somewhat: "If I am right, then 'truth' may, like 'proof', be a term with only intra-theoretic applications" (1970a, p. 266). The second thesis is that *the procedures of scientific investigation can be shown to be rational, and the appropriate sense of "rationality" can be explicated, only by drawing upon the substantive achievements of science.* One formulation of the thesis can be found in Chapter 1 of his book (1970), where he rejects the dichotomy of "the context of discovery" and "the context of

1. All references will be to the second edition of 1970, which contains "Postscript – 1969".
2. See, e.g., Shapere (1964) and (1971), the essays in Lakatos and Musgrave (1970), some of the essays in Suppe (1974), Scheffler (1967) and (1972), and Quay (1974). The criticisms of the last two authors in this list come closest of any I know to the comments of the present article.

justification" and the dichotomy between normative and descriptive disciplines. A more explicit formulation occurs as part of his answer to the charge of being an irrationalist:

I have not previously and do not now understand quite what my critics mean when they employ terms like 'irrational' and 'irrationality' to characterize my views. These labels seem to me mere shibboleths, barriers to a joint enterprise whether conversation or research. (1970a, p. 263)

. . . existing theories of rationality are not quite right and . . . we must readjust or change them to explain why science works as it does. To suppose, instead, that we possess criteria of rationality which are independent of our understanding of the essentials of the scientific process is to open the door to cloud-cuckoo land. (*ibid.,* p. 264)

Another statement, which occurred in a discussion rather than an essay, is very explicit and extends the thesis from scientific methodology to epistemology in general:

Although my professional identity is as an historian of science, what is on my mind when I get involved with the sort of thing I am doing here today is ultimately epistemology. I really want to know what sort of thing knowledge is, what it is all about, and why it is that it works the way it does. Now in order to do that, it seems to me the right move (I am glad somebody else said philosophy is an empirical enterprise) is to look around and try to see what is going on and what it is that people who have knowledge have got. If I then think that what I discover when I look gives me certain sorts of understanding of what goes on – makes it plausible that knowledge should be the sort of thing it is and should develop the way it does – then I can legitimately say that from the examination of scientific communities I am beginning to become a better epistemologist. (1974, pp. 512–3)

The two theses ought to be decoupled, since I shall argue that the second is partly correct while the first is thoroughly erroneous. If the first thesis were accepted, and truth were taken to be irrelevant to the progress of science, then the second thesis would have to be construed as saying that scientific achievements provide proximate justifications for methodological principles but it is vain to seek for ultimate justifications, for there are none to be found – a conclusion which might indeed deserve the charge of 'irrationalism'. On the other hand, if the first thesis is denied, and scientific progress is considered to be evidence that serious people immersed in research are collectively sensitive to the truth, then we can understand the second thesis in an entirely different way: we could acknowledge that sensitivity to the truth is very difficult to explain, and that the organons, systems of inductive logic, and theories of scientific inference proposed hitherto somehow all fall short of the mark, but we might hope that looking more closely than methodologists are accustomed to do at the substantive achievements of science will help to provide the desired explanation. In other words, we should turn to the history of science not for the purpose of seeking a surrogate for the approach to the truth, but rather for illumination on the question of how this approach is possible for human beings.[3]

The message that I am trying to convey in urging that Kuhn's two theses be decoupled has been said from time to time in different ways, for example in the following passages:

For that which the author had at heart throughout his studies of the history of science was to gain an understanding of the whole logic of every pathway to the truth. (Peirce, 1958, p. 175)

My suggestion is that Galileo's own view of his subtle, flexible, and discriminating method has in common with Plato two things: First, the reconciliation of a 'mathematical' and 'rational' conception of science with a full recognition of the difficulty of coming to know: a reconciliation that is only possible through a crucial emphasis upon the process of inquiry itself. Second, a conviction that inquiry as such is not a profitable subject for positive doctrinal exposition ('Discourse on Method', 'Rules for the Conduct of the Understanding'), foreshadowed in dramatic representation. In the *Two New Sciences,* on this reading, Galileo has offered us his view of *what science is,* in the form of a dramatised commentary upon the central work of his own scientific life. (Stein, 1974, p. 397)

I hope that there will be some value in reiterating and reformulating the message.

2

The first of Kuhn's arguments for the irrelevance of truth to scientific progress is essentially that for the purpose of explaining the phenomenon of progress we have no need of the hypothesis that science is drawing constantly nearer some fixed goal (1970, pp. 170–1). Scientific progress consists in unidirectionality with respect to certain desirable characteristics:

I believe it would be easy to design a set of criteria – including maximum accuracy of predictions, degree of specialization, number (but not scope) of concrete problem solutions – which would enable any observer involved with neither theory to tell which was the older, which the descendant. (1970a, p. 264)

3. Although my primary concern is with epistemology, it should be evident that the point of view which I am recommending also has consequences for the discipline of history of science. Specifically, human sensitivity to the truth should be taken seriously both as an *explanandum* and as an *explanans* in studies of scientific development. There is no question worthier of investigation by historians of science than one raised by Cohen: "Can we establish, by historical inquiry, what is the most likely social ensemble of human need, desire, conflict, prejudice, and training to produce a search for, and attainment and recognition of, true knowledge?" (1970, p. 233).

Unidirectionality of this kind no more depends upon the existence of a fixed goal than does the "steady emergence of more elaborate, further articulated, and vastly more specialized organisms", which, according to neo-Darwinian evolutionary theory, can be accounted for by natural selection without postulating orthogenesis (1970, pp. 172-3). A second argument, not included in the first edition of *The Structure of Scientific Revolutions,* is this:

There is, I think, no theory-independent way to reconstruct phrases like 'really there': the notion of a match between the ontology of a theory and its 'real' counterpart in nature now seems to me illusive in principle. (1970, p. 206)

I shall postpone for a while a consideration of Kuhn's first argument and start with an examination of his second, which is recognizable as a special version of a general type of skeptical reasoning that has frequently occurred in the history of philosophy. I do not know a commonly used name for this type of reasoning, but perhaps it could be called "arguments for the impossibility of transcendence", or more mundanely "trapped-in-a-box arguments". The box in which the subject is trapped varies from one version to another of the argument: the subject's own consciousness, the historical epoch in which the subject is born, or the cognitive peculiarities of the human race. In any case, transcendence is impossible, because in principle the only information accessible to the subject is contained within the box. An exemplary argument of this kind occurs in Hume's *Enquiry Concerning Human Understanding:*

It is a question of fact, whether the perceptions of the senses be produced by external objects, resembling them: how shall this question be determined? By experience surely: as all other questions of a like nature. But here experience is, and must be entirely silent. The mind has never anything present to it but the perceptions, and cannot possibly reach any experience of their connection with objects. The supposition of such a connection is, therefore, without any foundation in reasoning. (1955, Sect. XII, Part I)

The parallel between Kuhn's argument and that of Hume is evident. If, therefore, one accedes to the conclusion that "the notion of a match between the ontology of a theory and its 'real' counterpart in nature now seems to me illusive in principle," how can one consistently resist Hume's skepticism concerning knowledge of external objects? Conversely, and more constructively, Hume's skeptical argument may fail to be compelling because of a generic weakness in arguments for the impossibility of transcendence, and in this way an answer to Kuhn's version of skepticism can be discovered.

The response to Hume which I find most persuasive consists of two steps. The first, which is usually not made explicit, is to establish that indirect or hypothetico-deductive arguments are legitimate in epistemological investigations. To disallow such arguments would "block the way of inquiry" by ruling out of consideration *ab initio* a large spectrum of epistemological positions, whereas none are censored by allowing such arguments. Specifically, it is conceivable that phenomenalism would be found, in an investigation using indirect arguments, to have greater explanatory power than any alternative to it which is proposed; but if phenomenalism were accepted because indirect arguments were ruled inadmissible, then it would have been effectively insulated from the possibility of critical examination. The second step, which is developed in innumerable versions in the literature, consists in taking advantage of the legitimation of indirect arguments and in weighing the explanatory power of phenomenalism against that of critical realism, etc., the crucial evidence being the apparent persistence of physical processes when gaps occur in observation of them, the concurrence of observers with regard to events, the reliability and precision of certain laws formulated in terms of physical concepts, and the non-existence of laws of comparable reliability and precision formulated directly in phenomenalistic terms. If, in reply, a variant of phenomenalism is presented which is agnostic regarding the existence of external objects but admits that experience is organized as if such objects exist and bear causal relations to phenomena, then a critical realist could claim as decisive a settlement as one can reasonably expect of philosophical disputes: the variant is clearly parasitical upon critical realism, and the difference between them has been diminished to the point where the question of choice is a *Scheinproblem*.

The argument of the preceding paragraph misses many of the subtleties and important cognate issues that can be found in the literature on the existence of external objects. (At least I hope that this is so, for it would be depressing to think that such an immense corpus could be distilled without loss to so small a residue!) Nevertheless, I believe that the argument is fundamentally correct and that its solid structure will not be essentially changed by refinements. It is evidently applicable as a counterargument to any kind of "argument for the impossibility of transcendence", but the outcome of any particular application cannot be foreseen without examining the details; indeed, it is precisely the virtue of the kind of argument that I have sketched in answer to Hume that it moves freely by reflection and critical examination of evidence, thereby keeping the way of inquiry open.

What then should be said about Kuhn's particular version of the argument for the impossibility of transcendence? He surely is correct that what is "really there" is not directly accessible to us; but if indirect arguments are admissible, then one cannot say initially (whatever one may say at the conclusion of the analysis) that it is illusive in principle to consider "the match between the ontology of a theory and its 'real' counterpart in nature". Consider the hypothesis, which I shall call "the Hypothesis of Verisimilitude",[4] that *for well-established theories in a science in which high critical standards have been achieved such a match exists to some good degree of approximation.* Then if philosophical inquiry is to be open, the Hypothesis of Verisimilitude and its negation should initially both be on the same footing. An appeal to indirect argument consists in examining which of them has the greater explanatory power. It may be noted that if the Hypothesis of Verisimilitude turns out to be superior in explaining the relevant evidence, then an answer will also have been given to Kuhn's first argument, which had been temporarily set aside and which consisted essentially of the challenge, "Does it really help to imagine that there is some one full, objective, true account of nature and that the proper measure of scientific achievement is the extent to which it brings us closer to that ultimate goal?" (1970, p. 171).

There are two massive bodies of evidence which fall naturally into place upon the Hypothesis of Verisimilitude, and which are very mysterious indeed upon its negation. Kuhn neglects the first of these bodies of evidence almost entirely, but the other he recognizes and does try to fit into his scheme.

The first is the evidence that once a certain stage of maturity is reached in a science, a remarkable continuity of thought is maintained even across scientific revolutions, exemplary instances being the transition from classical mechanics to relativistic mechanics, from classical mechanics to quantum mechanics, from classical electro-magnetic theory to quantum electrodynamics, from special relativity to general relativity, from phenomenological thermodynamics to statistical thermodynamics, from phenomenological valence theory to quantum chemistry, from the Darwinian to the genetic theory of natural selection, and from the phenomenological theory of the gene to molecular biology. In each case the theory that was "overthrown" was in fact retired with honor: its domain of approximate validity was not only delimited but reaffirmed, and its continuity with the new theory, not just in approximate agreement with regard to a certain class of predictions but also with regard to conceptual structure, was recognized. In short, what Bohr called "the correspondence principle" holds, *mutatis mutandis,* between the new and the old theories in all these instances of scientific revolutions.[5] The phrase "once a certain stage of maturity is reached", which was used above, will evidently be problematic for historians of science, but the case which I am trying to make is

---

4. The terminology is borrowed from Popper (1963) but with some change of meaning.
5. See Tisza (1963), Shimony (1970), Post (1971), Koertge (1973), and Boyd (unpublished).

unaffected by the problem of dating, for even the foregoing short list of exemplary fulfillments of the correspondence principle suffices to constitute an impressive body of evidence for weighing the epistemological hypotheses in question. Nevertheless, the notion of maturity is important for the present considerations, for it suggests that a distinction ought to be made (without precluding the possibility of overlap and of intermediate cases) between those scientific revolutions which consisted primarily of the establishment of decisively new standards of investigation, and those revolutions which consisted primarily of conceptual generalization or of deepening the level of description.

The second body of evidence is just that which Kuhn cites in support of the unidirectional character of the development of science: that more and more problems are solved in ways that are considered to be satisfactory by the community of scientists, and the precision of solutions in general increases (1970, p. 171). Kuhn is so strongly impressed by the evidence of progress in the natural sciences in these and related respects that he cautions against pushing too far the conflation of the history of science with that of art or politics, towards which his argument had been drifting (*ibid.,* pp. 208–9).

It remains now to judge the Hypothesis of Verisimilitude and its negation as fairly as possible on the basis of the two bodies of evidence just summarized.

The first body of evidence consists of instances in which the correspondence principle has held after various sciences have reached maturity. If the Hypothesis of Verisimilitude is true, and if maturity is identified with the achievement of high critical standards, then it would follow that any theory which has satisfied those standards sufficiently well to be accepted as 'well-established' is a good approximation to the truth. Characterizing the approximation as 'good' imputes to the theory some validity within an appropriate domain, which is expected to be discernible from the standpoint of a theory yet closer to the truth. But this is just what the correspondence principle asserts. Thus, the Hypothesis of Verisimilitude accommodates the first body of evidence in an entirely natural manner.

My argument, however, is loose in texture, so that objections can reasonably be raised against its decisiveness. The source of the trouble is the occurrence of three vague expressions in the formulation of the Hypothesis of Verisimilitude: "well-established theories", "high critical standards", and "to some good degree of approximation". The effect of these vague expressions, it might be contended, is to make the content of the Hypothesis of Verisimilitude so ill-defined that any body of evidence drawn from the history of science would be compatible with it. In reply I would say that the vagueness should appear troublesome only to some one who conceives of methodology as a discipline to be developed entirely in advance of the sciences. According to Kuhn's own second thesis (to be defended, with certain qualifications, in the next section), the phrases "well-established theories" and "high critical standards" could be explicated by reflection upon exemplary scientific achievements. Similarly, exemplary instances of the correspondence relation will help to explicate the vague expression "good degree of approximation", though it may be wise to recognize the necessity for some reliance upon analogy in using this expression, just because the relation between any actual theory (a human artifact, however inspired) and the ideal theory which exactly matches the truth is an extrapolation from the relation holding between two actual theories.

That the second body of evidence falls into place naturally from the standpoint of the Hypotheses of Verisimilitude is almost a corollary of what has already been said. The displacement of a theory by one which is a better approximation to the truth will mainly preserve – possibly with refinements – the predictions and explanations achieved in the domain of validity of the former. One would expect, therefore, that revolutionary as well as normal development of a science subsequent to its maturity will, for the most part, increase the number of solved problems and the precision of solutions. To prevent misunderstanding it should be pointed out that eddies and retrograde motions do in fact complicate the unidirectional development of science, and that the explanation of unidirectionality offered on the basis of the Hypothesis of Verisimilitude can take these complications into account. A considerable amount of mathematical analysis or attention to fine physical details is sometimes required in order to correct the retrogression. (For example, Drude's classical treatment of conduction electrons yielded an explanation of Ohm's law and some remarkably good evaluations of electrical resistances; all of which seemed to be undercut by the quantum mechanical theorem of Bloch that electrons propagating in a periodic lattice experience no resistance; to which the correction was to take account of impurities and thermal fluctuations of the lattice.) Despite such complications, the central point is that the history of various sciences, after the achievement of certain critical standards, has been astonishingly progressive, and the Hypothesis of Verisimilitude removes some of the mystery from the progress without dispelling the sense of astonishment.

It is also interesting to consider a kind of scientific progress not emphasized by Kuhn: the typical history of experiments. A pioneer experiment is often delicate and hard to duplicate, so that its decisiveness is doubtful. In time, however, reliability is often increased to the point that the experiment is incorporated into the repertory of student laboratory work. The transformation is due to such improvements as the control of perturbations, the elimination of systematic errors, the correction of instabilities in the equipment, and the increase in sensitivity of detectors. From the standpoint of the Hypothesis of Verisimilitude, this kind of progress can be understood quite naturally: there is such a thing as the 'reality' to which the theory of interest refers, but the causal connections between this reality and the appearances in the laboratory are complicated and often masked by various factors, and the improvement of the experiment consists in the identification and control of these factors.

Both the evidence concerning the correspondence principle and that concerning the progress of science could only be regarded as fantastic networks of coincidences if the Hypothesis of Verisimilitude were false. The reason that this is so is virtually revealed by Kuhn's own analysis. He recognizes that scientific investigation is at least to some extent open to nature. Even his doctrine of the theory-ladenness of observation does not permit him to say that the full content of the observational report is under the control of the experimenter. But if nature plays a role in the results of experimental tests, how can either of the two bodies of evidence be explained unless nature is somehow cooperative? In a brief passage Kuhn says precisely this:

What must nature, including man, be like in order that science be possible at all? Why should scientific communities be able to reach a firm consensus unattainable in other fields? . . . It is not only the scientific community that must be special. The world of which that community is a part must also possess quite special characteristics. (1970, p. 173)

In the immediately following passage, however, Kuhn veers away from examining the epistemological consequences of the correct and important admission that he has just made:

That problem – What must the world be like in order that man may know it? – was not, however, created by this essay. On the contrary, it is as old as science itself, and it remains unanswered. But it need not be answered in this place. Any conception of nature compatible with the advance of science by proof is compatible with the evolutionary view of science developed here. (*ibid.*)

The final sentence is a sleight of hand. Presumably "the evolutionary view of science developed here" includes the proposition that there is no "permanent fixed scientific truth" which is the goal of scientific development (*ibid.*); but there is indeed a "conception of nature compatible with the growth of science" (*ibid.*) which contradicts this proposition, namely, the conception implicit in the Hypothesis of Verisimilitude. To readers interested in paradigms Kuhn has provided a splendid one: a paradigm of the abortion of a viable line of reasoning at exactly the moment that it becomes embarrassing to the author!

The case in favor of the Hypothesis of Verisimilitude and against Kuhn's dismissal of the role of truth can be summed up in the following way. Long ago Descartes compared the attempt to discover the laws of nature from reflection upon phenomena to the attempt to discover the rules by which a message has been encoded in a text. Suppose that we have such a text, and that after numerous conjectures the tentative decipherment has become more and more coherent. The success may be no more than a series of coincidences, so that the tentative decipherment is completely on the wrong track. But somehow it is more plausible that a good approximation to the correct rules of encoding has been found than that the long run of successes is coincidental. Kuhn's thesis that the truth plays no role in the progress of science is analogous to maintaining that progressive coherent decipherment could occur even though no such things as the initial message and the rules of encoding exist.

3

Kuhn's argument for the second thesis is far from explicit, but I shall try to reconstruct it without misrepresentation from a number of passages in his work.

(i) He makes a historical claim that the genesis of methodological principles is inseparable from substantive scientific discoveries. He says, for example, that "in learning a paradigm the scientist acquires theory, methods, and standards together, usually in an inextricable mixture" (1970, p. 109). It is evident from the discussion in Chapters 9–11 of his book that this assertion is intended to apply to the initial discovery as well as to later imitation and indoctrination. I do not deny that Kuhn's historical assertion is largely correct, but I wish to register disappointment that he does not present and analyze detailed historical evidence for it, which would be of great intrinsic interest and might throw some light upon the complex relations among scientific thought, the thought of pre-scientific civilized peoples, the reasoning of children, and "the mind of primitive man".

(ii) Kuhn is dubious about the possibility of applying explicit methodological principles, because of "the insufficiency of methodological directives, by themselves, to dictate a unique substantive conclusion to many sorts of scientific questions" (1970, p. 3). The most explicit justification that I have been able to find for step (ii) is on pp. 145–7 of Kuhn's book, where he elliptically discusses the possibility of judging a theory by either the methods of probabilistic verification or by falsification. The following is essentially all that he has to say about the former:

In their most usual forms, however, probabilistic verification theories all have recourse to one or another of the pure or neutral observation-languages discussed in Section X. One probabilistic theory asks that we compare the given scientific theory with all others that might be imagined to fit the same collection of observed data. Another demands the construction in imagination of all the tests that the given scientific theory might conceivably be asked to pass. Apparently some such construction is necessary for the computation of specific probabilities, absolute or relative, and it is hard to see how such a construction can possibly be achieved. If, as I have already urged, there can be no scientifically or empirically neutral system of language or concepts, then the proposed construction of alternate tests and theories must proceed within one or another paradigm-based tradition. Thus restricted it would have no access to all possible experiences or to all possible theories. As a result, probabilistic theories disguise the verification situation as much as they illuminate it. (1970, pp. 145-6)

The *demonstrandum* in this passage is the impossibility of computing specific probabilities in crucial situations of conflict between theories. Evidently, the details of the various probabilistic methods are irrelevant to Kuhn's demonstration, since not even an outline is presented of any one of them, and in fact the description of one of the methods ("Another demands etc.") is so perfunctory that I cannot decide which one he is referring to. Kuhn obviously considers the decisive point in the argument to be the theory-ladenness of observation, expressed in the sentence beginning "If, as I have already urged". The elliptical character of his argument then has some justification, for the details of the various probabilistic methods, and of any other method that might be considered, are irrelevant to this consideration.

It is worth pointing out, however, in anticipation of later analysis, that Kuhn's *demonstrandum* about the computation of probabilities could be reached in an entirely different way, by pointing out that numerical values are underdetermined in all extant formulations of probabilistic methods, whether because of the range of choice left open as a matter of principle in personalist theories, or because there are too many admissible c-functions, or because the principle of indifference is insufficiently precise, or because in frequency theories there are too many possible choices of reference classes. Here would be a reason for asserting "the insufficiency of methodological directives" which does take account of the detailed structure of extant methods. In the case of the method of falsification, a similar demonstration of 'insufficiency' could rest upon the difficulty of giving unambiguous criteria for "basic statements" (Popper, 1961, pp. 100-111). If it is then objected that no general proof has been provided for "the insufficiency of methodological directives" in all formulations of scientific method, including those which the ingenuity of future methodologists may devise, then an answer could be given which Kuhn's empiricist approach to epistemology should find congenial: the shortcomings of methods so far devised in this respect constitute inductive evidence that those of the future will suffer the same defect. The argument suggested in this paragraph has the evident advantage over the one extracted from Kuhn's text that it is unaffected by analyses which challenge the theory-ladenness of observation.[6] It also has the virtue of applying to 'normal science' as well as to 'scientific crises'. Nothing in Kuhn's own argument precludes the possibility of unequivocal calculations of the probability of a theory upon given evidence, so long as both the theory and the evidence are governed by a definite paradigm. But any one who has struggled with the problem of devising 'objective' procedures for specific probability calculations would be euphoric at achieving even this much!

(iii) Kuhn sometimes goes beyond step (ii) by asserting that the methodological content of a paradigm can never be completely abstracted and articulated. Scientists are able to

agree in their *identification* of a paradigm without agreeing on, or even attempting to produce, a full *interpretation* or *rationalization* of it. Lack of a standard interpretation or of an agreed reduction to rules will not prevent a paradigm from guiding research. (1970, p. 44)

He evidently is maintaining that the unabstracted residue of a paradigm compensates for the insufficiency of methodological directives which he had pointed out in step (ii). In order to elucidate the *modus operandi* of a paradigm, Kuhn refers (*ibid.*) to Polanyi's ideas on 'tacit knowing'; but in spite of admiration for Polanyi as a phenomenologist of scientific investigation, I do not think that these ideas will provide an answer to the criticisms which will be raised below.

The main criticism is that all passages which can be taken as formulations of Kuhn's second thesis are obscure on the following questions: Even if one grants that the genesis of methodological principles is inseparable from substantive scientific developments, might not some of them be rationally justified in abstraction from actual scientific discoveries (so that the context in which they originated would only be psychologically relevant, like the figures drawn in sand for the geometer)? If so, which ones can be justified in this way and how? The most informative quotation that I can find discusses "a preliminary codification of good reasons for theory choice" as follows:

These are, furthermore, reasons of exactly the kind standard in philosophy of science: accuracy, scope, simplicity, fruitfulness, and the like. It is vitally important that scientists be taught to value these characteristics and that they be provided with examples that illustrate them in practice. If they did not hold values like these, their disciplines would develop very differently. . . .

What I am denying then is neither the existence of good reasons nor that these reasons are of the sort usually described. I am, however, insisting that such reasons constitute values to be used in making choices rather than rules of choice. (1970a, pp. 261–2)

6. Cf. Shimony (1977).

Even from this passage I cannot determine whether the goodness of the reasons mentioned is supposed to consist only in their being abstracted from several successful paradigms, as suggested by the opinion cited earlier that "philosophy is an empirical enterprise", or whether there is some other basis for their rationality. There is no need, however, to worry the text for an answer to this question, since Kuhn could be asked directly to say which interpretation he intended. His work deserves censure on this point whatever the answer might turn out to be, just because it treats central problems of methodology elliptically, ambiguously, and without the attention to details that is essential for controlled analysis.

What, then, ought to be said about the rationality of methodological principles? Or, put another way, if Kuhn's second thesis is partly correct, how exactly should it be understood and qualified? The questions are evidently difficult, and I shall attempt only an outline in reply. Complete answers would require a fully satisfactory theory of scientific method, which we still have not achieved. One thing that does seem clear, however, is the indispensability of a variety of considerations which are organized so as to complement each other. Specifically, all of the following seem to me essential for a thorough understanding of the rationality of scientific procedures.

3.1. Deductive logic

The scientific method draws freely and sometimes crucially upon deductive logic. Whatever the basis of the principles of deductive logic may be, they are so reliable in comparison with other elements of the scientific method that in the present context there is no 'real and living doubt' about their rationality. Even the extreme empiricists, who maintain that all statements are ultimately to be judged according to their contribution to a coherent account of experience, do not usually propose to derive the principles of deductive logic from specific achievements of science.[7] Consequently, Kuhn's second thesis ought to be qualified at least to the extent of admitting that those principles of deductive logic which are appropriated by the scientific method stand in no need of justification from the history of science. The most that can be retained of Kuhn's second thesis at this juncture is the observation that substantive scientific achievements may indicate what parts of deductive logic are important for scientific method. Thus, Aristotle recognized the validity of inference by *modus tollens* but considered it to be inferior to affirmative demonstration

---

7. An exception is the proposal of Finkelstein and Putnam to change the logic of discourse because of quantum mechanical evidence. They are wrong, in my opinion, but the issue is too complex to discuss here. If they are correct, however, then Kuhn's second thesis would have wider applicability than I have allowed.

(*Posterior Analytics* I, 25, 87a); the reversal of his judgment by later methodologists may plausibly be attributed to the historical occurrence of exemplary eliminations of theories on the basis of observation (together, of course, with the revelation of weaknesses in the affirmations of Aristotelian physics).

3.2. Probability theory

Corresponding to the different senses of 'probability' there are various justifications of the axioms of probability, using arithmetic, measure theory, decision theory, etc. In no case is the justification based upon substantive scientific achievements. There is nothing to be retained here from Kuhn's second thesis except that scientific achievements suggested (or formed part of the intellectual milieu which suggested) the significance of the notion of probability for scientific investigation (see Hacking, 1975). There are still many problems concerning the role of probability in scientific method which are not completely solved. For example, if 'probability' is understood in the personalist sense, then the Dutch book argument of Ramsey and DeFinetti supplies a very convincing justification of the axioms, provided that it makes sense to bet upon the propositions to which probabilities are assigned, but the typical universal generalizations which occur in the natural sciences do not seem to be 'bettable'. There happens to be a surrogate for the Dutch book argument, namely the theorem of Cox, Good, and Aczél, but it depends upon premisses which are difficult to justify definitively, even though they are highly plausible (Shimony, 1970). Such problems, however, do not seem to me to lend any support for Kuhn's thesis, but only indicate that much remains to be clarified in decision theory when a game is played for truth rather than for monetary stakes (see Levi, 1967).

3.3. A posteriori considerations

Here is where Kuhn's second thesis is correct and important. His assertion of "the insufficiency of methodological directives" is nowhere better illustrated than by the notorious inadequacy of the axioms of probability theory, even with such reasonable supplements as the principle of indifference, to yield non-arbitrary numerical evaluations. The axioms and supplements also fail to yield a non-arbitrary rule for the tentative acceptance of a hypothesis – a circumstance which may be a virtue rather than a failing in practical decision-making, when utilities can be compared (Jeffrey, 1956; Carnap, 1963, p. 972), but which appears to be a serious inadequacy in the context of theoretical investigations. There may be no way of compensating for these insufficiencies that is justifiable in all possible worlds. But the counsel of wisdom may be that nothing of value is thereby lost, for surely we should be content with methods that work in the actual world. A moment's reflection, however, diminishes the appeal of this solution: if we knew what the actual world is like, the objectives of scientific inquiry would already be accomplished and there would be no need for the methods of science. Nevertheless, something can be salvaged from the suggestion: *we may know enough about the actual world, or at least have tentative suppositions upon which we may reasonably enough rely, to add specificity to a priori methodological principles.* Since the great achievements of science can reasonably be claimed to provide good approximations to the truth about the actual world (as argued in Section 2), we are justified in attempting to derive from them some of the specificity which is lacking in a priori principles. Thus, it is a commonplace that the prior probability of a simple hypothesis should be greater than that of a less simple one, but many methodologists have despaired of the utility of this rule just because of the difficulties of finding non-arbitrary a priori criteria for simplicity. If, however, serious attention is paid to Weyl's admonition that "we must let nature train us to recognize the true inner simplicity" (1949, p. 155), then it is possible to use such criteria as the order of differential equations and tensorial rank in making judgments of relative simplicity. The 'training' by nature consists of exemplary scientific discoveries in which the roles of differential equations, tensors, etc. are exhibited. Here, then, is strong support for Kuhn's assertion that paradigms guide research, and for part of his thesis that the procedures of scientific investigation can be shown to be rational only by drawing upon substantive scientific achievements. It must be emphasized, however, that the reason which has just been offered for the partial correctness of Kuhn's thesis is very different from Kuhn's own explanation that a scientist may have tacit knowledge of the methods implicit in a paradigm without being able to articulate them. Whether the scientist's knowledge is tacit or explicit, the crucial point is that it is warranted only by evidence that the actual world is constituted in a way that makes certain procedures appropriate.

3.4. Dialectic

The proposal to justify even part of the scientific method by a posteriori considerations raises serious problems about the structure of scientific knowledge. Unless one proceeds with some finesse, one might conclude that there are no rational grounds for the acceptance of those scientific achievements from which methodological principles are abstracted. But this conclusion is absurd (and therefore I credit Kuhn's protests against critics who attribute the opinion to him) for the obvious reason that some criteria, however rough, must be operative in the judgment that a theory is successful. It may be suggested at this point that those methodological principles which are drawn from deductive logic and probability theory will suffice to establish the criteria of success, so that nothing but a priori principles and experience are needed to identify the exemplary achievements upon which the a posteriori methodological principles are based. I very much doubt that the structure of scientific knowledge is as neat as this. Rather, it seems that some complex considerations of the interplay between the a priori and the a posteriori – which may appropriately be called "dialectic" – are indispensable. Much remains to be clarified about this dialectic. How, for example, are a posteriori guidelines to be used in prior probability evaluations at the same time that open-mindedness is maintained towards revolutionary hypotheses? Is it possible to formulate inductive arguments in support of inductive principles without vicious circularity? What role is played by vindicatory arguments, i.e., arguments of the form that nothing will be lost, and possibly something of value will be gained, if certain procedures are followed? (For a debate on the second and third questions see Teller (1975) and Shimony (1975).) The work of Imre Lakatos on the methodology of research programs (1970) is one of the important contributions to the study of this dialectic, even though he uses very different language to characterize what he is attempting to do.

3.5. Moral factors

There are principles of scientific methodology that are manifestly not so much intellectual as moral, for example that a scientist should not suppress data which are unfavorable to his own proposals and that the state should abstain from scientific censorship. Such principles can perhaps be regarded as special instances of a posteriori considerations, for they derive from our awareness of some of the evil proclivities of human beings. (They would not be articulated in the scientific method of the Houynhmhms, though perhaps they would be "tacitly known".) In a sense, however, all methodological principles are subsumed under moral ones:

The most vital factors in the method of modern science have not been the following of this or that logical prescription – although these have had their value too – but they have been the moral factors. First of these has been the genuine love of truth and conviction that nothing else could long endure. Given that men strive after the truth, and, in the nature of things, they will get it in a measure etc. (Peirce, 1958, p. 56)

Here, then, is an outline of the elements of a scientific methodology, drawing upon the contributions of many workers, which presumes to account for the rationality of scientific procedures. It makes no pretense to completeness or to freedom from conceptual difficulties, and indeed a number of difficulties were explicitly pointed out. Despite the unfinished state of methodology, however, enough has been accomplished, I think, to establish firmly the main contentions of this paper: (i) Kuhn is partly correct in his thesis that the rationality of scientific procedures derives from scientific achievements, but his thesis needs qualification and refinement; and (ii) the most important qualification is to acknowledge what his other thesis denies, that the community of serious investigators exhibits collectively a wonderful sensitivity to the truth.

Acknowledgments. I am very grateful to Prof. Martin Eger for his helpful comments, and to the National Science Foundation for support of research.

BIBLIOGRAPHY

Boyd, Richard, *Realism and Scientific Epistemology* (unpublished), Department of Philosophy, Cornell University.
Carnap, Rudolf, 1963, 'Replies and Systematic Expositions', in P. A. Schilpp (ed.), *The Philosophy of Rudolf Carnap,* Open Court, La Salle, Ill.
Cohen, Robert S., 1970, 'Causation in History', in W. Yourgrau and A. D. Breck (eds.), *Physics, Logic, and History,* Plenum, New York and London.
Finkelstein, David, 1969, 'Matter, Space and Logic', in *Boston Studies in the Philosophy of Science,* Vol. V (ed. by R. S. Cohen and M. Wartofsky), D. Reidel, Dordrecht and Humanities Press, New York.
Hacking, Ian, 1975, *The Emergence of Probability,* Cambridge University Press, Cambridge.
Hume, David, 1955, *Inquiry Concerning Human Understanding* (ed. by C. Hendell), Liberal Arts Press, Indianapolis.
Jeffrey, Richard, 1956, 'Valuation and Acceptance of Scientific Hypotheses', *Philosophy of Science* **23**, 237–246.
Koertge, Noretta, 1973, 'Theory Change in Science', in G. Pearce and P. Maynard (eds.), *Conceptual Change,* D. Reidel, Dordrecht.
Kuhn, Thomas, 1970, *The Structure of Scientific Revolutions,* second edition, University of Chicago Press, Chicago.
Kuhn, Thomas, 1970a, 'Logic of Discovery or Psychology of Research?' and 'Reflections on my Critics', in Lakatos and Musgrave (eds.) (1970).
Kuhn, Thomas, 1974, 'Second Thoughts on Paradigms', and contributions to discussions, in Suppe (1974).
Lakatos, Imre, 1970, 'Falsification and the Methodology of Scientific Research Programs', in Lakatos and Musgrave (eds.) (1970).

318   *Experience and reason*

Lakatos, I. and Musgrave, A. (eds.), 1970, *Criticism and the Growth of Knowledge,* Cambridge University Press, Cambridge.
Levi, Isaac, 1967, *Gambling with Truth,* Knopf, New York.
Peirce, Charles S., 1958, *Collected Papers,* Vol. VII (ed. by A. W. Burks), Harvard University Press, Cambridge, Mass.
Popper, Karl R., 1961, *The Logic of Scientific Discovery,* Science Editions, New York.
Popper, Karl R., 1963, *Conjectures and Refutations,* Harper and Row, New York and Evanston.
Post, Heinz, 1971, 'Correspondence, Invariance, and Heuristics', *Studies in History and Philosophy of Science* **2**, 213–255.
Putnam, Hilary, 1969, 'Is Logic Empirical?', in *Boston Studies in the Philosophy of Science,* Vol. V (ed. by R. S. Cohen and M. W. Wartofsky), D. Reidel, Dordrecht, and Humanities Press, New York.
Quay, Paul, 1974, 'Progress as a Demarcation Criterion for the Sciences', *Philosophy of Science* **41**, 154–170.
Scheffler, Israel, 1967, *Science and Subjectivity,* Bobbs-Merrill, Indianapolis.
Scheffler, Israel, 1972, 'Vision and Revolution: A Postscript on Kuhn', *Philosophy of Science* **39**, 366–374.
Shapere, Dudley, 1964, 'The Structure of Scientific Revolutions', *Philosophical Review* **73**, 383–394.
Shapere, Dudley, 1971, 'The Paradigm Concept', *Science* **172**, 706–9.
Shimony, Abner, 1970, 'Scientific Inference', in R. Colodny (ed.), *The Nature and Function of Scientific Theories,* University of Pittsburgh Press, Pittsburgh.
Shimony, Abner, 1975, 'Vindication: A Reply to Paul Teller', in *Induction, Probability, and Confirmation, Minnesota Studies in the Philosophy of Science,* Vol. VI (ed. by G. Maxwell and R. M. Anderson, Jr.), University of Minnesota Press, Minneapolis.
Shimony, Abner, 1977, 'Is Observation Theory-Laden? A Problem in Naturalistic Epistemology', in *Logic, Laws, and Life: Some Philosophical Complications* (ed. by R. Colodny), University of Pittsburgh Press, Pittsburgh.
Stein, Howard, 1974, 'Maurice Clavelin on Galileo's Natural Philosophy', *British Journal for the Philosophy of Science* **25**, 375–397.
Suppe, Frederick (ed.), 1974, *The Structure of Scientific Theories,* University of Illinois Press, Urbana.
Teller, Paul, 1975, 'Shimony's A Priori Arguments for Tempered Personalism', in *Induction, Probability, and Confirmation, Minnesota Studies in the Philosophy of Science,* Vol. VI (ed. by G. Maxwell and R. M. Anderson, Jr.), University of Minnesota Press, Minneapolis.
Tisza, Laszlo, 1963, 'The Conceptual Structure of Physics', *Reviews of Modern Physics* **35**, 151–185. Reprinted in L. Tisza, *Generalized Thermodynamics,* MIT Press, Cambridge, Mass., 1966.
Weyl, Hermann, 1949, *Philosophy of Mathematics and Natural Science,* Princeton University, Princeton.

PART E
Fact and value

12

On Martin Eger's
"A tale of two controversies"

Criticisms are presented against Eger's challenge to the demarcation between the natural sciences and ethics. Arguments are given both against his endorsement of the "new" philosophy of science and against his rejection of the fact–value dichotomy. However, his educational recommendations are reinforced rather than weakened by these criticisms.

Martin Eger's essay is extraordinarily rich in penetrating philosophical comments and in educational good sense. Nevertheless, I believe that there are serious errors in his fundamental philosophical theses, and much of this commentary will be devoted to exhibiting them. I shall then try to show that for the most part his educational recommendations are reinforced rather than weakened by my theoretical criticisms.

"NEW" AND "OLD" PHILOSOPHY OF SCIENCE

Eger challenges the demarcation of the natural sciences from the study of morals by questioning that they are different cognitively. The demarcation was clear and strict, he says, as long as an old, essentially positivist conception of the natural sciences was maintained and as long as the fact–value dichotomy was accepted. The "new philosophy of science," however, has profoundly criticized the old conception, and in spite of some reservations Eger on the whole assents to these criticisms. Likewise, he assents to arguments against the fact–value dichotomy, and indeed at one point he provides a very interesting argument of his own, which may be original while he makes no claim to this effect:

True, there is no consensus among *theoretical* experts on rules or principles, nor on frontier issues. . . . But in regard to exemplars – history's moral heroes – the situation is quite different. When it comes to these *practical* experts, a widely acceptable list can indeed be drawn up. And imitating exemplars, as Kuhn has shown, is at once surer and more flexible than acquaintance with rules. (Eger 1988, 315)

This work originally appeared in *Zygon* 23 (1988), pp. 333–40. © 1988 by the Joint Publication Board of *Zygon*. Reprinted by permission of the publisher.

Eger undervalues the "old" philosophy of science partly because he uses the term to refer to only a part of a diverse and complicated collection of methodological and epistemological doctrines. Positivism – which maintains that the total content of a scientific theory lies in its implications for human experience – was indeed very influential during the first half of the twentieth century. During this period there were also influential realists, who followed the tradition of Galileo Galilei, Isaac Newton, and John Locke of attempting to infer from experience the properties of an objective world which has an existence independent of human beings. (For example, Bertrand Russell in 1914 was a positivist, but by 1927 Russell had been converted to realism.) Eger also attributes to the old philosophy of science the use of algorithms of inference, but I have no idea to whom he may be referring. The formulations of scientific methodology in the best of the old philosophers of science (my personal favorites being Charles S. Peirce and Harold Jeffreys) are complex and sophisticated syntheses of diverse intellectual elements: hypothetico-deductive reasoning, probability theory, decision theoretical arguments, appeals to evolutionary biology to justify human skill in hypothesis formation, and appeals to the history of science for *a posteriori* refinements of method (see Shimony 1970). It should not be surprising that an adequate scientific methodology is complicated, if one considers the ambitiousness of the scientific enterprise: namely, to obtain good approximations to the objective truth about the universe at large, on the basis of experience which is very limited in space and time and constrained by the peculiarities of human faculties for gathering and processing information. It is true both that "Facts are stubborn things" and that "Nature loves to hide," and an adequate methodology must do justice to each of these divergent dicta.

I do not wish to deny that the new philosophy of science has made some real contributions. It has emphasized the indispensability of the history of science for a rich philosophy of science. Some of the innovators (notably Michael Polanyi and Thomas Kuhn) have pointed out that skilled scientific practitioners typically have much more "tacit knowledge" of their craft than they are able to articulate in explicit rules of scientific method. Some (notably Norwood Hanson) have drawn upon empirical psychology in order to carry out epistemological analyses of observation and other mental processes. For the most part, however, the great value of these insights has been debased by drawing from them relativistic and subjectivistic epistemological conclusions. The following are some of the important criticisms that should be made of the new philosophy of science.

First, the history of science need not be used as a surrogate for scientific methodology, as Kuhn maintains, with each historical epoch providing its paradigms which cannot be judged from a neutral standpoint. Instead, the history of science may be studied in order to provide *a posteriori* elements in scientific method, for there is no reason to believe that the human intellect is endowed *a priori* with all the methodological tools it needs for investigating the natural world. Experience is needed not only to learn substantive truths about nature, but also to learn how to learn (see Shimony 1976).

Second, the great virtue of the tacit knowledge of skilled investigators is not its tacitness but its knowledge. If methodologists eventually are able to articulate what these investigators know tacitly (as good athletic and musical coaches are able to do in their respective domains), then nothing is lost thereby and something is positively gained.

Third, the deployment of data from empirical psychology, especially from Gestalt psychology, in order to show that observations are "theory laden," provides *prima facie* evidence against the possibility of objective empirical assessments of competing scientific hypotheses. Yet a careful study of empirical psychology – such as the experiments on "cognitive dissonance" by Jerome Brunner and Leo Postman – reverses this judgment, and shows that human beings are capable of switching between integrative and analytic strategies of perception, and the latter is strikingly liberated from theory-ladenness (see Shimony 1977).

Finally, the occurrence of scientific revolutions is an insufficient reason to deny the meaningfulness of the concept of objective truth and to recommend that it be replaced by a concept of historically relativized truth, as Kuhn recommends (1970, 171–73). The most that one can legitimately infer is the unlikelihood that human beings can ever *achieve* the goal of objective knowledge of the universe. However, even this concession to relativism is excessive, for it fails to pay due attention to the detailed history of those scientific revolutions which have occurred since the seventeenth century. Typically in these revolutions the displaced theory is a good approximation to the displacing theory, with regard to empirical predictions and in some respects with regard to conceptual structure. In the terminology of Niels Bohr, there is a "correspondence principle" governing the relation between the old and the new theory. Consequently, the appropriate moral to be drawn from the occurrence of scientific revolutions is not relativism but a doctrine of successive approximations to the truth.

In summary I find a pervasive slovenliness of reasoning in the new philosophy of science. Its advocates have failed to use its excellent insights constructively by exploiting them for the refinement of scientific methodology and epistemology. This constructive enterprise requires hard work, which is evaded by their relativism and subjectivism, and of these Eger is too tolerant.

THE FACT–VALUE DICHOTOMY

The relation between facts and values is very subtle, and I make no pretense to professional expertise about it. However, I wish to present a few considerations which should make one resist any simple conflation of the natural sciences and ethics and to distrust the claim that rational criteria are the same in both domains.

At one level there is a universally admitted dichotomy: that what a person does and what a person ought to do in a given situation are not generally the same. What the person does is a fact, and perhaps what he or she ought to do is also a fact, but if so it is a fact of a different kind. The fact–value dichotomy would then be converted into a dichotomy within the domain of facts. But what, then, is the ontological status of the second type of fact, that is, of normative facts?

A possible answer is that desires are factual, and desires may or may not be achieved. Might not the ontological status of a normative fact be that of a desire? No, this suggestion is not sufficient for two reasons. Most people who are not morally nihilistic would make moral judgments not only about achievements but about desires themselves: that some desires are better than others. Secondly, in the same situation two different persons may have different desires as to what the actor should do. (The difference of opinion may depend upon whether the person judging is the same as the actor, but this is not the only crucial factor, for two people may disagree about what is desirable were their respective situations in the action to be exchanged.) Consequently, the identification of normative facts with desires would deprive normative facts of an objective or interpersonal ontological status.

It may be suggested that the ontological status of the normative fact is that of an authoritative prescription, with different versions of this point of view recognizing different authorities: God, society, the evolutionary history of the human race, and so on. I am skeptical that any of these appeals to authority can account for the motive force of a normative fact unless they endow the authority with a power of enforcement, to punish infractions and reward obedience. If the authority is so endowed, then its prescription is really a hypothetical imperative rather than a categorical imperative, for the motive force is effectively the desire of the subject to avoid pain or achieve gratification. In brief, despite the above-mentioned shortcomings of the identification of the normative fact with desire, it is unrealistic to neglect the role of desire in the analysis of value.

My own (tentative) position is to relate norms to desires, but to do so in the wise manner suggested by Aristotle's *Ethics,* which points out some remarkable facts about the structure of desire. He notes that all people agree verbally as to what is desirable: namely, happiness (Aristotle 1.4), but they have many different ideas as to what constitutes happiness. People also have subordinate desires – for example, for wealth or honor or learning – and the satisfaction of these may or may not lead to happiness. Furthermore, because a human being is neither a beast nor a god but a social animal, his or her happiness is bound up with the happiness of others, in the family or in the state (or – by a nonaristotelian extrapolation – in the human race as a whole). It is a fact about human nature that character – and hence what is desired by an individual – is to some extent plastic, and is formed by habituation and education. It is also a fact that not all modes of forming character are equally conducive to that universally desired but vague end, happiness. Most of Aristotle's *Ethics,* after the preliminaries of the first book, is devoted to the investigation of the moral and intellectual virtues, which he regards to be the true way to happiness. Even if one has followed Aristotle this far, however, one may suspect that he was unavoidably ethnocentric, because of the limitations of his social experience. One may draw upon the mass of evidence accumulated by anthropologists to suggest that there is not one but a plurality of ways to human happiness (e.g., a contemplative life, a kinaesthetic life, a ritually organized life). One is not thereby committed to a thoroughgoing cultural relativism, for not all cultures are equally satisfying to their practitioners in their own eyes; as Edward Sapir (1924) pointed out, some among the great variety of cultures are "genuine" and some are "spurious."

It should be clear why I resist Eger's attempt to narrow the gulf between the natural sciences and ethics. With regard to the former, I have argued, albeit briefly, that there is a domain of entities independent of human experience which are endowed with definite properties, and a scientific proposition is objectively true if it correctly characterizes this domain. Whatever the difficulties may be for human beings to discover on the basis of their limited experience the objective truth, it is, so to speak, "there" to be found out. In ethics, however, the ontological status of the normative facts is much more problematic. I do not wish to say that they are merely matters of convention or subjective opinion, for it is not the case that "anything goes"; the constitution of the human psyche and the social character of human beings set limits upon the range of life styles which will permit the achievement of happiness by the standards of the subjects themselves. Yet it is by no means clear that these constraints uniquely determine an optimum life style, and with it a unique set of norms.

PEDAGOGY

After these criticisms of Eger's philosophy of science and ethics I shall turn to his discussion of pedagogy, with which I largely agree. He is properly outraged about an authoritarian handling of the creation/evolution controversy combined with the slackness of the program of moral education in the public schools.

The passages cited by Eger against teaching creationism express anxiety about its corrosive effect upon the intellectual faculties of young students. I agree, of course, that intellectual faculties should be cultivated, and a primary way to do so is to perform experiments and demonstrations concerning phenomena which are simple enough to permit a fairly complete exercise in scientific methodology, without gaps and without the intrusion of authority. It must be quite confusing, and perhaps even demoralizing, to thoughtful students to be presented with fragmentary and authority-adulterated applications of the scientific method unless the instructor is candid about the lacunae in the reasoning. How is the student to know whether his or her vagueness of understanding is due to the incompleteness of the reasoning itself or to personal intellectual shortcomings? With this consideration in mind, how is a rational account of the theory of evolution to be presented to young students? Here is the most outstanding case in the history of science of a great theory which is confirmed globally, by an immense variety of taxonomic, zoogeographic, embryological, and paleontological evidence falling into place, rather than by the prediction of striking, unexpected phenomena. The long history of resistance to the theory of evolution (see Mayr 1982, 510–70) – because of genuine conceptual difficulties, not just because of stubborn dogmatism – shows how ill-suited this theory is for elementary instruction. What then would be the danger of a good, open debate by clever students about the creation/evolution issue? It is hard to imagine a creationist positively persuading a classmate who is not antecedently convinced to accept the Biblical account. The worst that is likely to happen, from the standpoint of an advocate of evolution, is that the class will forcefully feel the lacunae in the standard textbook presentations, and is this such a bad thing for the cultivation of intellectual faculties? I suspect (partly because of introspection) that people who object to permitting creationism to be discussed in the public schools fear that it somehow will be the opening wedge of a general anti-intellectual, authoritarian, fundamentalist, and fascist seizure of political power. However, I believe that the real danger of such a catastrophe lies in racial and economic tensions, which will not be assuaged by the prohibition of a debate on creationism.

The case against a curriculum of moral education which emphasizes "the critical attitude" and "choice among alternatives" can be built upon Aristotle's *Ethics*. A young person is not suited for lectures on ethics because of inexperience in the actions that occur in life (Aristotle 1.3). Furthermore, first principles in ethics (as opposed to those of the theoretical sciences, which are obtained by induction) are acquired by habituation (Aristotle 1.7), and inculcation of the moral virtues by habituation must precede the acquisition of the intellectual virtues by instruction (Aristotle 2.1). An important way in which the schools can contribute to this inculcation seems not to have been mentioned by Aristotle, namely, to capture the students' imaginations. For this purpose an exposure to the biographies of "history's moral heroes," in Eger's phrase, may be particularly efficacious.

At this point there is a major ideological conflict between Aristotle and the designers of the curriculum of moral education. Aristotle believes that ethics and politics are continuous, and that the state is responsible for the moral education of the child not just for the child's sake but for the good of the state as a whole. The designers of the curriculum of moral education, on the other hand, wish to develop the child's independence of judgment in order to be able to resist the authoritarian claims of the state.

How can one adjudicate this controversy? I would say, above all not *a priori*. Only on the basis of experience can one judge whether a person inculcated with moral virtues in childhood and only later exposed to ethical analysis is more self-confident, more judicious, more tolerant, and in general more rational than a person whose critical attitudes on moral matters is fostered in early childhood. Of course, the question is complex, and the answer depends crucially upon the mode of inculcation and upon the details of the relation between individuals and society. In appealing to empirical evidence, however, I do not mean to conflate these difficult ethical, political, and pedagogical questions with the problems of the natural sciences. Practical reason is not concerned with the aspects of the human mind which are genetically fixed, but rather with those which are plastic. Hence, the evidence which it must marshal has to be drawn from human history and from the experience of people who have struggled with the concrete problems of life.

REFERENCES

Aristotle, *Nicomachean Ethics.* Many editions, including 1941. *The Basic Works of Aristotle,* ed. Richard McKeon, 927–1112. New York: Random House.

Bruner, Jerome S. and Leo Postman. 1949. "On the Perception of Incongruity: A Paradigm." *Journal of Personality* 18: 206–23.

Eger, Martin. 1988. "A Tale of Two Controversies: Dissonance in the Theory and Practice of Rationality." *Zygon: Journal of Religion and Science* 23: 291–325.

Hanson, Norwood Russell. 1958. *Patterns of Discovery.* Cambridge: Cambridge Univ. Press.

Jeffreys, Harold. 1948. *Theory of Probability,* 2d ed. Oxford: Oxford Univ. Press.

Kuhn, Thomas. 1970. *The Structure of Scientific Revolutions,* 2d ed. Chicago: Univ. of Chicago Press.

Mayr, Ernst. 1982. *The Growth of Biological Thought.* Cambridge, Mass.: Harvard Univ. Press.

Peirce, Charles Sanders. 1932. *Collected Papers,* vol. 2, ed. Charles Hartshorne and Paul Weiss. Cambridge, Mass.: Harvard Univ. Press.

Polanyi, Michael. 1958. *Personal Knowledge.* Chicago: Univ. of Chicago Press.

Russell, Bertrand. 1914. *Our Knowledge of the External World.* London: Allen & Unwin.

Russell, Bertrand. 1927. *The Analysis of Matter.* London: Allen & Unwin.

Sapir, Edward. 1924. "Culture, Genuine and Spurious." *American Journal of Sociology* 29: 401–29.

Shimony, Abner. 1970. "Scientific Inference." In *The Nature and Function of Scientific Theories,* ed. R. G. Colodny. Pittsburgh: Pittsburgh Univ. Press.

Shimony, Abner. 1976. "Comments on Two Epistemological Theses of Thomas Kuhn." In *Essays in Memory of Imre Lakatos,* ed. R. S. Cohen, P. K. Feyerabend, and M. W. Wartofsky. Dordrecht: Reidel.

Shimony, Abner. 1977. "Is Observation Theory-Laden? A Problem in Naturalistic Epistemology." In *Logic, Laws, and Life: Some Philosophical Complications,* ed. R. G. Colodny. Pittsburgh: Univ. of Pittsburgh Press.

Index

329